Lecture Notes in Physics

Edited by J. Ehlers, München, K. Hepp, Zürich, and
H. A. Weidenmüller, Heidelberg
Managing Editor: W. Beiglböck, Heidelberg

33

Classical and Quantum Mechanical Aspects of Heavy Ion Collisions

Symposium held at the Max-Planck-Institut
für Kernphysik, Heidelberg, Germany,
October 2-5, 1974

Edited by H. L. Harney, P. Braun-Munzinger
and C. K. Gelbke

Springer-Verlag Berlin Heidelberg GmbH 1975

Editors:
Hanns Ludwig Harney
Peter Braun-Munzinger
Claus Konrad Gelbke
Max-Planck-Institut für Kernphysik
Postfach 103980
D–69 Heidelberg

Library of Congress Cataloging in Publication Data

Symposium on Classical and Quantum Mechanical Aspects
    of Heavy Ion Collisions, Max-Planck-Institut für
    Kernphysik, 1974.
    Classical and quantum mechanical aspects of heavy ion
collisions.

    (Lecture notes in physics ; 33)
    Bibliography: p.
    Includes index.
    1. Collisions (Nuclear physics)--Congresses.
2. Ions--Scattering--Congresses. 3. Quantum theory--
Congresses. I. Harney, Hanns Ludwig, ed. II. Braun-
Munzinger, P., ed. III. Gelbke, C. K., ed. IV. Title.
V. Series.
QC794.6.C6S94 1974   539.7'54   74-32179

ISBN 978-3-540-07025-2   ISBN 978-3-540-37313-1 (eBook)
DOI 10.1007/978-3-540-37313-1

P R E F A C E

In 1874, Gustav Robert Kirchhoff claimed that mechanics is a complete
and most simple description of movements. Fifty years later, in 1924,
Louis de Broglie established the theory of matter waves. Another fifty
years later, in 1974, the Symposium on "Classical and Quantum Mechanical
Aspects of Heavy Ion Collisions" especially dealt with the question
whether heavy ions may be described in terms of classical mechanics or
in terms of matter waves.

Due to a rapid development in experimental techniques a great deal of
precision data, especially on heavy ion elastic scattering and transfer reactions, is available by now. Various quasiclassical as well as
quantum mechanical approaches have been used for the interpretation of
these experiments. Until now it is not clear if the concepts of trajectories or diffraction are sufficient for the understanding of heavy ion
interactions or if full quantum mechanical treatments are necessary.
All three approaches, which might not only be complementary but perhaps
even controversial are currently applied for the interpretation of
transfer reactions.

At this moment it seemed appropriate to have an extensive discussion
on the physical foundation of the various concepts. The Symposium on
Classical and Quantum Mechanical Aspects of Heavy Ion Collisions held
in October 1974 at the Max-Planck-Institut für Kernphysik, Heidelberg,
tried to elucidate the similarities and point out the controversies of
the different approaches. In this book the invited talks of the Symposium are reproduced and give an account of the present state of the
art in the interpretation of heavy ion interactions.

In order to speed up the publication the manuscripts of the authors
have been reproduced photomechanically. The editors, therefore, do not
feel responsible for misprints.

P. Braun-Munzinger

C.K. Gelbke

H.L. Harney

Heidelberg, October 1974

<u>Symposium on Classical and Quantum Mechanical Aspects</u>

<u>of Heavy Ion Collisions</u>

Organized by

The Max-Planck-Institut für Kernphysik,

Heidelberg, Germany

<u>Organizing Committee</u>

P. Braun-Munzinger
P. Brix
C.K. Gelbke
H.L. Harney

<u>Program Committee</u>

R. Bock
D. Fick
W. von Oertzen
H.C. Pauli
D. Schwalm
H.J. Specht
H.A. Weidenmüller
J.P. Wurm

Program and Table of Contents

Note that contributed papers and discussions are not published in these
proceedings.

Acknowledgement

It is a pleasure to acknowledge the experienced and friendly help of Mrs. R. Häfner and Mrs. U. Spies in the organization of the symposium and the preparation of the present book.

Sponsors

We are endebted to

The Max-Planck-Gesellschaft zur Förderung der Wissenschaften, München,

   and

The BASF, Ludwigshafen,

   and

The BBC, Mannheim,

   and

The Knoll AG, Ludwigshafen

that sponsored the Symposium on Classical and Quantum Mechanical Aspects of Heavy Ion Collisions.

FOLDED POTENTIALS FOR THE DESCRIPTION OF HEAVY ION

ELASTIC SCATTERING AND TRANSFER REACTIONS*

C. B. Dover and J. P. Vary
Brookhaven National Laboratory**
Upton, New York, USA 11973

We present a simple model for the heavy ion optical potential based on the convolution of target and projectile densities with a suitably chosen nucleon-nucleon effective interaction. This model is shown to be appropriate for a number of low energy peripheral processes, such as elastic scattering and one and two particle transfer reactions. Application of the method to cluster states in light nuclei and relativistic heavy ion interactions is also considered.

The optical potential is of central importance for heavy ion interaction processes, since it is widely used to describe elastic scattering as well as more complicated reactions through the DWBA formalism. In principle, one would like to relate the optical potential for composite particle scattering to the fundamental nucleon-nucleon (NN) interaction, in an approach that systematically includes many body corrections. Several such microscopic models are available for nucleon-nucleus scattering [1-3]. However, some phenomenological adjustments are usually necessary in order to obtain good fits to experimental data. We recognize that a quantitative microscopic theory of nucleus-nucleus interactions is somewhat beyond our reach at present. Hence, we propose a compromise which retains the more important physical features of a microscopic approach, but is flexible enough to fit data. We are thus able to preserve much of the predictive power of a fundamental theory. The essential ingredients are

1) the neutron and proton density distributions of the interacting nuclei, and

2) the nucleon-nucleon effective interaction in the two nucleus medium.

We present here a sample of results from a folding model which relates this input information to the optical potential $V_{opt}(r)$:

$$V_{opt}(r) = \int \rho_A(\underset{\sim}{r}_1)\rho_B(\underset{\sim}{r}_2)G(\underset{\sim}{r}+\underset{\sim}{r}_1-\underset{\sim}{r}_2)d\underset{\sim}{r}_1 d\underset{\sim}{r}_2 \qquad (1)$$

where ρ_A and ρ_B are the total densities of nuclei A and B, and G is the NN effective

* Invited paper for the Symposium on Classical and Quantum Mechanical Aspects of Heavy Ion Collisions, Heidelberg, Germany, October 2-5, 1974; presented by C. B. Dover.

** Supported by the U. S. Atomic Energy Commission.

interaction. In Eq. (1), we assume for simplicity a local interaction G(r) which is averaged over spin and isospin. Various models which are similar in form to Eq. (1) have also been employed in the discussion of atom-atom collisions and hadron-hadron interactions at very high energy [4]. Several similar treatments of composite particle processes also exist [5], but these involve the folding of a nucleon-nucleus optical potential with the density of the target (or the projectile). By this means, one hopes to phenomenologically include the effects of some higher order correlation terms, but the target and projectile are not treated symmetrically. We prefer to use the symmetric form (1), which is the first term in a consistent density expansion of the optical potential. This form of the folding model enables us to identify G as the underlying NN amplitude. Note that G is an amplitude rather than an NN potential; i.e., G includes rescattering corrections to all orders with Pauli restrictions and other many body effects. Thus Eq. (1) does not correspond to treating the NN potential in perturbation theory, which would be quite inadequate [6].

In fitting experimental elastic scattering data, we parametrize G in terms of an adjustable complex strength constant $\bar{f}$ and a fixed range parameter r_0. Thus, higher order corrections to Eq. (1) are absorbed by suitably modifying $\bar{f}$. As we shall see later, at higher energies one obtains good agreement between the phenomenological value of $\bar{f}$ and the value obtained from a microscopic estimate. In addition, we show that with an adjustable $\bar{f}$, Eq. (1) remains a useful model for the potential for peripheral processes at low energies.

The main features of the potential model of Eq. (1) are the following: a) it relates the geometry of the nucleus-nucleus potential to the geometries of the projectile and target nuclei. It thus represents a very strong theoretical prejudice as to the radius and effective diffuseness of the heavy ion optical potential. b) by construction, the folded potential has no continuous geometrical ambiguities. One can also show that there are no discrete ambiguities (in potential depth). c) the fixed geometry of the model greatly reduces the amount of computing time necessary to obtain a fit.

Some comments on the validity of the folding model are in order. Such an expansion is expected to converge at high energies [1] or low densities. The latter condition is the key to the present situation. For heavy ion elastic scattering or relatively simple particle transfer processes, the interaction is primarily restricted to the surface region, where the effective densities are low. If the two nuclei interpenetrate appreciably, more complicated many particle transfer and breakup channels will be populated. The folding model would be quite inappropriate for a particular complicated channel (5 particle transfer, say). However, the total reaction cross section σ_R should be well described by the folding model, or any other reasonable model for that matter, if it is close to the geometric limit.

The applicability of the folding model at low energies is thus restricted to peripheral processes, which are sensitive to the surface and tail region of the optical potential $V_{opt}(r)$. In the tail region, the folded potential should be closer to the truth than phenomenological Woods-Saxon potentials. However, the present elastic scattering data do not provide a sufficient constraint to determine the rate of fall-off of $V_{opt}(r)$ in the tail region, so one cannot reject phenomenological potentials. Of course, the folding model is quite wrong in the nuclear interior, since it neglects Pauli restrictions and correlation corrections. However, this fact is largely irrelevant, since the processes which we consider are known to be insensitive to the form of the real potential for small r [6]. One could probably obtain better fits to some of the transfer data (particularly two particle), while preserving the quality of the elastic fit, by using a phenomenological ansatz (a Woods-Saxon, say) for the interior region, and joining it smoothly to a folded potential in the surface region.

We now outline the scope of our efforts. In previous work, the use of the folding method has been restricted mainly to elastic and inelastic scattering [4,6]. However, for heavy ion processes, elastic and inelastic scattering place a rather weak constraint on the form of the potential. For instance, elastic scattering only determines the value of the real potential near a critical radius $r \approx 1.5 \ (A^{1/3}+B^{1/3})$ fm., where A and B are the projectile and target mass numbers [6]. Hence, to see if our model has non-trivial content, it is necessary to test it for a wide range of phenomena at various energies and for different nuclei. The crucial test is whether the folded shape, with the strength determined by a fit to elastic scattering, also reproduces one and two particle transfer data in the DWBA. We provide several illustrative examples. We also mention briefly the application of the model to cluster states in light nuclei [7], which tests the extrapolation of the method from continuum to bound state problems, and also to relativistic heavy ion interactions [8]. The details pertaining to the various ramifications of the model are to be found in refs. [7-11].

We now discuss the ingredients of our calculations. For nuclear densities, two prescriptions were tested, in order to determine the sensitivity of the potential to the tail region of the densities: a) proton densities $\rho_p(r)$ taken from electron scattering analyses, and neutron density $\rho_n(r)$ assumed proportional to proton density; b) $\rho_p(r)$ and $\rho_n(r)$ obtained from Hartree-Fock or shell model calculations; total density obtained from $\rho(r) = \sum_i W_i |\varphi_i(r)|^2$, where i runs over all occupied neutron and proton bound states, $\varphi_i(r)$ is the single particle wave function and W_i is the statistical weight (2j+1 for filled shell). To obtain $\varphi_i(r)$ we used Hartree-Fock results [12] or wave functions corresponding to Woods-Saxon potentials which give a best fit to empirical single particle binding energies [13] for a range of nuclei. This latter recipe is expected to provide a reasonable representation of $\rho(r)$ in the tail region, since the large r behavior will be dominated by the least bound orbits whose binding energies are well known. On the other hand, electron scattering analyses do not provide a sufficiently accurate density for large r. We find in fact that the quality of our elastic scattering fits depends fairly sensitively on having an adequate description of the tail region of the density. Fits obtained with prescription b) above are consistently better than those obtained using a). We show later that the use of proper densities is essential to establish contact between the best fit and theoretical values of the potential strength.

For the parametrization of the NN effective interaction G, we use the ansatz

$$G = \bar{f}(Ne^{-r^2/r_0^2})$$ (2)

where $\bar{f}$ is a complex depth parameter and N is a normalization constant chosen so that $N\int d^3\underset{\sim}{r}\, \exp(-r^2/r_0^2) = -2\pi\hbar^2/M$, where M is the nucleon mass. This choice of N enables us to relate $\bar{f}$ to the usual NN forward scattering amplitude. If we neglect many-body effects, the simplest theoretical estimate for $\bar{f}$ is

$$\bar{f} = \sum_\ell \bar{f}_\ell$$ (3)

where $\bar{f}_\ell$ is a spin and isospin averaged partial wave amplitude. For example, for s-waves we have

$$\bar{f}_0 = (1-\xi/2)\,\bar{f}_{1_{S_0}} + 3\xi/2\,\bar{f}_{3_{S_1}}$$

$$\bar{f}_{S\ell J} = \frac{2\ell+1}{k}\,\exp(i\delta_{S\ell J})\sin\delta_{S\ell J}$$ (4)

where $\delta_{S\ell J}$ is the free space NN phase shift for spin S, orbital angular momentum ℓ and total spin J. The factor ξ (1/2 for N = Z projectiles) yields the proper spin-isospin average, and k is the lab NN wave number, evaluated at the incident lab energy per particle. A range parameter $r_0 = 1.4$ fm was chosen. This value is consistent with what we expect from the long range part of the NN potential; it also corresponds to a sharp minimum in the χ^2 function as a function of r_0 for a number of α scattering cases which we examined.

We now present some typical numerical results. For elastic scattering, the two parameter fits with a complex $\bar{f}$ are of uniformly good quality. For α scattering, we only try to fit the data in the region of strong diffraction oscillations, consistent with the limitations of a first order model. The high quality of the fits in the diffraction region is illustrated in Fig. 1 for the $\alpha + {}^{208}$Pb reaction. Also shown in Fig. 1 is a typical fit to heavy ion elastic scattering, in this case ^{18}O + ^{60}Ni. For heavy ion processes, we employ the entire angular range in the fit, since the measured data only extend down to $\sigma/\sigma_R \approx 0.01$, which still corresponds to a peripheral process. It is not clear that a local single channel optical model description remains physically meaningful in the region far below $\sigma/\sigma_R \approx 0.01$. The fits to the heavy ion data are comparable to those obtained with conventional Woods-Saxon potentials with 4-6 parameters. This indicates that the folding model provides a succinct description of those features of the geometry of the optical potential which are important for peripheral processes. This is perhaps its greatest merit for

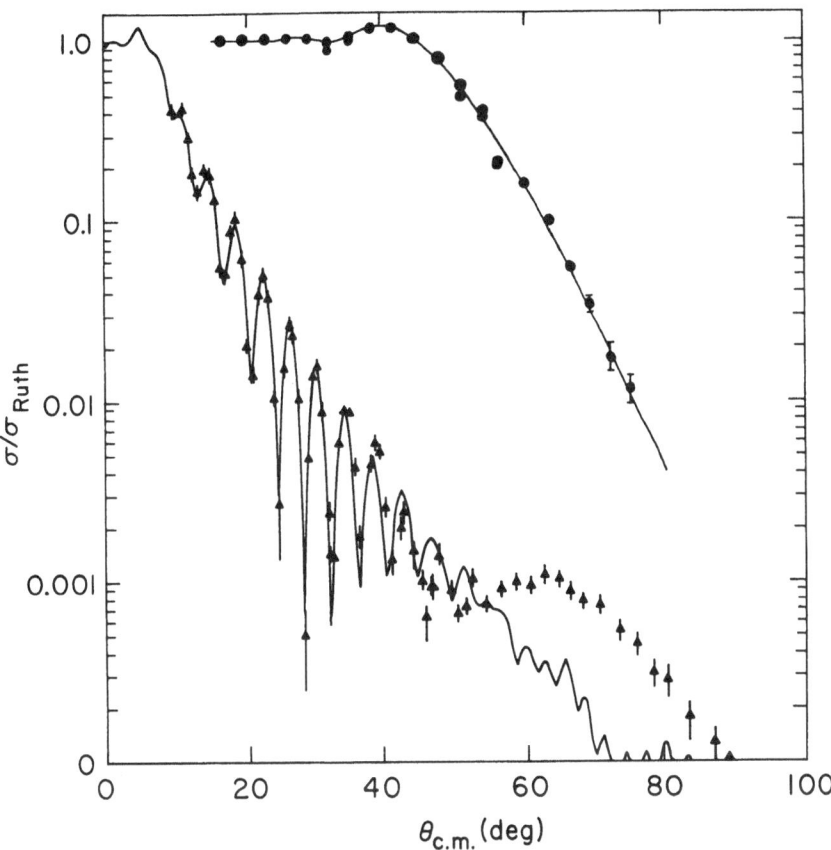

Fig. 1   The top curve shows the fit to the Brookhaven $^{18}O + ^{60}Ni$ elastic data [14]
at 62.92 MeV with $\bar{f} = 1.27 + 0.9i$ fm.   The lower curve displays the fit to the
Maryland $\alpha + ^{208}Pb$ data [16] at 139 MeV with $\bar{f} = 1.79 + 1.21i$ fm.

low energy applications.

The real and imaginary parts of the folded potential corresponding to the
$^{18}O + ^{60}Ni$ fit are shown as solid lines in Fig. 2. A Woods-Saxon (WS) potential used
to fit the transfer data is also shown as a set of dashed lines [14].   Several observations are in order: The real part of the folded and WS potentials agree well in
the region of the critical radius $r \approx 1.5\ (A^{1/3} + B^{1/3})$, indicated by an arrow in Fig.
2.   This is the part of the potential which is determined by elastic scattering.   The
behavior of the two potentials is quite different in the interior; the folded potentials are characteristically very deep near $r = 0$.   The imaginary potentials also
differ in the surface region, so it is clear that the elastic scattering data is much
less sensitive to the details of the absorption.   Particle transfer reactions, on the
other hand, are more sensitive to the magnitude of the absorption.   It is only with
the aid of such reactions that we are able to begin to sort out different potential
models.

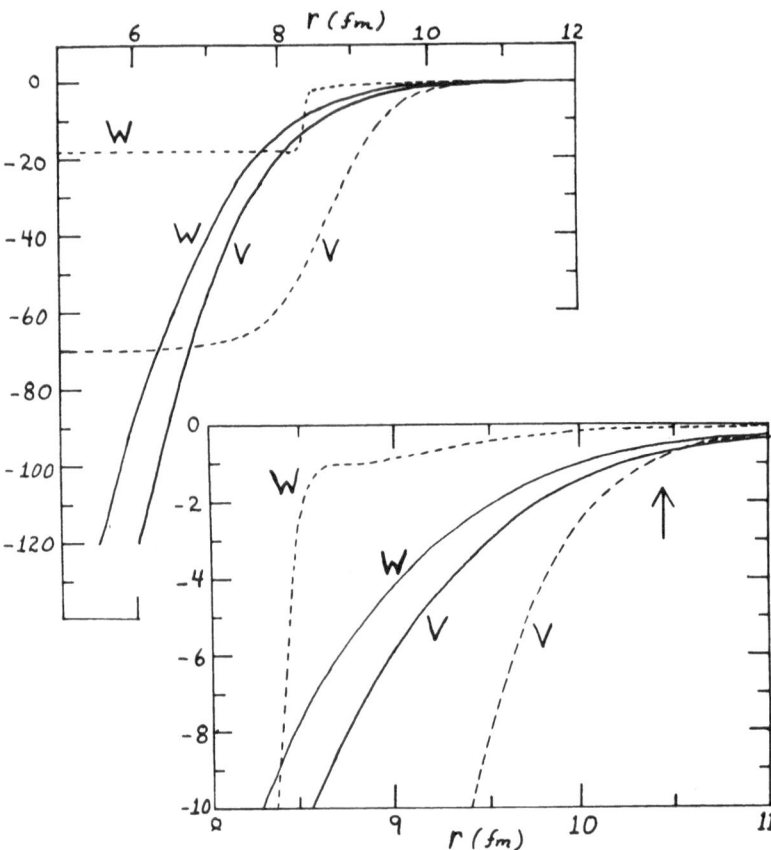

Fig. 2   The solid curves V and W are the real and imaginary parts of the folded potential corresponding to the $^{18}O + ^{60}Ni$ fit of Fig. 1.   The dashed curves
correspond to Woods-Saxon potentials of LeVine et al. [14] which yield a
good fit to the $^{60}Ni(^{18}O, ^{16}O)^{62}Ni$ reaction.

Another observation is that the <u>shape</u> of a folded potential differs from that
of a WS potential, in that it cannot be characterized by a single diffuseness parameter in the crucial surface region. At any radius r, we can define an "equivalent"
WS potential by matching the folded potential to the tail $V_0 \exp(R/a)\exp(-r/a)$ of a
WS potential. This procedure generates the <u>ambiguity plot</u> shown in Fig. 3. Such a
plot can be used to generate a respectable first guess to a WS potential which will
fit the elastic data. The main point here is that the effective diffuseness "a" of
the folded potential is <u>radius dependent</u>. However, the folding model gives values
of a for heavy targets which are in the usual range $0.4 \lesssim a \lesssim 0.6$ fm, when the matching to a WS form is done in the region of the critical radius for elastic scattering.
For bound state problems, such as the $\alpha + ^{16}O$ cluster states in ^{20}Ne [7,9], the effective diffuseness is considerably larger ($\approx$ 1 fm) in the region where the α cluster is
localized. This large diffuseness enables one to obtain close to the correct energy

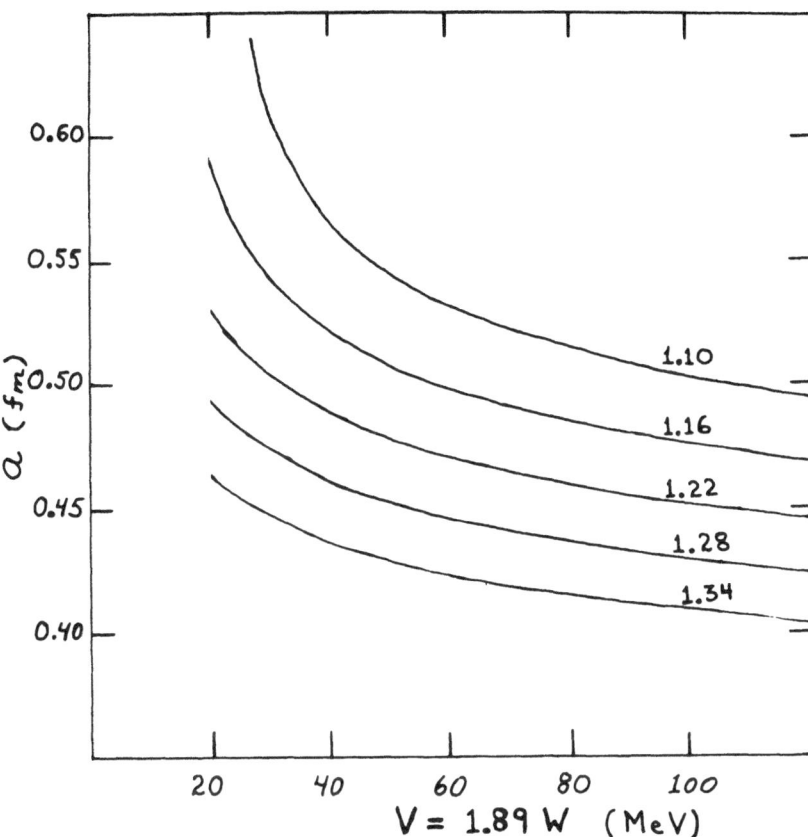

Fig. 3   Ambiguity plot from folding model analysis of 166 MeV Orsay data for
$\alpha + {}^{40}Ca$ [5]. Families of equivalent Woods-Saxon parameters for the real
depth V, the imaginary depth W and the diffuseness a are shown. Each curve
is labelled by r_0, where $R = r_0(A^{1/3} + B^{1/3})$ is the radius parameter of the
well.

splittings within the ground and first excited rotational bands with a <u>single</u>

strength parameter $\bar{f}$ for each band, whereas WS potentials with a conventional dif-

fuseness of 0.6fm require a considerable renormalization of the depth for <u>each</u>

state.

Another crucial test is the energy dependence. For the case of α particle

scattering, there is sufficient data (from 40-166 MeV) to study empirically the

<u>energy dependence</u> of the real and imaginary parts of the folded potential. For

heavy ions, it is difficult to make such a systematic study at the present time,

since most of the data are restricted to the low energy region of 5-10 MeV per par-

ticle. Higher energy data would be most welcome.

In Fig. 4, we show as dots the real and imaginary parts of $\bar{f}$, empirically

adjusted at each energy to provide a best fit to α elastic scattering data. The

real part of the potential is found to <u>decrease</u> with increasing energy, as is familiar

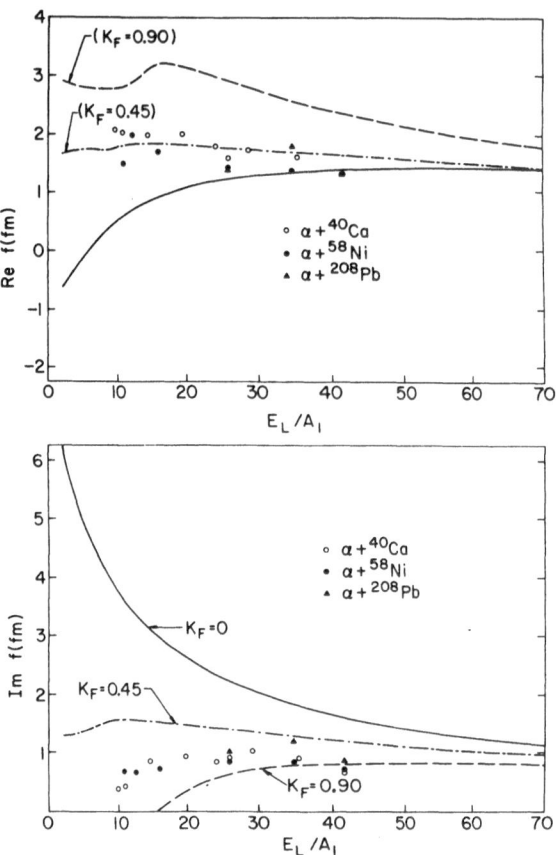

Fig. 4   Energy dependence of the real and imaginary parts of the depth parameter $\bar{f}$.
Empirical values of $\bar{f}$ determined by best fits to α scattering data are shown
as dots, circles and triangles, as a function of the lab energy/particle
E_L/A_1. The theoretical curves labelled by k_F are calculated as described
in the text.

from proton-nucleus scattering analyses [15]. Correspondingly, the imaginary depth
<u>increases</u> with energy, as expected in general from arguments based on density of
states. The fits shown in Fig. 4 refer to three different target nuclei, ^{40}Ca, ^{58}Ni,
and ^{208}Pb. No significant target dependence of $\bar{f}$ is seen, which reassures us that
the folding of densities correctly accounts for the variation of cross sections with
target at the same energy.

The comparison of these results with a theoretical calculation of $\bar{f}$ is provided
by the solid curves in Fig. 4. These theoretical estimates are obtained in a simple
interacting Fermi gas model, starting from the free space NN amplitude of Eq. (3).
We pass from the free NN amplitude to an <u>effective</u> amplitude in the medium by applying
the following series of corrections:

a) we first perform the <u>spin-isospin average</u> as in Eq. (4), including s,p, and
d waves in Eq. (3).

b) we next perform a <u>Fermi average</u> of the forward scattering NN amplitude over the internal motion of nucleons in the projectile and target, both being characterized by the same Fermi momentum k_F. This average assumes the form

$$< \bar{f} > = (16/\rho^2) \sum_{\substack{|\underset{\sim}{k}| < k_F \\ |\underset{\sim}{k}'| < k_F}} \bar{f}(\underset{\sim}{k} + \underset{\sim}{p}, \underset{\sim}{k}'; \underset{\sim}{k} + \underset{\sim}{p}, \underset{\sim}{k}') \qquad (5)$$

in the LAB system. Here $\rho = 2k_F^3/3\pi^2$ is the total density and p is the lab momentum corresponding to the projectile energy per particle.

c) Using $<\bar{f}>$, we now apply an __off-shell__ __correction__ which takes account of the fact that nucleons propagate in a single particle potential in the nucleus, and hence the NN scattering takes place off the energy shell. One may also think of this as a binding correction. It is well established that the effective interaction G depends smoothly on the off-shell energy. Including the off-shell effect via a Taylor expansion about the on-shell point, we obtain a corrected amplitude f_R given by

$$f_R(E) = < \bar{f}(E) > - (<u_1> + <u_2>) \left. \frac{d}{d\omega} < \bar{f}(\omega) > \right|_{\omega = E} \qquad (6)$$

where E is the lab kinetic energy per particle and $<u_1>$ and $<u_2>$ are average potential energies for particles 1 and 2. We take $<u_1> = <u_2> \approx -15$ MeV for surface nucleons. This correction is not very large and, like the Fermi motion average, tends to smooth out the energy dependence of the amplitude.

d) Using $f_R(E)$, we now supply a correction for the __Pauli__ __principle__. Our final approximation for the amplitude $\bar{f}$ is then

$$\bar{f}(E) = f_R(E)/(1 - f_R(E)F(E, k_F)) \qquad (7)$$

where F is complicated function [11] which ensures that the two nucleons always propagate above the Fermi sea in intermediate states. A detailed discussion of the many body and Pauli corrections to the NN amplitude is given in ref. [11].

The results shown in Fig. 4 display the sensitivity of the calculated $\bar{f}$ to the choice of k_F. Typical values characteristic of local densities in the nuclear overlap region are $k_F = 0.45 \text{fm}^{-1}$ and 0.90fm^{-1}. The values of $\bar{f}$ obtained using the free space amplitude of Eq. (3) ($k_F = 0$) are also shown for comparison.

From Fig. 4, we see that the calculated real part of $\bar{f}$ agrees well with the empirical values for $k_F = 0.45 \text{fm}$. No __single__ value of k_F reproduces both the real and imaginary parts of $\bar{f}$. Our calculation does reproduce the general __trends__ of $\bar{f}$ as a function of energy, however. Note that the free space amplitude ($k_F = 0$) is a rather __poor__ approximation to $\bar{f}$, particularly at low energies. The many body effects, in particular the Pauli principle corrections, are crucial in obtaining any sort of reasonable agreement with empirical values of $\bar{f}$. At low energies (below 40 MeV/ particle), there is considerable spread in the theoretical predictions, depending on which k_F is chosen. The small spread in the best fit values, however, and the high quality of the fits suggests that a more careful treatment of the many body effects would be worthwhile. The quality of the fits demonstrates that the folding model, even in its present simple form, is useful in the low energy region. Note that above 40 MeV per particle, the many body corrections become small, and the calculated and empirical values of $\bar{f}$ agree well, in both the real and imaginary parts. Thus our microscopic model should have considerable predictive power for intermediate and high energy heavy ion processes. In this region, there is very close contact between the underlying effective NN amplitudes and the nucleus-nucleus potential.

Before proceeding to discuss a variety of applications, it is worthwhile to dwell for a moment on the simplicity of our approximation to the lowest order potential. The only dependence on the projectile and target (except ξ) is through their ground state mass distributions. The entire energy dependence and the influence of many-body effects are incorporated in $\bar{f}$. While it may be desirable to eliminate some of these approximations later, this simple form yields a <u>single</u> theoretical curve for $\bar{f}(E)$ which expedites a comparison between the empirical and theoretical results for a variety of projectile-target situations. We note in passing that such a comparison is only meaningful when high quality fits to the data are obtained. This requirement demands a careful treatment of the densities and the finite range of the effective NN force.

We now show some examples of transfer reactions, calculated with distorted waves generated by the folded potential which fits elastic scattering. We thank A.J. Baltz for assistance in these computations. We have chosen non-trivial examples [14] which exhibit some forward peaking, in order to provide a proper challenge to the model. Semiclassical bell-shaped angular distributions do not provide a very strong constraint on the theory. In Fig. 5, we show some results for one particle transfer [14]. The folded model provides a good account of the data, both for the ground and excited state transitions. In this and several other cases, the results are comparable in quality to those obtained with Woods-Saxon potentials [14]. The fine details could be improved by weakening the absorption a bit, since this would have much less influence on the elastic than on the transfer data, and would enhance the forward peaking for the transfer reaction.

An example of two neutron transfer [14] is shown in Fig. 6. Here again the folding model provides a reasonable description of the data, if we weaken the absorption somewhat from the elastic best fit value. As we transfer more and more particles, we start to probe the interior of the potential, where the folding model is expected to break down. We see some evidence for this if we compare Figs. 5 and 6; the fit to one particle transfer is considerably more quantitative than that for two neutron transfer. Some of this effect may be due to the limitations of the DWBA.

One can also look for bound states of a folded potential [7,9]. For example,

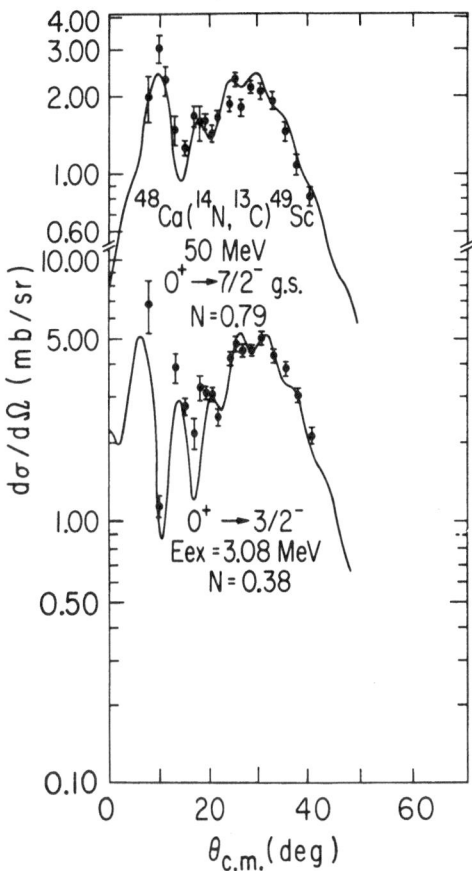

Fig. 5   One particle transfer cross sections [9] obtained using the folded potential
which best fits elastic scattering. Data is from Brookhaven [14]; the DWBA
code was supplied by A. J. Baltz.

bound states or single particle resonances of an α cluster in the potential generated
by an ^{16}O core can be identified with members of the ^{20}Ne ground or excited state rotational bands. The folding model with a single $\bar{f}$ per band yields close to the experimental energy splittings as well as the correct α widths for continuum states and
B(E2) values for transitions within the ground state band. The folding model can
also be used to predict the positions of high spin states populated in three particle
transfer reactions [17]. In many cases, the spins of such levels are not well established, so the folding model performs a useful predictive function. The folding
model approach to cluster states will be treated in detail by B. Buck at this conference, so we omit further discussion here.

Very high energy heavy ion cross sections ($\gtrsim$ 1 GeV/particle) have also been
calculated using an impact parameter representation for the scattering amplitude and
a folding model to generate the potential [8]. The folded potential with a free NN
estimate for $\bar{f}$ should be valid at high energy, since the many-body corrections to $\bar{f}$

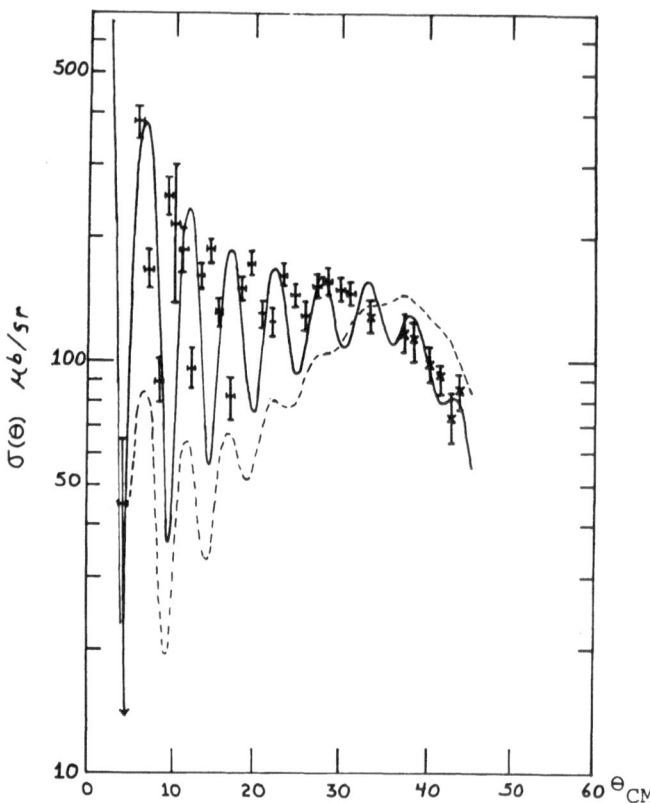

Fig. 6 Two particle transfer cross section(dashed line) for the ^{60}Ni$(^{18}$O$,^{16}$O$)^{62}$Ni
reaction at 62.92 MeV obtained using the best fit folded potential of Fig. 1.
The data is from LeVine et al. [14]. Numerical densities obtained from a
single particle model [13] were used. Similar calculations using electron
scattering densities are shown in ref. [9]. The solid line shows the effect
of decreasing the absorption by 50%; a reasonable fit to the elastic data is
maintained.

are small. The folding model yields a set of predictions for high energy cross

sections and factorization ratios based on methods of standard nuclear theory. We

thus obtain a useful set of predictions to be compared with those arising from more

sophisticated particle physics concepts. These results have been reviewed in detail

recently [10].

In summary, the folding model for nucleus-nucleus interactions correlates a

wide range of data within a single framework capable of being extrapolated to make

predictions in new situations. The model is still partly phenomenological in the

low energy region ($\lesssim$ 40 MeV/particle), since the strength of the potential is not

reliably given by our crude form of many body theory, and must be adjusted empirically

in order to fit data. However, the model gives a good account of one and two particle

transfer data as well as elastic scattering. It is expected to work well for any

peripheral process, since such processes are insensitive to the potential in the

nuclear interior. For intermediate and high energy reactions, the folding model provides an optical potential with no free parameters, assuming the nuclear densities ρ are given; the data may then place constraints on the behavior of ρ in the tail region.

We feel that there is some hope for a basic theory in which one tries to utilize one's knowledge of NN interactions and the geometrical properties of nuclei to derive a potential for nucleus-nucleus scattering. In this way, one tries to achieve a consistent description of as wide a class of phenomena as possible. In any case, the folding model may indicate a more economical way of parametrizing the potential; phenomenology is best tempered with some guidance from theory.

A number of other applications of the method are under study: a) α and heavy ion inelastic scattering [18]; b) use of the folded potential as an input into a coupled channel formalism; c) systematic predictions of the energies of high spin states populated in three and four particle heavy ion transfer reactions; d) calculation of multiplicities of particles produced in very high energy nucleus-nucleus collisions [10]; e) calculation of fusion barriers. We are also examining a number of refinements to the model, such as a more careful treatment of the density dependence of the effective interaction G. In general this will lead to somewhat different geometries for the real and imaginary parts of the optical potential. We may also parametrize the interior of the potential separately, and join on to the folded potential in the surface region; this should enable us to improve the description of the transfer data, while retaining a high quality fit to elastic scattering.

REFERENCES

1. A.K. Kerman, H. McManus and R.M. Thaler, Ann. Phys. (N.Y.) $\underline{8}$ (1959) 551.
2. J. Hüfner and C. Mahaux, Ann. Phys. (N.Y.) $\underline{73}$ (1972) 525.
3. G.W. Greenlees, G.H. Pyle and Y.C. Tang, Phys. Rev. $\underline{171}$ (1968) 1115.
4. T.T. Chou and C.N. Yang, Phys. Rev. Letters $\underline{20}$ (1968) 1213; Phys. Rev. $\underline{170}$ (1968) 1591.
5. B. Tatischeff and I. Brissaud, Nucl. Phys. $\underline{A155}$ (1970) 89; A.M. Bernstein and W.A. Seidler, Phys. Letters $\underline{34B}$ (1971) 569; P. Mailandt, J.S. Lilley and G.W. Greenlees, Phys. Rev. Letters $\underline{28}$ (1972) 1075; Phys. Rev. C $\underline{8}$ (1973) 2189; R.S. MackIntosh, Nucl. Phys. $\underline{A210}$ (1973) 245; L. West, S. Cotanch and D. Robson in Proceedings of Munich Conference (1973), Vol. I, p. 383.
6. G.R. Satchler, Invited talk at International Conference on Reactions Between Complex Nuclei, Nashville, Tennessee June 1974.
7. B. Buck, C.B. Dover and J.P. Vary (to be published).
8. S. Barshay, C.B. Dover and J.P. Vary, Phys. Letters $\underline{51B}$ (1974), 5 and BNL preprint 18972, May 1974.
9. J.P. Vary and C.B. Dover, Phys. Rev. Letters $\underline{31}$ (1973) 1510.
10. J.P. Vary and C.B. Dover, invited paper for the Second High Energy Heavy Ion Summer Study, Lawrence Berkeley Lab (July 15-26, 1974).
11. C.B. Dover and J.P. Vary (to be published).
12. J. Negele, private communication.
13. D.J. Millener and P.E. Hodgson, Nucl. Phys. $\underline{A209}$ (1973) 59.
14. E.H. Auerbach et al., Phys. Rev. Letters $\underline{30}$ (1973) 1078; M.J. Schneider et al., Phys. Rev. Letters $\underline{31}$ (1973) 320; C. Chasman, S. Kahana and M. Schneider, Phys. Rev. Letters $\underline{31}$ (1973) 1074; P. Bond et al., Phys. Letters $\underline{47B}$ (1973) 231; M.J. Levine et al., to appear in Phys. Rev. Comments.
15. W.T.H. van Oers and J.M. Cameron, Phys. Rev. $\underline{184}$ (1969) 1061.
16. D.A. Goldberg and S.M. Smith, Phys. Rev. Letters $\underline{29}$ (1972) 500; D.A. Goldberg, S.M. Smith and G.F. Burdzik, University of Maryland Technical Report No. 74-107, June (1974) and private communication.
17. K. Nagatani et al., Phys. Rev. Letters $\underline{31}$ (1973) 250.
18. C.B. Dover, P. Moffa and J.P. Vary (in preparation).

SEMI-MICROSCOPIC APPROACH TO THE IMAGINARY
PART OF THE HEAVY ION OPTICAL POTENTIAL

Christian Toepffer
Department of Physics
University of the Witwatersrand
Johannesburg, South Africa

The absorption in heavy ion scattering is treated by considering the Pauli-forbidden regions of phase space for the scattering of bound nucleons. An energy dependent imaginary potential is derived which is both surface transparent and strongly absorptive in the interior region. The model is sufficient to explain the main features of the ^{16}O - ^{16}O excitation functions up to 80 MeV.

1. Introduction

In heavy ion reactions there are usually many channels open and the original structure of the colliding nuclei may be completely destroyed. A strong coupling allows for large amounts of matter, energy, linear and angular momentum to be transferred from the relative motion, a collective degree of freedom, to the intrinsic degrees of freedom. This leads to a depletion of probability in the elastic channel which can be described by introducing an absorptive imaginary part into the nucleus-nucleus potential. The very qualitative nature of these considerations should be emphasized, a microscopic, self-consistent theory of heavy ion reactions does not yet exist.

The data have been usually fitted by ad-hoc potentials of Woods-Saxon type, the radius and diffuseness parameters were found to be in general agreement with nuclear parameters obtained from other experiments. The depth of the absorptive potential had to be linearly dependent on the energy at least[1].

When it was discovered that the ^{18}O - ^{18}O system has, in marked contrast to the ^{16}O - ^{16}O system, no prominent gross structure oscillations in its excitation function[2], two nuclear structure arguments were brought forward as a basis for an angular momentum dependent absorptive potential:

1) The amount of angular momentum which can be carried away by the reaction channels at a given energy may vary from system to system[3].

2) The density of the (pre-) compound states at a given energy depends on the

angular momentum[4].

Recently the angular distribution of the ^{12}C - ^{20}Ne system has been studied. It does not show the gross structure oscillations observed in ^{16}O - ^{16}O scattering. Similar values of angular momentum are brought into the same compound nucleus in both cases. It is therefore argued that entrance channel effects are responsible for the presence or absence of gross structure in these systems[5].

Thus in order to explain certain anomalies in heavy ion scattering the ad-hoc potentials are modified by taking into account specific details of the nuclear structure, which are supposed to be important. Such a procedure is rather arbitrary. By introducing a ℓ-dependent absorption the ^{16}O - ^{16}O excitation function could be fairly well reproduced up to the maximal energy E_{cm} = 35 MeV which was available at the time[3]. However, this potential fails completely at higher energies, where recently experiments have been performed by the Oak Ridge group[6].

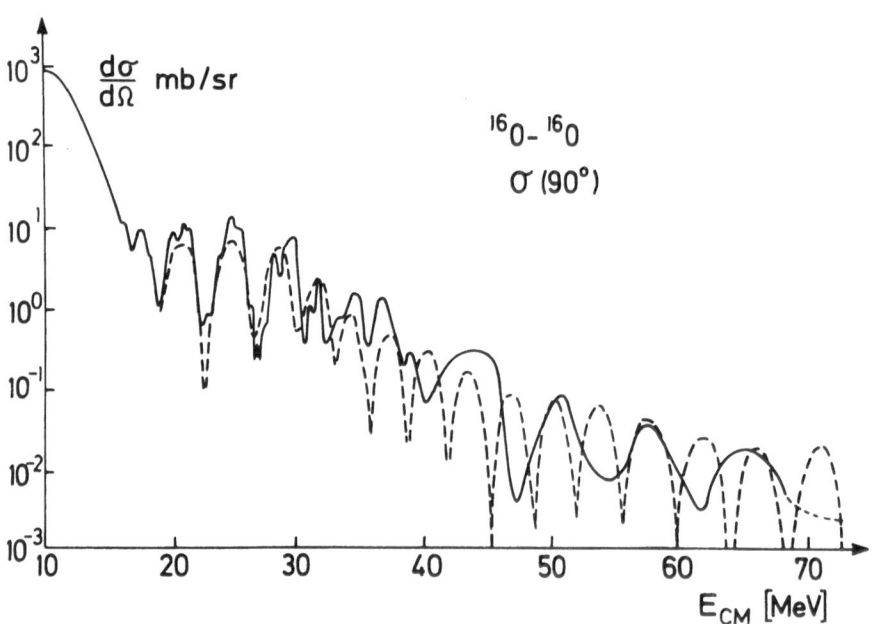

'Fig. 1. In the high energy region the excitation function calculated with a ℓ-dependent absorption[3] (dashed line) oscillates too fast in comparison with the experimental curve[6].

Of course one should not conclude from this that angular momentum matching, the

compound state density and entrance channel effects are unimportant for elastic heavy
ion scattering. What is required, however, is a derivation of optical potentials from
very general principles incorporating common properties of all nuclei and then to
superimpose the more specific features.

2. The dynamic absorption model

A suitable starting point is the "frivolous scattering model" [7] of Goldberger
which successfully describes the absorption of free nucleons in nuclei [8]. It can be
extended to describe the absorption in the collision of two complex nuclei [9].

Using quasi-classical methods the imaginary potential W can be related to the
absorption coefficient κ

$$W = h^2\kappa/(2\mu) \sqrt{\kappa^2/4 + 2\mu(E_{cm}-V)/h^2} \qquad (1)$$

where E_{cm} is the centre-of-mass energy, V the real potential and μ the reduced mass
of the nuclei. In order to calculate κ or its inverse, the mean free path λ, one
considers the colliding nuclei in phase space where they occupy limited six-dimensional
regions separated by the distance vector R and the relative momentum K per nucleon.
Fermi-type mass distributions are assumed and the intrinsic momentum distributions
are idealized by Fermi spheres the radii of which are determined by the compound mass
density $\rho(\underset{\sim}{r}, R)$

$$k^2_F(\underset{\sim}{r}, R) = (3/2\pi^2\rho)^{2/3} + 5/12 \; \xi \; (\nabla\rho/\rho)^2 \qquad (2)$$

Here the usual inhomogeneity correction with $\xi = 1/9$ is added [10].

When the colliding nuclei start to overlap elementary collisions between the nucleons lead to the creation of 2p - 2h states in momentum space. These states describe
inelastic excitations as well as transfer processes and are therefore the doorway states
for all nonelastic channels. In particular, thermalization leads to compound states.
The absorption is therefore determined by the formation cross section $\bar{\sigma}$ of these states.
If one folds the formation cross section $\bar{\sigma}(\underset{\sim}{r}, R, K)$ with the densities of the scattering
nuclei over all possible positions $\underset{\sim}{r}$ of elementary collisions, i.e. the overlap volume,
one obtains κ as a function of K and R

$$\kappa(R, K) = \int d^3r \; \rho_1(\underset{\sim}{r}) \; \rho_2(R - \underset{\sim}{r}) \; \bar{\sigma}(\underset{\sim}{r}, R, K) \qquad (3)$$

$\bar{\sigma}$ can be calculated by averaging the cross section for elementary collisions of bound
nucleons $\sigma_b(k)$ over the intrinsic momentum distributions at the position $\underset{\sim}{r}$ of the

scattering event (see eq. (2)).

$$\bar{\sigma}(\underset{\sim}{r}, \underset{\sim}{R}, \underset{\sim}{K}) = (V_F(\underset{\sim}{r}, \underset{\sim}{R}))^{-2} \int\limits_{F_1(\underset{\sim}{r},R)} d^3k_1 \int\limits_{F_2(\underset{\sim}{r},R)} d^3k_2 \; 2k/K \; \sigma_b(k) \qquad (4)$$

Here V_F is the volume of the Fermispheres F_1 and F_2 which are separated by a momentum $\underset{\sim}{K}$ per nucleon and $\underset{\sim}{k} = \frac{1}{2}(\underset{\sim}{k}_1 + \underset{\sim}{K} - \underset{\sim}{k}_2)$ is the relative momentum of the scattering nucleons. By the factor $2k/K$ care is taken of the relative motion of the scattering centres.

The cross section $\sigma_b(k)$ for the scattering between nucleons of target and projectile systems is smaller than the free nucleon-nucleon cross section. It can be calculated by integrating the free nucleon-nucleon cross section $\sigma_f (\underset{\sim}{k}, \underset{\sim}{k}')$ over all those directions of the final relative momentum $\underset{\sim}{k}'$ which lead to states allowed by the Pauli principle

$$\sigma_b(k) = \int\limits_{\Omega p} \sigma_f (\underset{\sim}{k}, \underset{\sim}{k}') \; d\Omega_{\underset{\sim}{k}'} \qquad (5)$$

Ωp is the region of solid angle allowed by the Pauli principle. It is obtained by a geometrical construction. Because of energy and momentum conservation the endpoint of k' has to fall on a sphere R with radius k which cuts two spherical cones out of F_1 and F_2. The states on the intersection of these cones are already occupied in both nuclei. Therefore scattering is forbidden into this intersection and, because the colliding nucleons cannot be distinguished, into the region which one obtains by reflection at the origin of R. (Fig.2)

3. Results and Conclusions

Collecting formulae (1) to (5) and averaging over protons and neutrons the imaginary potential can be derived from the free nucleon-nucleon cross sections and the density distributions of the colliding nuclei. The model has been applied to the $^{12}C-^{12}C$(see ref.9) and $^{16}O-^{16}O$ excitation functions (Fig.3). The average slope and the peak to valley ratios of the gross structure oscillations of the experimental excitation function are well reproduced. This indicates that both the energy and the radial dependence of the imaginary potential are reasonably well derived in this model.

The physical origin of the energy dependence of the absorption is obvious: If K increases, the Pauli-forbidden overlap region between F_1 and F_2 shrinks. Therefore more phase space will be available after the elementary collisions and the absorption becomes stronger.

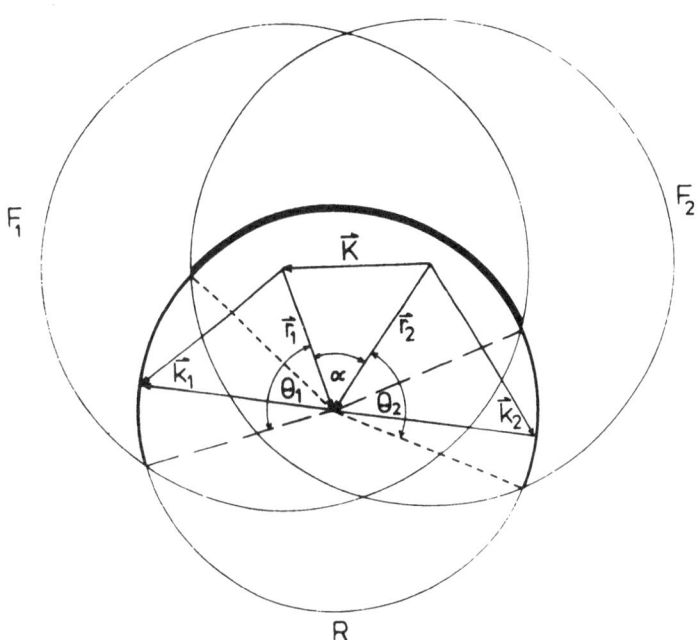

Fig. 2.   The Fermi spheres F_1 and F_2 are separated by the momentum K per nucleon and
intersected by the auxiliary sphere R. θ_1 and θ_2 are the~angles of the
spherical cones cut out of F_1 and F_2.   The Pauli forbidden region is marked
in heavy black on the circumference of R.

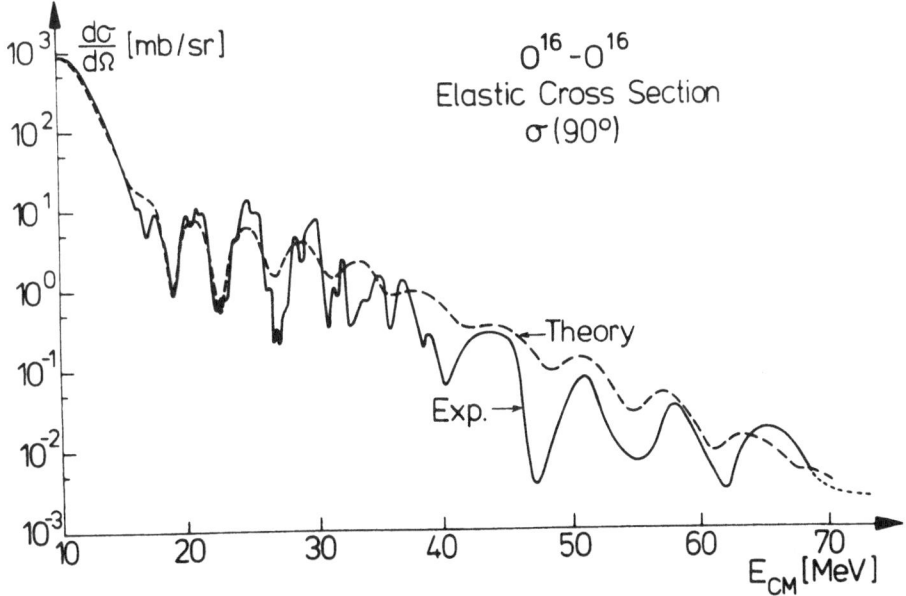

Fig. 3.   Full line: Experimental excitation function [1,6];
Dashed line: Theoretical excitation function according to the dynamic
absorption model discussed here.   Compare with Fig. 1.

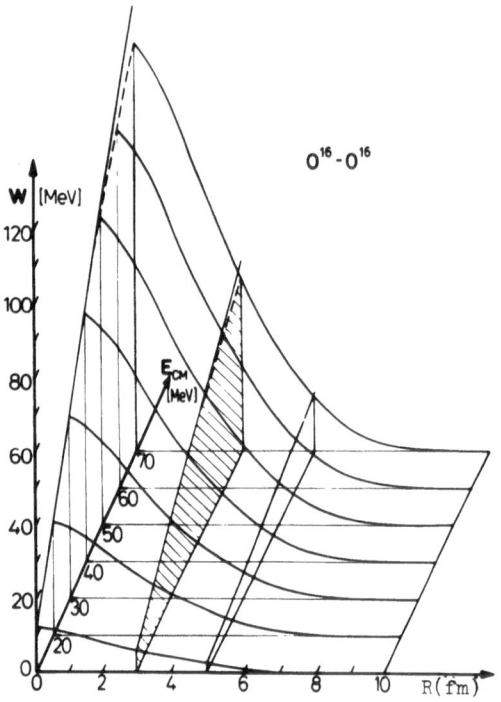

Fig. 4.   The absorptive potential W (R, E_{cm}) for a ^{16}O - ^{16}O collision.   The energy
dependence is almost linear in this energy region, although with a different
slope at different distances.

The radial dependence of the imaginary potential is mainly determined by the
overlap density in eq. (3).   W turns out to be both surface transparent and strongly
absorbing in the interior region.   Such absorptive potentials favour the formation
of quasi-molecular states, indeed a very similar type of potential has been used in
the Regge-pole analysis of ^{16}O - ^{16}O scattering [11].   (Fig. 5.)

The form of the overlap density depends on the radii and diffusivities of the individual density distributions of the colliding nuclei.Both the ^{18}O - ^{18}O and the
^{12}C - ^{20}Ne systems have larger (aggregated) radii and are more diffuse than the
^{16}O - ^{16}O system.   This can be translated into angular momentum space using quasi-
classical arguments.   The diffraction model [13] then predicts a stronger absorption
and damping of the gross structure as has been observed in these systems.

This,of course, does not presuppose that other arguments based on the individual
structure of the colliding nuclei are not important too.   The dynamic absorption
model presented here incorporates some very general aspects of heavy ion scattering
and may thus be a useful starting point for the more detailed considerations.

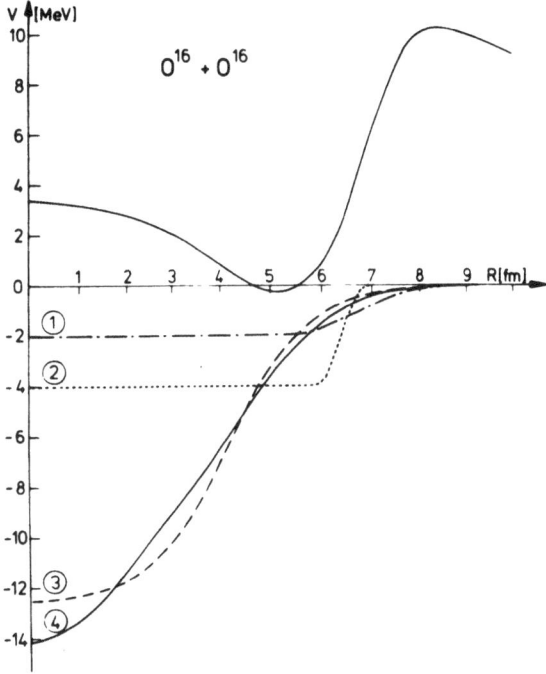

Fig. 5.   Comparison of the absorptive potential W_4 discussed here with the ad-hoc
potentials W_1 [1)] and W_2 [12)] and the potential W_3 of McVoy [11)].   The upper
part shows the real potential V [1)].

On the other hand some shortcomings should be noticed, which are typical for the
present models in heavy ion physics:

1)   Only in the elementary collisions the nucleons are properly treated as
     indistinguishable particles.   The use of the distance and momentum coordin-
     ates R and K presupposes, however, that the nuclei retain some of their identity
     also in the overlap region.   This, again, may be offset by the fact that the
     compound density is used for determining the local Fermi momentum $k_F(r, R)$
     in eq. (2).

2)   The treatment is not self-consistent.   The real potential and the prescrip-
     tion how to calculate the compound density have to be taken from elsewhere.
     Fortunately the latter point does not seem to be very crucial as far as the
     absorption is concerned.   If, for example, one assumes that a uniform
     compression takes place during the collision, then the overlap density
     as well as the Fermi momentum k_F and with it the Pauli-forbidden region
     (Fig. 2) become larger than in the adiabatic case.   Both effects tend to
     cancel each other.

4. Transport equation

To derive the dynamic absorption model the significant considerations are made in momentum space. One would like to have a theory which gives both the spatial and momentum probability distributions as functions of time. Boltzmann's transport equation which can be derived from the usual equation of motion for Green's functions serves this purpose. The elementary collisions between the nucleons are described by the collision integral in this equation. Introducing Laudau's quasi-particle model the transport properties of nuclear matter can easily be derived [14]. One finds, for example, $\eta = 1.95 \cdot 10^{-18}$ (kT)$^{-2}$ MeV sec/fm^3 and $\tau = 1.04 \cdot 10^{-18}$ (kT)$^{-2}$ sec for the viscosity and the relaxation time as functions of the nuclear temperature T, respectively. A comparison with the single particle time $\tau_{sp} = 10^{-22}$ sec shows that in actual nuclei finite size effects will be dominant.

Although it is a classical picture, the description of heavy ion reactions (and fission) in terms of transport equations may be much more fruitful than the use of optical potentials. Instead of calculating the loss of probability in selected channels one obtains positive answers on the amounts of energy and momentum which are transferred to the intrinsic degrees of freedom.

The dynamic absorption model was developed together with B. Fink. The excitation functions have been calculated by T. Morovič using an optical model code of H.J. Fink.

References

1) J.V. Maher, M.W. Sachs, R.H. Siemssen, A. Weidinger and D.A. Bromley, Phys. Rev. 188 (1969) 1665.

2) R.W. Shaw, R. Vandenbosch and M.K. Mehta, Phys. Rev. Lett. 25 (1970) 457.

3) R.A. Chatwin, J.S. Eck, D. Robson and A. Richter, Phys. Rev. C1 (1970) 795.

4) G. Helling, W. Scheid and W. Greiner, Phys. Lett. 36B (1971) 64, and H.J. Fink, W. Scheid and W. Greiner, Nucl. Phys. A188 (1972) 259.

5) R. Vandenbosch, M.P. Webb and M.S. Zisman, Proc. Int. Conf. on Reactions between Complex Nuclei, Nashville, June 10-14, 1974, Vol. 1, p. 9, North-Holland, Amsterdam (1974).

6) M.L. Halbert, C.B. Fulmer, S. Raman, M.J. Saltmarsh, A.H. Snell and P.H. Stelson, Bull. Am. Phys. Soc. 18 (1973) 1387 and Phys. Lett. 51B (1974) 341

7) D.A. Bromley, Proc. Int. Conf. on Nuclear Reactions Induced by Heavy Ions, Heidelberg 1969, p. 27ff, North-Holland, Amsterdam (1970).

8) M.L. Goldberger, Phys. Rev. 74 (1948) 1269.

9) B. Fink and C. Toepffer, Phys. Lett. <u>45B</u> (1973) 411.

10) K.A. Brueckner, J.R. Buchler, S. Jorna and R.L. Lombard, Phys. Rev. <u>171</u> (1968) 171.

11) K.W. McVoy, Phys. Rev. <u>C3</u> (1971) 1104.

12) R.H. Siemssen, J.V. Maher, A. Weidinger and D.A. Bromley, Phys. Rev. Lett. <u>20</u> (1968) 175, A. Gobbi, Proc. Symp. on Heavy Ion Scattering, Argonne Nat. Lab. ANL - 7837 (1971), p.63.

13) W.E. Frahn and R.H. Venter, Ann. of Phys. <u>24</u> (1963) 243.

14) C. Toepffer, 12th Int. Winter Meeting on Nucl. Phys., Villars, Switzerland (Jan. 1974) and Proc. Int. Conf. on Reactions between Complex Nuclei, Nashville, June 1974, Vol. 1 p. 129, North-Holland, Amsterdam (1974), G. Wegmann, Phys. Lett. <u>50B</u> (1974) 327.

FUSION BARRIERS

Hans J. Krappe

Hahn-Meitner-Institut für Kernforschung Berlin
Bereich Kern- und Strahlenphysik
Berlin-West, Germany

Abstract
Various models for the calculation of heavy-ion interaction barriers
in the sudden approximation are compared. Their relation to experimental reaction thresholds and optical potentials is discussed. The relation between interaction, fusion, and fission barriers is explained
within the frame of a multidimensional model.

1. Introduction

The theory of heavy-ion fusion reactions has been approached in the
general frame of fission theory or by means of the general formalism
of reaction theory. A key element in the theory of fission is the potential-energy surface and in particular the fission barrier. Reaction
theory on the other hand is built on such concepts as optical potentials, transmission coefficients and transfer form-factors. An important role plays the interaction barrier i.e. the maximum of the real
part of the ion-ion interaction potential. The relation between these
barriers and their bearing on experimentally observed threshold energies for heavy-ion fusion reactions shall be discussed in the following.

2. Phenomenological Optical Potentials

In contrast to fission theory in the conventional formalism of reaction theory each partial wave is usually treated as a one-dimensional
problem. The price for this simplification of the original many-body
problem is well known. If one wants to have a smooth energy dependence of the potential to allow a simple parametrisation it becomes non-
hermitian and allows at best the derivation of energy-averaged cross
sections.

    Generally this optical potential is then parametrised in a way
supposed to be reasonable and the parameters are fitted to elastic

scattering data. From the very construction of this potential it is clear, that its real part does not have an independent physical meaning irrespective of what the shape and strength of the imaginary potential is. The same is of course true for the semiclassical[1-3] and classical versions of the theory[4,5]. In the latter case it has been shown that the friction coefficient plays the role of the imaginary part of the optical potential[6,7]. Again a fit of experimental data defines friction and potential together as a unity and not as independent physical quantities.

3. One-dimensional Potential Models

With few exceptions[8-11] heavy-ion interaction potentials have only been calculated in one-dimensional models. The distance between the centers of mass of the two ions is used as the dynamical variable. Clearly this variable loses its meaning where the density distributions strongly overlap. In order to continue the one-dimensional description into this region often a linear sequence of shapes is more or less arbitrarily specified along which the system is supposed to move towards a fused compound state. Because of this arbitrariness the calculated potentials are of limited usefulness in this range of short distances.

Various proposed models will be discussed in the following roughly in the order of decreasing sophistication and increasing numerical expedience.

3.1. Models based on an effective two-particle interaction

Only very recently constrained truely self-consistent Hartree-Fock calculations have been used to evaluate the adiabatic interaction potentials[12,31] as well as potentials for hot nuclei[12]. Only fairly light systems of equal ions (^{4}He and ^{16}O) have been investigated so far.

On the basis of the Thomas-Fermi approximation in the form developped by Brueckner et al.[13] Ngô et al.[14] recently calculated interaction potentials and barrier heights. They use density distributions for the separate ions derived from Hartree-Fock calculations in conjunction with the "frozen density approximation", sometimes called the "sudden" approximation. Of course in experiments which yield information on the barrier height the energy will be fairly close to the barrier where the sudden approximation does not seem to be realistic. Therefore the potentials may only be reliable outside the strong overlap region. The same restriction applies to constrained Hartree-Fock calculations because of a certain arbitrariness in the choice of the constraining operator in the overlap region.

Effective two-particle interactions generally used in these calculations are not specifically designed and should not be used for nuclear densities much lower than the average nuclear saturation density. That has been shown in particular for the Skyrme force[32]. One may therefore doubt the reliability of the nuclear interaction potential in the tail region. Recently effective two-particle potentials have been derived which should be appropriate in the low-density region[15]. But still there remain some doubts about the applicability of the local density approximation in the nuclear surface where the de Broglie length of the nucleons is larger than the diameter of the interaction area.

3.2. Simple folding models

In order to relate the real part of the nucleon-nucleus optical potential to the real part of the α-nucleus potential it has long been common practise simply to fold the α density-distribution into the real nucleon-nucleus potential[16,17].

The same prescription has more recently been used to generate a real heavy-ion interaction potential[4,17-20]. Again in the overlap region serious errors are expected because of the total neglect of antisymmetrisation between target and projectile nucleons and the assumption of "frozen" densities and potentials.

A subtle problem is the use of the free nucleon-nucleus scattering-potential to describe the scattering of bound nucleons. Their binding energy of 6-10 MeV is usually much larger than the kinetic energy of relative motion per particle at the barrier. Lacking any accurate knowledge of the off-energy shell behaviour of the optical nucleon-nucleus potential one may estimate the uncertainty by the decrease in the potential strength by a 10 MeV increase of the scattering energy on shell. Taking the energy dependence of the optical potential from reference[33] one gets about 5 % in the overall strength.

In order to determine the tail of the interaction potential one has to know the tail of the mass distribution and of the optical potential. In contrast to the charge distribution these two functions are not too well known in the tail region. At best the strength, the radius and the surface thickness are determined unambiguously. Therefore at best the same parameters of the folded distribution can be calculated[20]. The far tail end of the interaction potential is consequently not very reliably determined. The same remark applies to the models to be discussed in the next paragraph.

3.3 Theories based on the liquid-drop model

The liquid-drop model itself as well as its droplet-model generalisation[34] does not yield any nuclear interaction between two nuclei as long as they do not overlap with their equivalent sharp surfaces. Furthermore the surface energy for strongly necked-in configurations is overestimated. These deficiencies originate in the asymptotic expansion underlying the liquid-drop model. Here the surface thickness and the range of the forces are supposed to be very small compared to any geometrical parameters of the configuration under consideration. This expansion breaks down close to the scission line.

Two fairly similar models have been proposed to overcome these limitations of the liquid-drop model without destroying its numerical simplicity and flexibility[28,29]. In ref.[29] the following ansatz is made for the nuclear contribution to the deformation energy of a non-spherical drop

$$E_{def} = -\frac{a_s(1-\kappa_s I^2)}{8\pi^2 a^4 r_o^2}\left[\int_{deformed} d^3r\, d^3r'\, \frac{\exp(\frac{-|r-r'|}{a})}{\frac{|r-r'|}{a}} - \int_{sphere} d^3r\, d^3r'\, \frac{\exp(\frac{-|r-r'|}{a})}{\frac{|r-r'|}{a}}\right] \quad (1)$$

where a_s is the surface energy constant, κ_s the surface symmetry energy constant, $I = (N-Z)/(N+Z)$ the neutron excess, and a is a constant of the dimension of a length which has to be added to the usual liquid-drop parameters to account for the finite range of the nuclear force and the finite surface thickness. The first integral is to be taken in both variables over the volume of the equivalent sharp-surface drop, the second over the volume of the equivalent sharp sphere with the radius parameter r_o. The limit $a \to o$ yields the usual liquid-drop model.

The ansatz has been shown to have the following properties[29]. Choosing the parameters of the model appropriately one gets essentially the same potential energy surface as in the usual liquid-drop model for those parts of the deformation-parameter space which are explored in nuclear fission up to the saddle point. At the scission line the derivatives of the potential are no longer discontinuous. Instead there appears an attractive nuclear potential between two non overlapping spherical nuclei of the form

$$V_{int}(r) = -4\left(\frac{a}{r_o}\right)^2 a_s(1-\kappa_s I^2)\, g\left(\frac{R_1}{a}\right) \cdot g\left(\frac{R_2}{a}\right) e^{-r/a}\left(\frac{r}{a}\right)^{-1} \quad (2)$$

with

$$g(x) = x \cosh x - \sinh x.$$

R_1, R_2 and r are the equivalent sharp radii of the two ions and the distance between their centers, respectively. The length parameter a of the model is seen to determine the range of the nuclear proximity force. As the model does not contain further length parameters besides a and r_0 only the two lowest moments of the interaction-potential function are determined.

Assuming r_0 to be derived from electron scattering and suitable assumptions about the neutron skin[20] the remaining three parameters a_s, κ_s, and a can in principle be determined from a fit of ground-state binding energies and fission-barrier heights. No such fit has been made so far. Instead experimental interaction-barrier heights have been used besides fission barriers to determine the following parameter set

$$a_s = 24.7 \text{ MeV}, \quad \kappa_s = 4, \quad a = 1.4 \text{ fm}, \quad r_0 = 1.16 \text{ fm} \tag{3}$$

Only the combination $a_s(1-\kappa_s I^2)$ for the neutron excess of heavy nuclei has been determined accurately rather than a_s and κ_s independently. But comparison with earlier fits of the surface energy constant to ground state masses[35] shows that the parameter set (3) may not have to be revised substantially after a more careful fit.

In contrast to Greiner's model[28] ansatz (1) does not allow for compression degrees of freedom and does not contain shell corrections to the smooth energy surface. But it is nevertheless assumed to be a reasonably simple extension of the liquid-drop model into and beyond the scission region.

Even simpler extensions of the liquid-drop model have been proposed in order to derive a nuclear proximity force. A method containing far more arbitrariness has been used by Bondorf et al.[5]. They simply interpolate between the liquid-drop energy of two separate spherical nuclei at infinity and that of the fused spherical compound nucleus by means of a Woods-Saxon function. The parameter set proposed is

$$V_0 = 17 \left[A_1^{2/3} + A_2^{2/3} - (A_1 + A_2)^{2/3} \right] \text{ MeV}$$
$$r_0 = 1.3 \text{ fm} \tag{4}$$
$$a = 0.9 \text{ fm.}$$

By simply guessing the shape of the fusion-barrier configuration Kalinkin and Petkov derived a barrier height from the liquid-drop model[22]. At least for lighter systems the shape at the barrier is certainly much less compact than a spheroid assumed to be the fusion-barrier configuration by these authors.

Considerably more physical information is incorporated in models which are based on a general theorem which relates the strength F of the attractive force between two touching spherical drops with radii R_1 and R_2 to their surface-tension constant γ:

$$F = 4\pi\gamma \frac{R_1 R_2}{R_1 + R_2}. \tag{5}$$

The theorem has been used for quite a while in the theory of coargulation in aerosol physics[23-24] and was recently rediscovered in nuclear physics independently by Swiatecki[25], Wilczyński[26], and Bass[27]. In order to get an explicit expression for the proximity force some short-range functional form has to be assumed which involves a strength, a range, and a radius parameter and is most conveniently taken to be an exponential

$$V(r) = V_o \exp\left(\frac{R_1 + R_2 - r}{a}\right) \tag{6}$$

where $R_1 = r_o A_1^{1/3}$, $R_2 = r_o A_2^{1/3}$ with some suitable radius parameter r_o. The strength V_o can be determined from eq. (5) for $r = R_1 + R_2$ and the range parameter a remains to be fitted to experimental interaction-barrier heights. Clearly (6) is only valid for $r > R_1 + R_2$.

In the derivation of equation (5) the assumption of "frozen" density distributions is made. Moreover it represents only the first term of an expansion in powers of a/R. For macroscopic droplets the zeroth order term is sufficient for all practical purposes. But in nuclear physics considerable errors are introduced by a restriction to this term. The interaction potential (2) being properly related to the liquid-drop model of course yields eq. (5) in the limit $a/R \rightarrow 0$. It also shows that a light system like $^{16}O + ^{16}O$ has a contact force which is smaller by a factor of more than 3 than the estimate (5). Bass has partly corrected for that deficiency by chosing an extremely small radius parameter ($r_o = 1.07$ fm). Thus his potential is shifted towards smaller r-values compared to the potential (2). He also uses the somewhat smaller surface energy constant of ref.[36] ($a_s = 17.94$, $\kappa_s = 1.78$). The range parameter used by Bass, a = 1.35 fm is very close to the parameter set (3).

Instead of the exponential function (6) Wilczyński uses a decay function related to the square of the density[26]

$$\frac{dV}{dr} = c \frac{(R_1 - R_2 + r)(R_2 - R_1 + r)}{r} \left(1 + \exp\frac{r - R_1 - R_2}{a}\right)^{-2} \tag{7}$$

with

$$R_1 = 1.11 \ A_1^{1/3} \ fm; \ R_2 = 1.11 \ A_2^{1/3} fm; \ a = 0.54 \ fm,$$

$$c = -6.22 \ \{2-1.78(I_1^2+I_2^2)\} \ MeV.$$

The function (7) has a considerably shorter range than either Bass' or Krappe and Nix's function (2). Though Wilczyński makes use of equation (7) at the contact point $r = R_1+R_2$ only, one may somewhat unfairly integrate the equation to get a proximity potential for the purpose of comparison with other models.

4. Quantitative Comparison of Calculated and Experimental Interaction-Barrier Heights

The largest contribution to the interaction barrier comes from the Coulomb field about which there is little disagreement. It is the deviation of the interaction from the Coulomb potential which shows the nuclear proximity force. In figs. 1-3 the trivial Coulomb potential at the contact point $z_1z_2e^2/1.16(A_1^{1/3}+A_2^{1/3})$ has therefore been subtracted from calculated and experimental interaction-barrier heights. Plotted is this difference against the target mass for nuclei along the line of β-stability and for three frequently used projectiles. The curve labelled Wilczyński in fig. 2 is calculated from a nuclear potential, which follows from integrating eq. (7). Because of the short range of the force the resulting attractive potential is very weak, the barriers are correspondingly high. The curves labelled Bass and Krappe et al. are derived from the potentials (6) and (2) respectively. For the heaviest systems (Krypton on targets with A > 90 and A > 210 for (6) and (2) respectively) there is no maximum in the range of $r \geq 1.16 \ (A_1^{1/3}+A_2^{1/3})$, where the "frozen" model can at best be believed. The interaction barriers agree in the two models within 3 MeV.

The effect of the deformation of actinide and rare earth nuclei is taken into account in the extended liquid-drop model in a way described in ref.[29]. It is seen to lower the barriers substantially.

Barriers derived from a Woods-Saxon potential with parameters (4) are shown by the lowest curve in fig. 2 and seem to correspond to a nuclear attraction which is too strong all over the periodic table. From a couple of barrier heights given in ref.[14] the curve labelled Ngô et al. has been constructed as an example of a Thomas-Fermi calculation. It agrees quite well with the extended liquid-drop model (2).

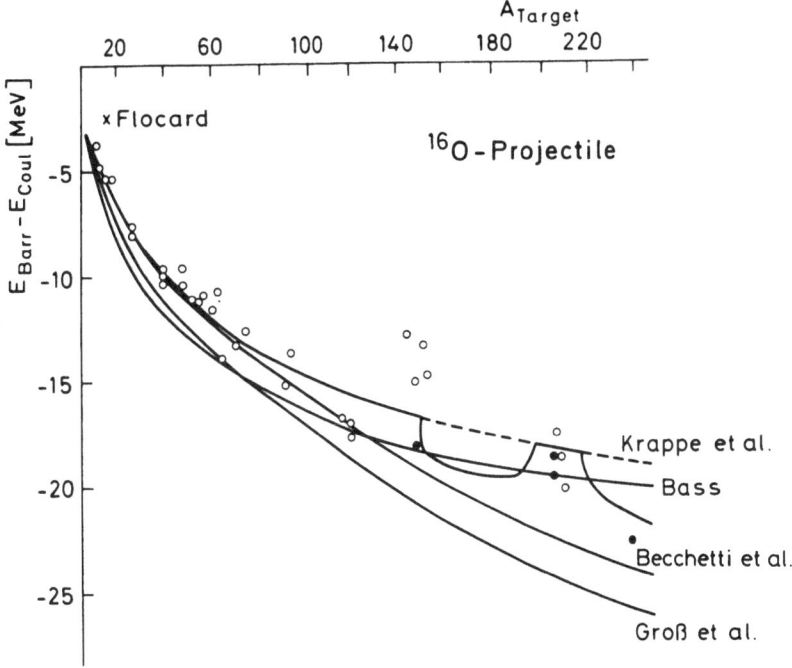

Fig. 1: Calculated interaction-barrier heights of oxygen projectiles
on various targets (full lines). Barriers derived from optical potential fits of elastic scattering data (open circles).
Reaction barriers derived from a barrier-penetration analysis
of reaction excitation functions (full dots). The trivial Coulomb potential $z_1z_2e^2/(R_1+R_2)$ is always subtracted.

As an example of the folding typ potentials I will take the one
used by Gross et al.[4]. The density distribution and the optical potential are assumed to have Woods-Saxon shapes with parameters

$$V_0 = 50 \text{ MeV} \quad , \quad a_v = 0.65 \text{ fm} \quad R_v = 1.25 \, A_2^{1/3} \text{ fm}$$

$$a_\rho = 0.54 \text{ fm} \quad , \quad R_\rho = (1.12 \, A_1^{1/3} - 0.86 \, A_1^{-1/3}) \text{ fm}.$$

The curves in figs. 1-3 were actually calculated by means of a simple
approximation to the folding integral[21]

$$V_{Nucl}(r) = \left[\sum_{i=0}^{2} \alpha_i (r-R_{12})^i\right]\left[\frac{r-R_{12}}{a} - \ln(1+\exp\{\frac{r-R_{12}}{a}\})\right] \quad r > R_{12}$$

with

$$R_{12} = 1.3 \, (A_1^{1/3} + A_2^{1/3}) \quad , \quad a = 0.61$$

$$\alpha_0 = 33 \quad , \quad \alpha_1 = 2.0 \quad , \quad \alpha_2 = 3.0 \, .$$

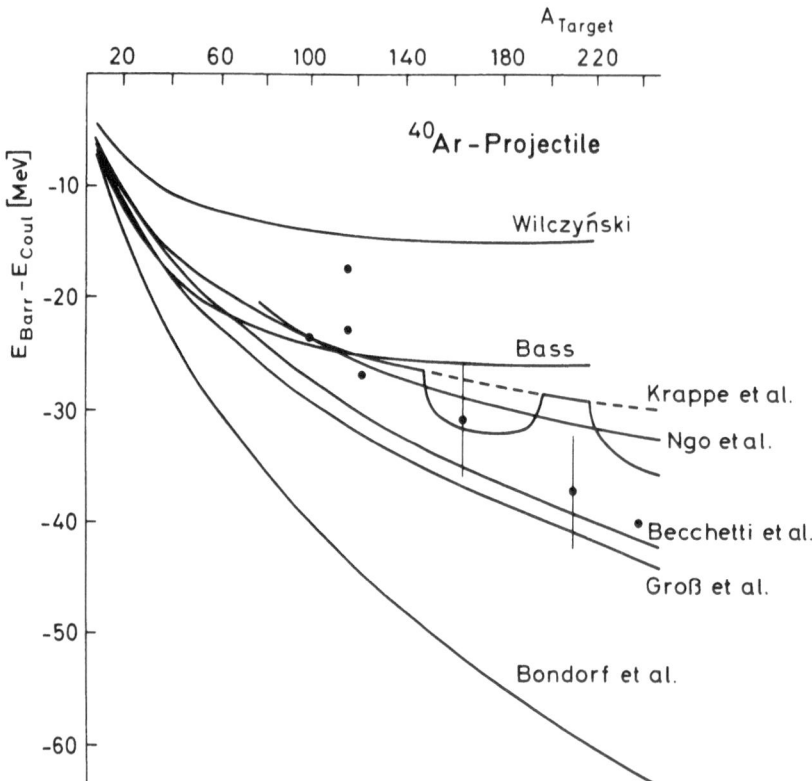

Fig. 2: Same as fig. 1 for argon projectiles.

Using the parameter set proposed by Myers[20, 34)]

$$V_o = (49 + 0.4 \ z_2 \ A_2^{-1/3}) \ \text{MeV}, \qquad a_v = 0.65 \ \text{fm},$$

$$R_v = (1.16 \ A_2^{1/3} + 0.45 - \frac{1.69}{1.16A_2^{1/3} + 0.45}) \ \text{fm}$$

$$a_\rho = 0.54 \ \text{fm}, \qquad R_\rho = (1.16 \ A_1^{1/3} - 0.86 \ A_1^{-1/3}) \ \text{fm}$$

one gets only slightly lower barriers, the difference being 3 MeV for the heaviest systems. But there is a more drastic difference between the extended liquid-drop model predictions and the results from the folding model. Particularly for heavy targets the liquid-drop barriers are higher by about 12 MeV for ^{40}Ar on ^{208}Pb and 23 MeV for ^{84}Kr on ^{208}Pb. The difference is particularly large for close distances between the equivalent sharp surfaces. The reason may be the lack of properly accounting for the saturation properties of nuclear matter in the folding procedure. The same conclusion has been reached in ref.[20)]. It is also revealed by a failure of the folded potential to satisfy eq. (5) for a reasonable surface-tension constant γ in the limit $A_1 = A_2 \to \infty$. The folding potential can therefore not be connected with the liquid-drop energy surface in the interior region.

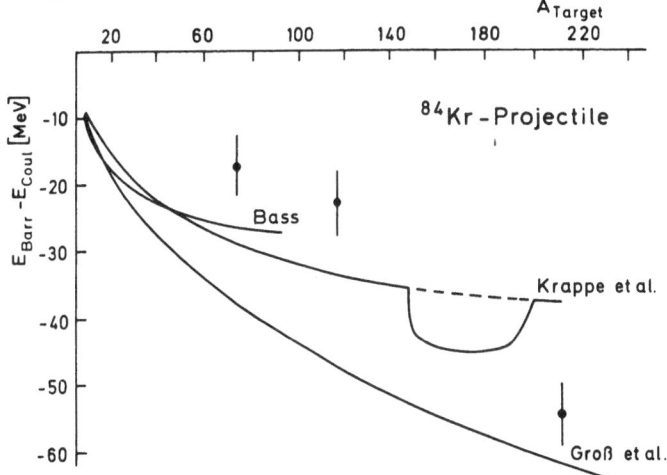

Fig. 3: Same as fig. 1 for krypton projectiles.

The full dots (with error bars in those cases where they are given by the authors) represent experimental reaction-barrier heights. The word experimental needs some qualification in this context. The values are derived from excitation functions of the sum of transfer, fission and compound formation cross sections by means of a parabolic barrier-penetration formula[43]. Using this formula the rather strong model assumption has been made that the crossing of the one-dimensional barrier is the necessary and sufficient condition for a reaction to happen. If the assumptions are weakend by only requiring a finite, smoothly energy-dependent reaction probability beyond the barrier one may still derive threshold energies (but no longer reaction radii) from the penetration formula. But if the barrier turns out to be in a region of considerable density overlap the concept of a one-dimensional barrier penetration may become invalid.

Data were taken from refs[37-42] and the compilation by Vaz and Alexander[30]. For lack of more complete experimental data in some cases partial rather than total reaction cross sections have been used. In fig. 1 maxima of optical potentials are also included which are fitted individually to elastic scattering data of oxygen on various target nuclei[44-55] (open circles). Despite the fact that deep potentials are included as well as shallow ones and 3-, 4-, and 6-parameter fits have been used the resulting potential maxima seem to be determined fairly unambiguously. The data for Neodymium and Samarium

targets from ref.[48] result from rough estimates of optical-model parameters which may explain that they do not follow the general trend.
The curves labelled Becchetti et al. in figs. 1 and 2 were calculated
with optical potential parameters fitted specifically to the system
$^{16}O+^{40}Ca$ [46]. The resulting function follows the trend of folded potentials rather than the liquid-drop estimate. The use of optical-model parameters far away from the region where they are fitted seems to
require some caution.

5. Multidimensional Energy Surfaces

If the barrier occurs at a distance where the densities already overlap appreciably the one-dimensional theory has to be extended into a
multidimensional scheme. The expression (1) can be used as a suitable
basis. The next problem is the choice of the relevant collective degrees of freedom. So far this has always been a matter of guessing and
the following discussion will not be better in that respect.

In order to investigate the influence of the neck-formation degree of freedom Nix's parametrisation of the shape by three smoothly
joined quadratic surfaces of revolution[56] has been used. Fig. 4 gives
a contour plot of the energy around the scission configuration as a
function of the distance d between the centers of mass of the two fragments and the smallest cross section area of the neck. Negative values
of the area parameter β correspond to separated fragments facing each
other with little noses as remnants of the ruptured neck. The lower
boundary of the diagram corresponds to two spherical nuclei as long
as they do not overlap. The remaining shape parameters are constrained
in such a way that the volume of the left half corresponds to a bismuth
nucleus the left one to krypton. The quadrupole moment of each half is
roughly kept zero so that the neck formation is essentially an octupole mode.

In the entrance channel the system enters the strong interaction region at the lower right corner of fig. 4. From 15 fm on inwards
the energetically most favourable configuration at a given distance
has already a neck with about 16 fm^2 cross section. But the valley of
two spheres is separated by a small mountain ridge from the "neck valley". For comparison the point of contact of the equivalent sharp spheres occurs at a distance of 11.8 fm. There is not too much energy to be
gained by forming the neck and the inertia connected with an octupole
mode is fairly large. It is therefore doubtful whether this degree of
freedom will be faster than the motion in the radial direction. Clearly a dynamical calculation is necessary to determine the actual trajectory.

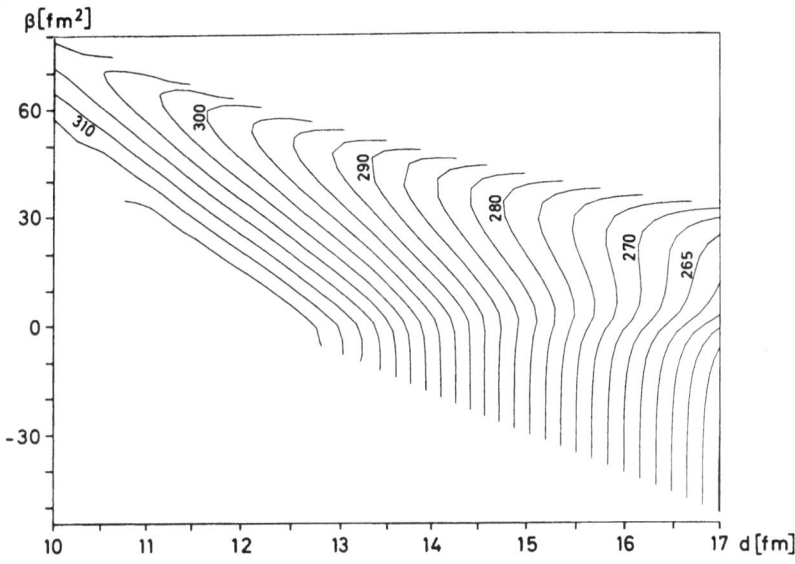

Fig. 4: Contour plot of the potential energy of the system krypton plus bismuth as a function of the distance d between the two centers of mass and the neck-size parameter β. Negative values of β correspond to separated fragments. Distance between contour lines is 2.5 MeV.

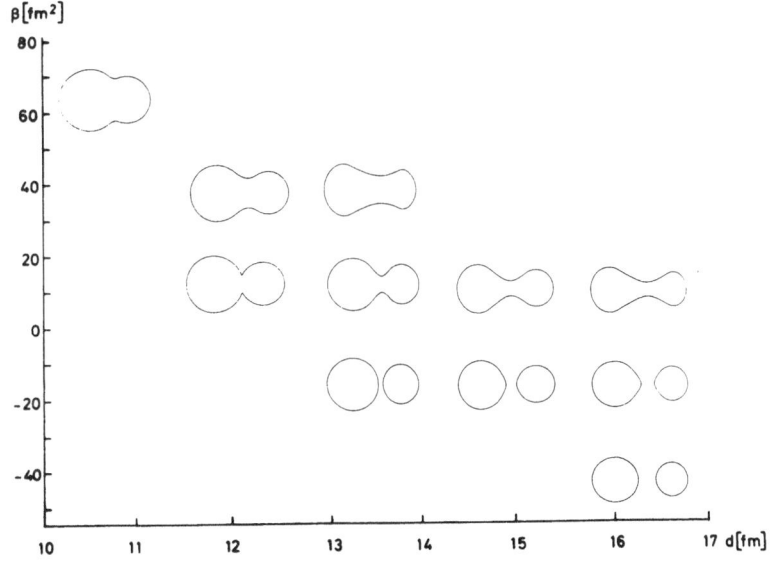

Fig. 5: The class of shapes in the range of d and β values used in fig. 4.

The system considered here has a fissility larger than 1 so
that there is no fission barrier in the liquid drop model. The fission
valley starts right at the spherical configuration its bottom falling
with increasing quadrupole deformation. The spherical configuration
corresponds to a somewhat lower energy than spherical krypton and bismuth drops just touching each other. The system therefore enters the
strong overlap region on the wall of the fission valley.

Besides the octupole or neck-formation degree of freedom the
quadrupole deformation of both fragments and the asymmetry or mass and
charge exchange degrees of freedom have to be considered. For the
krypton plus bismuth system energy is gained by allowing a prolate deformation and a decrease of the asymmetry which would bring it to the
bottom of the fission valley. A calculation of energy surfaces as
function of these degrees of freedom is now in progress.

For lighter systems a fission barrier exists dividing the intrinsic region around the spherical compound state and the outside
area. The height of the fission saddle-point is by definition the
lowest energy at which the inside region can be reached from outside.
But there is no reason for fusion trajectories just to hit the saddle
point. Therefore compound-formation thresholds will always be larger
than fission barriers and in general determined by dynamical considerations rather than derived from static potential-energy surfaces
alone [11].

6. Summary

There seems to be reasonable agreement between one-dimensional Thomas-
Fermi and extended liquid-drop theories of the interaction barrier.
Folded potentials on the other hand yield systematically lower barriers for heavier systems (roughly $z_1 z_2 > 600$), where the barrier lies
within 2 fm between the equivalent sharp surfaces. In classical one-
dimensional dynamical calculations folded potentials turn out to do
better than liquid-drop potentials if used together with a strong friction force[58]. They seem to describe the effective potential required
in such calculations quite well although they are derived in a different frame. In view of the fact that the neck-formation degree of freedom may become important already at distances of 3-4 fm it is not surprising that one-dimensional models have to be modified at shorter
distances.

References

1) J. Knoll and R. Schaeffer, Extended Seminar on Nuclear Physics, Trieste 1973, to be published; S. Levit, U. Smilansky, and D. Pelte, preprint Rehovot 1974
2) R. A. Malfliet, private communication (1974)
3) R. A. Broglia, S. Landowne, R. A. Malfliet, V. Rostokin, and A. Winther, Phys. Lett. 11C (1974) 1
4) D. H. E. Gross and H. Kalinowski, Phys. Lett. 48B (1974) 302
5) J. P. Bondorf, M. I. Sobel, and D. Sperber, Phys. Lett. C, to be published
6) R. Beck and D. H. E. Gross, Phys. Lett. 47B (1973) 143
7) B. Giraud, J. Le Tourneux, and E. Osnes, International School of Physics Enrico Fermi, Varenna, Course 62 (1974)
8) R. Beringer, Phys. Rev. Lett. 18 (1967) 1006; J. Maly and R. Nix, Suppl. J. Phys. Soc. Jap. 24 (1968) 678; P. W. Riesenfeldt and T. D. Thomas, Phys. Rev. C2 (1970) 711
9) H. Holm, D. Scharnweber, W. Scheid and W. Greiner, Z. Physik 231 (1970) 450
10) A. S. Jensen and C. Y. Wong, Nucl. Phys. A171 (1971) 1
11) J. R. Nix and A. J. Sierk, Physica Scripta, to be published
12) P. G. Zint, K. H. Paßler, and U. Mosel, Verh. Dt. Phys. Gesellschaft VI 9 (1974) 98
13) K. A. Brueckner, J. R. Buchler, and M. M. Kelly, Phys. Rev. 173 (1968) 944
14) C. Ngô, B. Tamain, J. Galin, M. Beiner and R. J. Lombard, Nucl. Phys., to be published; preprint IPNO/TH 74-19
15) C. Dover, contribution to this conference
16) C. J. Batty, E. Friedman, and D. F. Jackson, Nucl. Phys. A175 (1971) 1
17) S. G. Kadmenskii, V. E. Kalechits, S. I. Lopatko, V. I. Furman, and V. A. Khlebostroev, Sov. Jour. Nucl. Phys. 10 (1970) 422
18) R. A. Broglia and A. Winther, Phys. Lett. 4C (1972) 153
19) D. M. Brink and N. Rowley, Nucl. Phys. A219 (1974) 79
20) W. D. Myers, Nucl. Phys. A204 (1973) 465
21) H. Kalinowski, private communication (1974)
22) B. N. Kalinkin and I. Z. Petkov, Acta Phys. Pol. 25 (1964) 265
23) R. Bradley, Phil. Mag. 13 (1932) 853
24) B. Deryaguin, Koll. Z. 69 (1934) 155
25) J. Randrup, W. J. Swiatecki, and C. F. Tsang, Nuclear Chemistry Annual Report 1973, Berkeley LBL-2366 p. 143
26) J. Wilczyński, Proc. Third IAEA Symp. on Physics and Chemistry of Fission, Rochester 1973, vol. II p. 269, Vienna 1974
27) R. Bass, Phys. Lett. 47B (1973) 139; Proc. Int. Conf. Reac. between Complex Nuclei, Nashville 1974 vol. I p. 117; North Holland Publ. Comp. 1974
28) H. Holm and W. Greiner, Phys. Rev. Lett. 24 (1970) 404
    H. J. Fink, W. Scheid, and W. Greiner, Nucl. Phys. A188 (1972) 259
29) H. J. Krappe and R. Nix, Proc. Third IAEA Symp. on Physics and Chemistry of Fission, Rochester 1973, vol. I p. 159, Vienna 1974
30) L. C. Vaz and J. M. Alexander, preprint Stony Brook 1973
31) H. Flocard, Phys. Lett. 49B (1974) 129
32) J. W. Negele, Proc. Intern. Workshop II on Gross Properties of Nuclei and Nuclear Excitations, Hirschegg AED-Conf -74-025-000, p.4, T. H. Darmstadt 1974
33) M. P. Fricke, E. E. Gross, B. J. Morton, and A. Zucker, Phys. Rev. 156 (1967) 1207; C. B. Fulmer, J. B. Ball, A. Scott, and M. L. Whiten, Phys. Rev. 181 (1969) 1565
34) W. D. Myers and W. J. Swiatecki, Ann. Phys. 55 (1969) 395
35) P. A. Seeger, Proc. Fourth Intern. Conf. on Atomic Masses and Fundamental Constants, Teddington 1971 p. 255
36) W. D. Myers and W. J. Swiatecki, Ark. Fys. 36 (1967) 343

37) H. Gauvin, R. L. Hahn, Y. Le Beyec and M. Lefort, Proc. Int. Conf. Reac. between Complex Nuclei, Nashville 1974, vol. I, p. 109; North Holland Publ. Comp. 1974

38) F. Hanappe, C. Ngô, J. Peter and B. Tamain, Proc. Third IAEA Symp. on Physics and Chemistry of Fission, Rochester 1973, vol. II p.289, Vienna 1974

39) M. Ishihara, R. Broda, G. B. Hagemann, B. Herskind, S. Ogaza, and H. Ryde, Proc. Int. Conf. React. between Complex Nuclei, Nashville 1974, vol. I, p. 111, North Holland Publishing Comp. 1974

40) Y. Le Beyec, M. Lefort and M. Sarda, Nucl. Phys. $\underline{A192}$ (1972) 405

41) R. Bimbot, D. Gardès, and M. F. Rivet, Phys. Rev. $\underline{C4}$ (1971) 2180

42) Y. Le Beyec, M. Lefort, and A. Vigny, Phys. Rev. $\underline{C3}$ (1971) 1268

43) C. Y. Wong, Phys. Rev. Lett. $\underline{31}$ (1973) 766

44) A. W. Obst, D. L. McShan, and R. H. Davis, Phys. Rev. $\underline{C6}$ (1972)1814

45) B. C. Robertson, J. T. Sample, D. R. Goosman, K. Nagatani, and K. W. Jones, Phys. Rev. $\underline{C4}$ (1971) 2176

46) F. D. Becchetti, P. R. Christensen, V. I. Manko, and R. J. Nickles, Nucl. Phys. $\underline{A203}$ (1973) 1

47) J. Orloff and W. W. Daehnick, Phys. Rev. $\underline{C3}$ (1971) 430

48) A. M. Friedman, R. H. Siemssen, and J. G. Cuninghame, Phys. Rev. $\underline{C6}$ (1972) 2219

49) A. Gobbi, R. Wieland, L. Chua, D. Shapira, and D. A. Bromley, Phys. Rev. $\underline{C7}$ (1973) 30

50) M. C. Bertin, S. L. Tabor, B. A. Watson, Y. Eisen, and G. Goldring, Nucl. Phys. $\underline{A167}$ (1971) 216

51) K. O. Groeneveld, L. Meyer-Schützmeister, A. Richter, and U. Strohbusch, Phys. Rev. $\underline{C6}$ (1972) 805

52) G. Delic, IKDA 74/6 preprint, Darmstadt 1974

53) N. R. Fletcher, L. West, and K. W. Kemper, Bull. Am. Phys. Soc. $\underline{16}$ (1971) 1149

54) E. H. Auerbach, and C. E. Porter, Proc. Third Conf. Reac. between Compl. Nuclei Asilomar 1963, p. 19

55) M. C. Mermaz, Extended Seminar on Nuclear Physics, Trieste 1973, to be published

56) J. R. Nix, Nucl. Phys. $\underline{A130}$ (1969) 241

57) L. G. Moretto, D. Heunemann, R. C. Jared, R. C. Gatti, and S. G. Thompson, Proc. Third IAEA Symp. on Physics and Chemistry of Fission, Rochester 1973, vol. II p. 351, Vienna 1974

58) D. H. E. Gross, H. Kalinowski and J. De, Contribution to this conference

USE OF SEMICLASSICAL QUANTITIES IN TRANSFER REACTIONS

H. J. Körner
Physik-Department, Techn. Universität
München
8046 Garching, W.-Germany

Semiclassical quantities are used to discuss several features of transfer and inelastic scattering reactions induced by heavy ions. Experimental data on absorption phenomena in elastic scattering and the relative importance of quasielastic channels are presented. Transfer probabilities and the enhancement of two-nucleon transfer cross sections
due to correlated nuclear structure are discussed. The question of two-
step, simultaneous and successive transfer reactions is investigated
for $^{16,18}O$ induced two neutron transfer reactions, and the "exotic"
$(^{16}O,^{15}C)$ reaction.

1. Introduction

Recent investigations[1] have provided an appreciable amount of
new information on heavy ion induced reactions. Transfer reactions were
studied in order to explore the importance of recoil effects, multistep
processes, and the nature of forward oscillations in angular distributions. Inelastic scattering processes were investigated in order to
understand the various aspects of interference phenomena in target and
projectile excitations. The new data demonstrate the progress in experimental techniques achieved during the last years: Improvements in energy
resolution and particle identification methods due to magnetic spectrograph techniques, time-of-flight methods, and the capability to handle
extremely high counting rates.

The basic motivation for these experiments is: a) To understand
the details of the reaction mechanism quantitatively, and b) to explore
the effects which are specific to, or more pronounced in heavy ion
induced reactions than in light ion induced reactions.

The quantitative analysis of transfer reactions is generally based
on DWBA calculations; semi-classical approximations in these are used
to simplify and accelerate the numerical calculations. The quality of

the analysis is judged from a comparison with corresponding light ion
induced reactions, and from the fits to the shape of the measured angular distributions. The selectivity of a certain reaction is established
by comparison with a corresponding light ion induced reaction[2].

Some basic problems seem to emerge from these procedures: a) An
optical potential is needed in the DWBA calculations. The analysis of
elastic scattering data does not appear to be a unique prescription to
determine such a potential. One example for such difficulties will be
discussed in the following paragraph. b) Additional channels such as
inelastic excitations and successive transfer processes have to be
included in the quantitative analysis. Full DWBA calculations therefore
become extremely lengthy and complicated; c) The interpretation of the
imaginary part of the optical model potential seems to be less clear
in heavy ion induced reactions than in light ion induced reactions.
Experiments at incident energies above the barrier indicate that an
appreciable fraction of the reaction cross section can be due to quasielastic processes, and not to absorption, i.e. formation of a compound
nucleus system[3].

In the following paragraphs we shall investigate how simple classical or semiclassical concepts can be used to understand the various
phenomena, and to estimate the importance of more complicated processes.
Furthermore, we shall try to demonstrate how the improved experimental
techniques can be used to study some of the questions raised above.
Since forward oscillation phenomena are most pronounced at incident
energies well above the Coulomb barrier, and will be covered in the
talk by Garret we shall especially be interested in the behaviour of
cross sections below and slightly above the Coulomb barrier. In addition to transfer reactions we discuss some relevant features of
inelastic scattering reactions because the physics of these two processes is intimately correlated.

2. Ambiguities in Optical Potentials: An Example

The analysis of elastic scattering data often yields optical model
potentials which do not fit the transfer angular distributions, especially in the case of mismatched transitions. This deficiency is then repaired by allowing rather arbitrary modifications in the parameters of
the potential used. Two procedures have been discussed to resolve these
problems: a) A simultaneous analysis of elastic scattering, inelastic
scattering, one and more nucleon transfer data. The argument is that
these various processes probe rather different regions of the ion-ion

potential; b) an investigation of forward-angle cross sections in transfer reactions at several incident energies well above the Coulomb barrier. The argument is that this cross section is due to inner trajectories deflected towards forward angles by the attractive (real) ion-ion potential, and thus becomes more and more sensitive to the interior region of the potential, as the incident energy is raised.

The second suggestion will be discussed by Garret. In the following we will investigate the first procedure[4]. Fig. 1 shows an elastic scattering angular distribution for 56 MeV ^{16}O ions incident on ^{48}Ca.

The data are measured down to $\sigma_{el}/\sigma_R \sim 1/300$. The various curves represent fits to these data with 11 different sets of optical model parameters taken from the literature. The fit values deviate only insignificantly from the literature values, and are summarized in table 1. Although the parameters of these potentials vary over a large range of values all potentials describe the elastic scattering data extremely well; differences in the various calculations only show up for the largest scattering angles. Consequently the transmission coefficients are more or less identical for all these fits, and the reaction cross section (last column of table 1) is determined with a very high degree of accuracy.

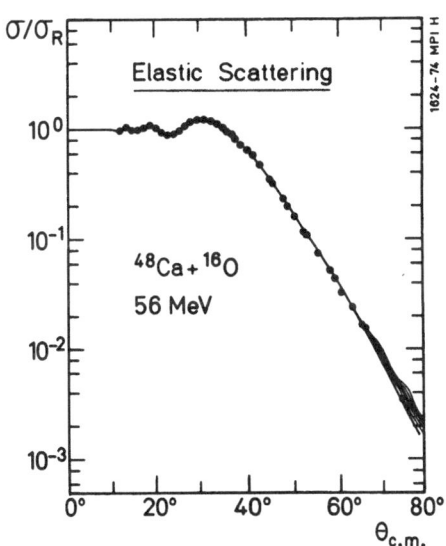

Fig. 1   Elastic scattering angular distribution for 56 MeV ^{16}O ions incident on ^{48}Ca. The solid curves are fits with the potentials summarized in table 1

Figs. 2 and 3 show calculated one proton transfer angular distributions for the same sets of potentials. The full-recoil code LOLA (ref. 5) was used in these calculations for the $^{48}Ca(^{16}O, ^{15}N)^{49}Sc$ reaction to the $1f_{7/2}$ ground state (L = 4) and $2p_{3/2}$ 3.08 MeV excited state (L = 2) in ^{49}Sc. Again, no appreciable difference is found among all these calculations, and it would hardly be possible to decide about one or the other potential from a comparison with experimental data.

V	R_r	a_r	W	R_i	a_i	σ_R
312.78	1.208	.408	104.92	1.337	.188	1.281
149.06	1.168	.496	11.09	1.233	.566	1.313
100.10	1.200	.500	24.04	1.207	.482	1.290
78.02	1.218	.497	11.25	1.232	.571	1.313
68.50	1.190	.561	17.04	1.280	.457	1.295
65.56	1.241	.490	12.32	1.266	.455	1.281
43.50	1.259	.520	47.98	1.203	.458	1.306
33.75	1.344	.424	110.15	1.274	.280	1.286
22.15	1.342	.476	6.24	1.380	.318	1.263
15.90	1.341	.508	8.18	1.330	.517	1.327
12.73	1.363	.510	8.00	1.363	.510	1.370

Table 1   Optical model parameters and reaction cross section obtained from fits to the elastic scattering data shown in fig. 1

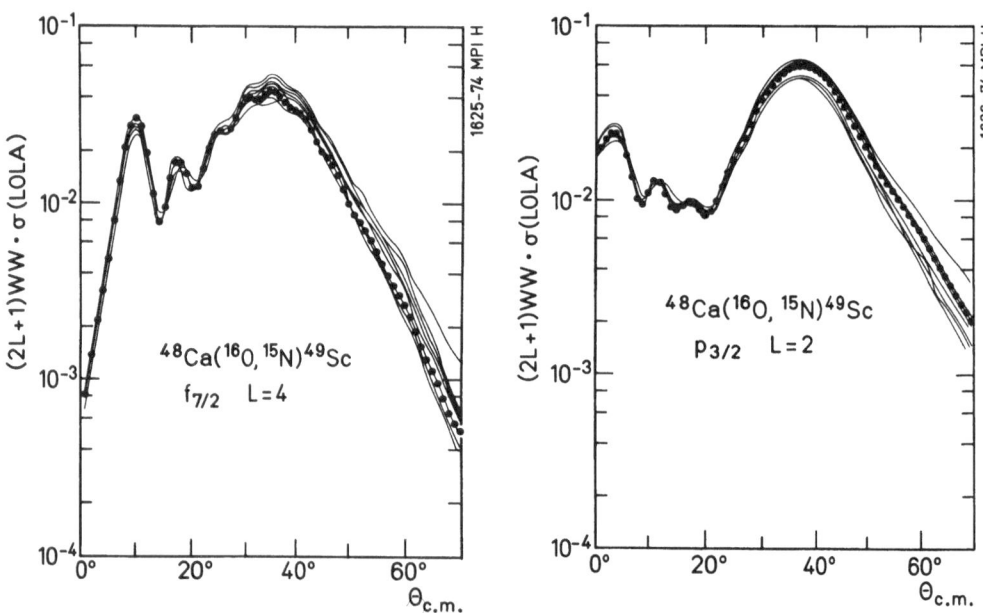

Fig. 2 Calculated angular distributions for the $^{48}Ca(^{16}O, ^{15}N)^{49}Sc$ g.s. reaction at 56 MeV for the optical potential parameters given in table 1

Fig. 3 Same as fig. 2, for the $^{48}Ca(^{16}O, ^{15}N)^{49}Sc$ reaction to the 3.08 MeV excited state in ^{49}Sc

However, although the data[6] for the well-matched L = 4 g.s. transition fit well to these calculated angular distributions, those for the mismatched L = 2 transition to the $2p_{3/2}$ excited state don't. This problem may be repaired by a modification of the optical potential geometry, for instance a variation of the diffuseness parameter in the exit channel. Similar problems arise for other one-proton transfer reactions on ^{48}Ca, for instance the $(^{15}N, ^{14}C)$ and $(^{19}F, ^{18}O)$ reactions, whenever a transition is badly matched.

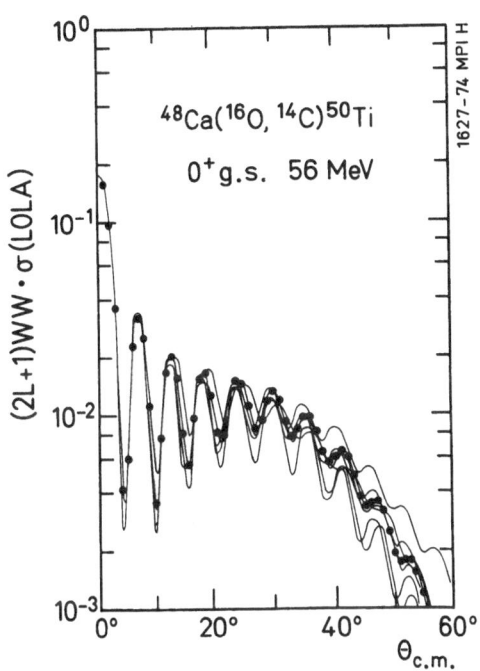

Fig. 4 Same as fig 2, for the ^{48}Ca$(^{16}O, ^{14}C)^{50}$Ti g.s. reaction at 56 MeV incident energy

Finally, we show (see fig.4) corresponding calculations for the ^{48}Ca$(^{16}O, ^{14}C)^{50}$Ti L = 0 two-proton transfer reaction to the ^{50}Ti ground state. Again, these curves only deviate from each other at larger angles, where the cross section becomes small and more and more featureless, and there appears to be little hope to differentiate between the various potentials from a comparison with the experimental data.

Thus, we conclude, that the simultaneous analysis of elastic scattering and transfer data obtained at one incident energy will not help to distinguish different optical potentials. The question whether data for different incident energies would help has still to be explored.

3. Semiclassical Quantities

Consider a beam of (classical) particles incident on an ensemble of target nuclei. Each particle follows a _trajectory_ specified by an _impact parameter ρ._ The cross section for elastic scattering is then

$$\frac{d\sigma}{d\Omega} = \frac{\rho}{\sin\theta}\left|\frac{d\rho}{d\theta}\right|$$

Thus the classical problem is to find the scattering angle θ for each
impact parameter ρ, or, more general, the <u>deflection function</u> θ(ρ) of
the problem[7]. Alternatively, we could use the <u>distance of closest</u>
<u>approach D</u> for a given trajectory, and have then to find a connection
of D, θ and angular momentum.

The classical picture relies on the fact that the wave length of
the heavy ions is small as compared to the distance of closest approach,
or the nuclear dimensions; this localization in r-space allows us to
use the trajectory concept. The whole approach is one of classical
geometric optics, with the geometrical dimensions of "apertures" being
large as compared to the wave length of the radiation. The connection
between the quantum-mechanically correct scattering amplitude and the
classical or semiclassical approximation for the cross section has first
been studied by Ford and Wheeler[7].

Although the classical picture was made up only for elastic
scattering we may try to use it for inelastic scattering and transfer
reactions as well. Using quantum mechanics one can calculate the excitation or transfer amplitudes for any time t during the collision process.
In the semiclassical approach the time dependence of the interaction
between the two colliding nuclei is determined by the classical trajectory for the relative motion. The cross section for populating a given
final state is then equal to the elastic scattering cross section times
the probability P that an excitation or transfer occurred during the
collision process:

$$\frac{d\sigma}{d\Omega} = P \cdot \frac{\rho}{\sin\theta} \cdot \left|\frac{d\rho}{d\theta}\right|$$

This concept is applicable if the transfer of energy, mass, charge, and
angular momentum is small enough, so that the classical trajectory is
not disturbed appreciably. In general we have to expect a dependence of
the cross section on differences between trajectories in the entrance
and exit channels. Secondly, we may face the situation, especially in
two- and more-nucleon transfer reactions, that the approximation of
classical optics is no longer applicable, i.e. the transfer process is
confined to such a narrow region of the nuclear surface (or in L-space)
that wave aspects of the interaction process become important and even
dominant.

4. Elastic and Inelastic Scattering

4.1. GENERAL FEATURES

For sub-Coulomb incident energies the relation between the scattering angle θ and the impact parameter ρ is well defined for all values of θ. For larger incident energies more than one impact parameter may lead to the same scattering angle. The details of this behaviour depend on the strength of the nuclear potential[8]. For scattering angles near to the (Coulomb) rainbow angle contributions from two trajectories to the elastic and inelastic scattering have been discussed[9]. At backward angles we have a unique correspondence between impact parameter and scattering angle if we exclude spiraling phenomena. In the following we shall discuss some data in the context of the trajectory picture.

In fig. 5 we reproduce elastic scattering data obtained[10] with carbon and oxygen projectiles at incident energies above the Coulomb barrier for a number of targets in the mass regions of A~50 and A~90. The data are plotted as σ/σ_R versus a normalized classical distance of closest approach, $d=D(\theta)/(A_1^{1/3}+A_2^{1/3})$, calculated under the assumption of pure Coulomb scattering. Two points emerge from these plots: a) A rather uniform general behaviour of all the data within the two mass regions, with a uniquely defined breakoff from the Rutherford cross section at $d_0 \sim$ 1.68 fm. The solid curves in fig. 5 are fits to the data, with a parametrization $\sigma/\sigma_R = 1$ for $D \geq D_0$, and $\sigma/\sigma_R = \exp |(D - D_0)/\Delta|$ for $D < D_0$. It is seen that most of the data are reproduced by these fits to within $\pm$ 30 %. b) An apparently systematic difference between the A ~ 50

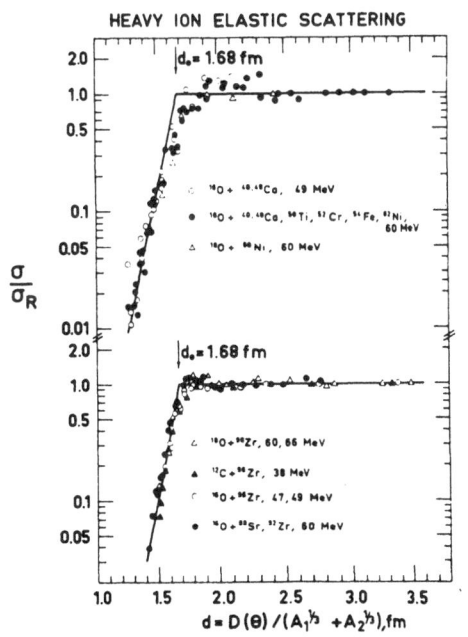

HEAVY ION ELASTIC SCATTERING

$d_0 = 1.68$ fm

$^{16}O + {}^{40,48}Ca$, 49 MeV
$^{16}O + {}^{40,48}Ca$, ^{48}Ti, ^{52}Cr, ^{54}Fe, ^{62}Ni, 60 MeV
$^{16}O + {}^{58}Ni$, 60 MeV

$d_0 = 1.68$ fm

$^{16}O + {}^{90}Zr$, 60, 65 MeV
$^{12}C + {}^{90}Zr$, 38 MeV
$^{16}O + {}^{90}Zr$, 47, 49 MeV
$^{16}O + {}^{88}Sr$, ^{91}Zr, 60 MeV

$\dfrac{\sigma}{\sigma_R}$

$d = D(\theta)/(A_1^{1/3} + A_2^{1/3})$, fm

Fig. 5 Elastic scattering cross sections for various combinations of targets, projectiles, and energies. The solid lines are best fits as explained in the text

and A ∼ 90 regions; the breakoff from the Rutherford cross section
appears to accur much more abruptly for target nuclei with A ∼ 90.

In the following we investigate whether this second observation
can be substanciated. We thereby concentrate on incident energies near
to the Coulomb barrier and backward angle data, where the distance of
closest approach is uniquely defined for a given incident energy. The
purpose of this is to deduce the relative surface behaviour of different
projectiles, and to check the extent to which the results depend on a
specific target nucleus. We find:

a) A projectile dependence: Fig. 6 reproduces results obtained by
Eisen et al.[11] for the elastic scattering of ^{16}O and ^{18}O ions from
^{52}Cr at backward angles. The elastic cross section never exceeds σ_R,
consistent with the assumption of only one contributing trajectory. A
significant difference between the two sets of data is observed, with
a much earlier and smooth
deviation of σ from the Rutherford value for the ^{18}O projectiles, i.e. a deviation from
pure Coulomb scattering at
much larger values of D than for
the ^{16}O projectiles.

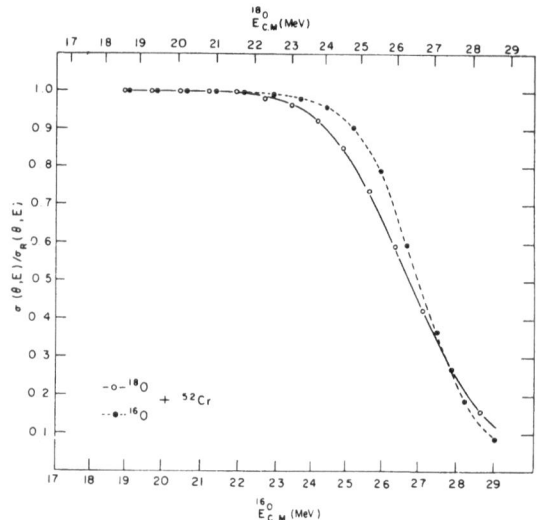

Fig. 6 A comparison between ^{16}O
(dotted line) and ^{18}O (solid
line) scattered from ^{52}Cr
at θ 180°. The solid lines
are best fits as explained
in ref. 11

b) A target-mass dependence:
Fig. 7 shows backward angle
(θ ∼ 180°) excitation functions[12]
for the elastic and inelastic
scattering (target excitation)
of ^{18}O projectiles incident on
^{60}Ni and ^{92}Mo targets. The data
are plotted vs d_o, a normalized
distance of closest approach, with
$Z_1 Z_2 e^2 / E_{CM} = d_o (A_1^{1/3} + A_2^{1/3})$.
The elastic scattering data re-
produce the effect already indicated in fig. 5: A much earlier
and gradual deviation from the
Rutherford cross section for the
^{60}Ni target, as opposed to the
rather abrupt breakoff for the ^{92}Mo target.

Correlated to this behaviour we observe an earlier deviation from
the Coulomb excitation cross section in the ^{60}Ni inelastic excitation,
with a shallow destructive interference minimum followed by the rise

of the normalized cross section due to nuclear excitation. For the ^{92}Mo target the interference minimum is missing, and the rise of the normalized inelastic excitation cross section correlates with the deviation of the elastic cross section from the Rutherford value.

It should be added that the smooth deviation of σ/σ_R from unity for the ^{18}O + ^{60}Ni system is accompanied by the gradual onset of one- and two-neutron transfer processes. Actually, an appreciable fraction of the missing elastic cross section is due to these transfer channels and inelastic scattering.

Excitation functions measured at backward angles are useful to scan the radial dependence of the form factor in the inelastic scattering process, without the complications introduced by contributions from more than one trajectory. Theoretical calculations[12] indicate that the energy dependence of the cross section is extremely sensitive to the geometry of the form factor.

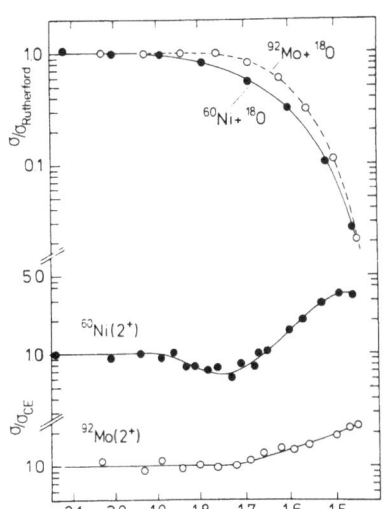

Fig. 7 Elastic and inelastic scattering data obtained for ^{60}Ni+^{18}O, and ^{92}Mo+^{18}O at $\theta \sim 180°$. The distance of closest approach d_0 is defined in the text. The solid lines are eye-guides only

4.2. CONTRIBUTION OF TWO TRAJECTORIES

This phenomenon has been discussed by Malfliet et al.[9], and more recently by Broglia[13]. The arguments are presented together with data (see fig. 8) for the elastic and inelastic scattering (projectile excitation) of 72 MeV ^{18}O ions incident on ^{120}Sn (ref. 12). The incident energy is high enough so that the nuclear potential does influence the trajectories, and may cause a multivalued deflection function for scattering angles near to the rainbow angle. We shall adopt the argument that two trajectories contribute to the cross section; these two originate from both sides of the rainbow. All possible other trajectories are assumed to be damped out by absorption.

Consequently, cross sections have to be calculated by a coherent summation of two contributions. The oscillations of σ/σ_R around unity then reflect the interference of the elastic scattering amplitudes due to the two trajectories. If we ascribe cross sections σ_R and σ_N to the outer (larger impact parameter) and inner trajectories, respectively, we get

$$\sigma = \sigma_R \left| 1 + (\sigma_N/\sigma_R) + 2\sqrt{\sigma_N/\sigma_R} \right| \text{ for } \theta = 62°.$$

Since $\sigma/\sigma_R \sim 1.2$ at this angle, we obtain $\sigma_N \sim 0.01\ \sigma_R$ in this case. At more forward angles the oscillations get damped out quickly, because of the increase of σ_R.

Inelastic scattering allows to study these interference effects more directly. Due to the short range of the nuclear form factor the outer trajectory contributes Coulomb excitation only, and the cross section due to this trajectory varies slowly with scattering angle. The inner trajectory contributes Coulomb <u>and</u> nuclear excitation, and is responsible for the

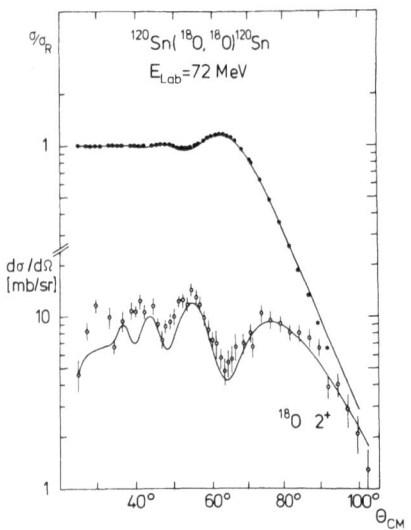

Fig. 8 Elastic and inelastic (projectile excitation) scattering of ^{120}Sn + 72 MeV ^{18}O. The solid lines are fits to the elastic (ABACUS) and inelastic (collective model form factor, DWUCK) data[12]

rapid oscillations observed in the inelastic scattering angular distributions. Due to the dominant nuclear scattering amplitude, which is of opposite sign as the Coulomb scattering amplitude, one obtains a phase rule[9]: Maxima in σ/σ_R for elastic scattering line up with minima in the inelastic scattering angular distributions, and vice versa. Complications due to the Q-value dependence, possible reorientation effects, and the phase of the inelastic scattering form factor[13] will not be discussed here. The data shown in fig. 8 provide an example for this phase rule, and thus support the concept of two contributing trajectories to the elastic and inelastic scattering processes.

Three points should be added at this stage: 1) The earlier explanation[14] of the Coulomb nuclear interference phenomena in terms of only one important trajectory and a minimum in the radial form factor does not describe the data on inelastic scattering <u>angular distributions</u> obtained at incident energies above the Coulomb barrier, whereas the two-trajectory-concept does. Therefore, such data carry a different type of information as those on backward angle excitation functions discussed earlier. 2) The inelastic scattering angular distributions for target excitation are in general quantitatively described by theoretical calculations which use a collective model form factor and deformation parameters derived from B(E2) values. This is important in order to judge the reliability of calculations for transfer reactions which include the influence of additional inelastic excitation processes. 3) Similarly, as discussed in section 2, the analysis of inelastic scattering data does in general not allow to differentiate between various optical model parameter sets which fit accurate data on the elastic scattering equally well.

4.3. CONTRIBUTIONS DUE TO QUASIELASTIC PROCESSES AND FUSION

Fig. 9 shows preliminary data on backward angle excitation functions for the system $^{18}O + {}^{118}Sn$ obtained with the Argonne time-of-flight set up[15]. These data are one example for the new kind of information that can be obtained with the improved experimental techniques: A complete separation of elastic, inelastic scattering and transfer reactions was possible in this case. The transfer channels are mainly neutron transfer at the lower incident energies. In the spirit of the semiclassical approximation all the flux going to a certain solid angle is proportional to the elastic scattering; thus the cross sections for the various processes are successively summed up, and are normalized to the Rutherford cross section at this angle. It is evident that the cross sections for quasielastic processes (inelastic scattering and transfer) are

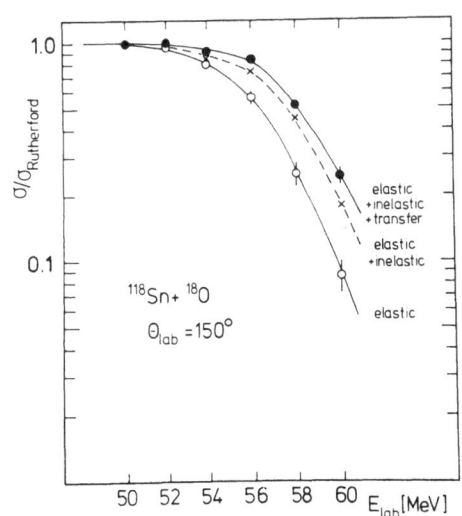

Fig. 9 Excitation functions for the $^{18}O + {}^{118}Sn$ system. The solid lines are eyeguides, only

appreciable: Between 56 and 60 MeV incident energy they add up to ~ 20% of the Rutherford cross section at the corresponding energy.

These results are not specific to incident energies near to the Coulomb barrier, and to ^{18}O projectiles, as can be seen from a comparison of the total reaction cross sections and the total cross sections for quasielastic processes obtained by an integration of differential cross sections. Data obtained by time-of-flight techniques[16] for the $^{18}O + ^{120}Sn$ system at 72 MeV, and for the $^{16}O + ^{122}Sn$ system at 74 MeV incident energy indicate that even at these higher incident energies the quasielastic cross sections are an appreciable fraction of the total reaction cross sections. Total reaction cross sections are ~ 1000 mb in these cases, and the quasielastic cross sections contribute ~ 30 % to these cross sections. Similar conclusions were obtained by von Oertzen et al.[17] for ^{16}O projectiles incident on $^{206,208}Pb$ target nuclei at energies slightly above the Coulomb barrier.

A complementary information is obtained from measurements of fusion cross sections. As an example we reproduce (see fig.10) data obtained with ^{32}S projectiles by Gutbrod et al.[18]. The solid lines in the lower part of the figure represent the total reaction cross sections as calculated from fits to the elastic scattering data. The comparison with the measured fusion cross sections shows that in these systems the main part of the reaction cross section is due to fusion. This conclusion is supported by data obtained at Munich[19] for the inelastic scattering and transfer reactions in the $^{32}S + ^{40}Ca$ system. At 100 MeV incident energy (E_{CM} = 55.6 MeV) the integrated quasielastic cross sections are smaller than 10 % of the total reaction cross section. Since the $^{32}S + ^{40}Ca$ system is especially stable against inelastic excitation

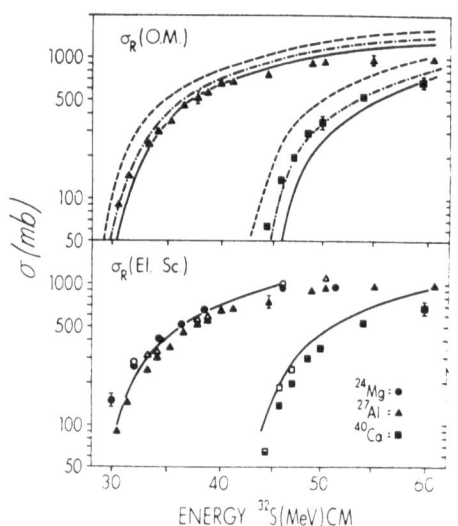

Fig.10 Fusion cross sections for ^{32}S ions incident on ^{24}Mg, ^{27}Al, and ^{40}Ca. The solid lines represent total reaction cross sections as explained in ref. 18

and transfer processes, it is not clear whether this result is a general one for ^{32}S induced reactions; the data shown in fig. 10 for ^{24}Mg and ^{27}Al targets indicate, however, that this behaviour may be common to ^{32}S ions incident on lighter target nuclei.

5. Transfer Reactions

5.1. GRAZING-TYPE ANGULAR DISTRIBUTIONS

Fig. 11 shows angular distributions for the two-neutron stripping and pickup reactions ^{18}O + ^{120}Sn $\rightleftharpoons$ ^{16}O + ^{122}Sn leading to the ground states and first excited states in the Sn nuclei, respectively[16]. The incident energies are well above the Coulomb barrier. These data will be used to discuss some of the features of heavy ion induced transfer reactions.

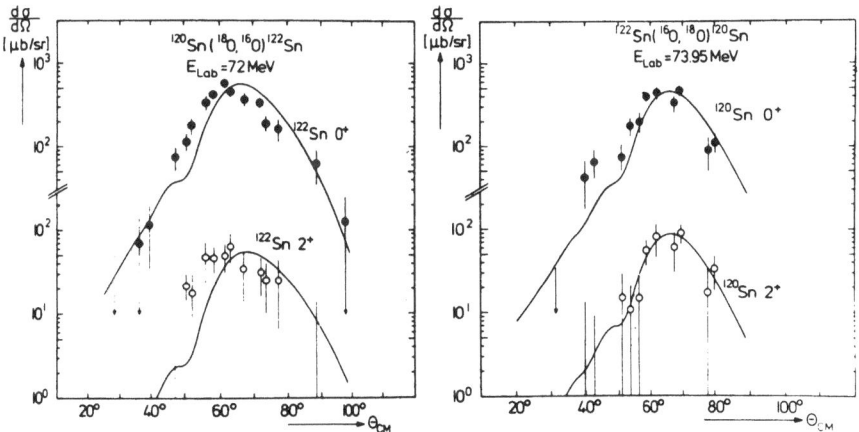

Fig. 11   Angular distributions for the ^{18}O + ^{120}Sn $\rightleftharpoons$ ^{16}O + ^{122}Sn transfer reactions leading to the g.s. and first excited states. The solid lines are calculated with the DWBA program LOLA[5].

All transitions show grazing-type bell-shaped angular distributions; the angular width indicates that many L-values contribute to the two-neutron transfer process. The grazing angle ($\theta_{CM} \sim 65°$) correlates with the rainbow angle; this is where the nuclear interaction becomes of dominant importance. The cross section at larger angles decreases because of absorption <u>and</u> flux conservation as particles with lower impact parameters are either absorbed or scattered to the forward direction. Consequently, the cross section forward to the grazing peak

is due to the larger partial waves <u>and</u> to flux scattered to the forward
direction, i.e. particles with lower partial waves.

5.2. LOCALIZATION IN L-SPACE

For the same example, the $^{120}Sn(^{18}O,^{16}O)^{122}Sn$ g.s. reaction at
72 MeV incident energy, we may also test the applicability of the
classical trajectory picture. From DWBA calculations we find that the
dominant partial wave is L = 38 in this case, and the half width of
the transfer amplitude distribution is ΔL = 11. The corresponding
diffraction angular spread would be $\Delta\theta_d$ = $1/\Delta L \sim 5°$, which is small as
compared to the position of the grazing peak, and the width of the
measured angular distribution. As has been discussed in the literature[20]
different situations may exist in systems like ^{16}O + ^{32}S, and incident energies well above the Coulomb barrier, where the diffractive
angular spread may be as large as 30°. In such cases the details of a
calculated deflection function become meaningless.

5.3. TRANSFER PROBABILITIES

A definition of the transfer probability for a well-matched reaction
has already been given above. In cases of Q-value-mismatch (however, not
too large mismatch, because semiclassical considerations would fail in
these cases) the differential cross section can be written as[10]

$$\frac{d\sigma}{d\Omega} = \sqrt{(\sigma/\sigma_R)_i \, (\sigma/\sigma_R)_f} \cdot \overline{\sigma_R} \cdot P_{tr}(\theta) \cdot F(Q)$$

where $\overline{\sigma_R}$ is an averaged Rutherford scattering cross section. The Q-value
dependence of the transfer cross section is described by the function
$F(Q)$. In order to extract the transfer probability $P_{tr}(\theta)$ at the grazing
angle or distance of closest approach one has to estimate this Q-value
dependant factor. For $(^{16}O,^{14}C)$ reactions von Oertzen et al.[21] have
shown that semiclassical and DWBA estimates of F(Q) agree to within 20%.

Fig. 12 shows an analysis of data for the $^ASn(^{18}O,^{16}O)^{A+2}Sn$ g.s.
reactions obtained[22] at 60 MeV incident energy for A = 112 to 124. The
calculations were performed with the DWBA program LOLA (ref. 5). These
data are especially suited to test a calculated Q-value dependence for
neutron transfer reactions because all these g.s. transitions should be
of equal strength, and the Q-values vary over a range of $\sim$ 5 MeV. A
two-neutron cluster was assumed in the calculations, and optical model
parameters which fit the 72 MeV ^{18}O + ^{120}Sn and 74 MeV ^{16}O + ^{122}Sn
elastic and inelastic scattering data[12,16]. The experimental data
shown in fig. 12 are normalized by the corresponding relative (t,p)

cross sections[23]; this correction is 30 % at most. The calculations (solid curve, arbitrarily normalized by a factor 1.4) indicate that the Q-value dependence is correctly described. Furthermore, the same normalization is obtained from the analysis of the $^{18}O + {}^{120}Sn \rightleftharpoons {}^{16}O + {}^{122}Sn$ g.s. transition data shown in fig. 11.

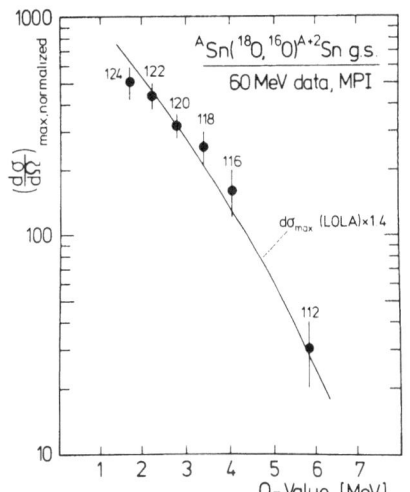

Fig. 12 Experimental and calculated maximum cross sections for two-neutron transfer reactions on even A Sn nuclei, plotted vs. Q-value.

Two examples for transfer probabilities will be quoted: (a) Von Oertzen's analysis of the $(^{16}O,^{14}C)$ two proton transfer ground state transitions for N = 82 target nuclei [21], which yields $P_{tr}(2p) \sim 10^{-3}-10^{-2}$. A comparison with the one-proton transfer cross sections for the same target nuclei indicates that $P_{tr}(2p)$ is approximately 25 times the squared one-proton transfer probability, i.e. $P_{tr}(2p) \sim EF \cdot |P_{tr}(1p)|^2$, with an enhancement factor EF $\sim$ 25. Thus the two-proton transfer cross sections on N = 82 target nuclei are enhanced quite appreciably.

(b) Similar results are obtained for the $^{18}O + {}^{120}Sn \rightleftharpoons {}^{16}O + {}^{120}Sn$ two-neutron transfer reactions to the corresponding ground states (see fig. 11). Using the experimental information[16] on the one- and two-neutron transfer cross sections we obtain $|P_{tr}(1n)|/|P_{tr}(2n)| \sim 5$ at $\theta_{CM} \sim 65°$, and an enhancement of the two-neutron cross section EF $\sim$ 25. Furthermore, in the $^{A}Sn(^{18}O,^{16}O)^{A+2}Sn$ reactions[16,22], the ground state transitions are enhanced by factors 30 - 50 over the transitions to the first excited 2^+ states in the Sn nuclei.

5.4. TWO-STEP, SIMULTANEOUS AND SUCCESSIVE TRANSFER PROCESSES.

Glendenning and Ascuitto's calculation[24] of the influence of inelastic excitations on the $^{18}O + {}^{120}Sn \rightleftharpoons {}^{16}O + {}^{122}Sn$ two-neutron transfer process has stimulated experimental investigations of this reaction, as well as of other two-particle transfer reactions[25], like $(^{16}O,^{14}C)$ and $(^{12}C,^{14}C)$. As discussed above the inelastic excitation

of Sn nuclei by ^{16,18}O projectiles itself is described quantitatively
by theoretical calculations. According to the coupled channel calculations these inelastic excitations would not influence the g.s. to g.s.
two-neutron transfer differential cross sections, but would have two
effects on the angular distributions for transitions to the first
excited 2^+ states: a) A destructive interference (i.e. reduced cross
section) near to the grazing angle for the (^{18}O,^{16}O) stripping reaction,
and a constructive interference in the same angular range for the
(^{16}O,^{18}O) pickup reaction. b) A forward rise and oscillations in the
angular distribution for the stripping reaction. This second effect
is calculated to be most pronounced at incident energies $\geq$ 100 MeV,
and barely evident in the calculations for 80 MeV incident energy.

The data shown in fig. 11 are consistent with these predictions.
The ^{120}Sn(^{18}O,^{16}O)^{122}Sn reaction at 72 MeV incident energy does not
show a forward rising cross section for the 2^+ state. A comparison of
the maximum cross sections for stripping and pickup to the 2^+ states
indicates, however, a systematic difference which is of the predicted
order of magnitude and not due to a Q-value dependence: The reduced
(i.e. Q-value corrected) cross section for stripping is only one third
of the cross section for pickup. A crude semiclassical estimate (two-
neutron g.s. to g.s. transfer probability $\sim 10^{-3}$; inelastic excitation
cross sections $\sim$ 10 mb/sr) indicates that the two-step contribution to
the 2^+ cross section could be as large as 10 µb/sr, which is of the
right order-of-magnitude to explain the observed difference between
the stripping and pickup cross sections.

Data obtained at higher incident energies, and on other reactions
which actually show the predicted forward rise and oscillations in the
2^+ cross section will be discussed by Scott[26].

The question whether two or more nucleons are transferred in one
step (simultaneously, as a cluster) or in two steps (sequentially, via
a number of intermediate states) has been discussed recently[27]. According to these calculations both processes are of comparable magnitude
in the ^{94}Mo(^{18}O,^{16}O)^{96}Mo g.s. and ^{208}Pb(^{16}O,^{14}C)^{210}Pb g.s. reactions.
Experimental evidence for the importance of multistep processes cannot
easily be extracted from the data, because of the presence of the
one-step cluster transfer, and the difficulty to calculate correct
absolute cross sections for multi-nucleon transfer reactions. Therefore,
one should either compare calculated and measured _ratios_ of cross
sections, or investigate a case in which the one-step cluster transfer
is not possible.

Examples for the first type of information might be the (^{16}O,^{18}O) and $(^{18}O, ^{20}O)$ two-neutron pickup reactions, for instance the $^{48}Ca(^{16}O, ^{18}O)^{46}Ca_{g.s.}$, $^{120}Sn(^{18}O, ^{20}O)^{118}Sn_{g.s.}$, and $^{122}Sn(^{16}O, ^{18}O)^{120}Sn_{g.s.}$ reactions. In these cases[15,16] the cross sections for processes in which the light particle (^{18}O or ^{20}O) is left in its first excited 2^+ state are much larger than expected from the corresponding $^{16,18}O(t,p)$ reactions. Therefore, it would be useful to have a theoretical estimate of these cross sections, or ratios of these and the corresponding g.s. transition cross sections, for simultaneous and successive transfer of two neutrons.

An example for the second type of information is provided by the $^{48}Ca(^{16}O, ^{15}C)^{49}Ti$ reaction studied at Argonne[28]. This reaction in which two units of charge and one unit of mass are transferred cannot be regarded as a single-step cluster transfer process. Whatever the details of the reaction mechanism may be, the experimentally determined cross sections provide a measurement of the strengths of transfer processes other than the one-step processes usually considered.

An (angle-integrated) energy spectrum for this reaction is shown in fig. 13. The two most-energetic groups result from the population of the $1f_{7/2}^{-1}$ ^{49}Ti g.s., with ^{15}C in its $2s_{1/2}$ g.s., and $1d_{5/2}$ first excited state, respectively. The angular distributions for these transitions are forward peaked, and the differential cross sections are of the order 5 - 50 μb/sr. These cross sections are of comparable magnitude as those for conventional cluster transfer reactions on ^{48}Ca, like the $(^{16}O, ^{13}C)$, $(^{16}O, ^{14}N)$, and $(^{16}O, ^{18}O)$ reactions to low-lying final states.

The observed strength of the $^{48}Ca(^{16}O, ^{15}C)^{49}Ti$ reaction is consistent with a crude semiclassical estimate. Consider the transitions to be the $1f_{7/2}^{-1}$ ^{49}Ti g.s. via the two steps $(^{16}O, ^{17}O)$ and $(^{17}O, ^{15}C)$, where the intermediate states are those of ^{47}Ca (the $1f_{7/2}^{-1}$ g.s.), and ^{17}O (the $1d_{5/2}$ g.s., and the $2s_{1/2}$ first excited state). If one assumes that the neutron configurations

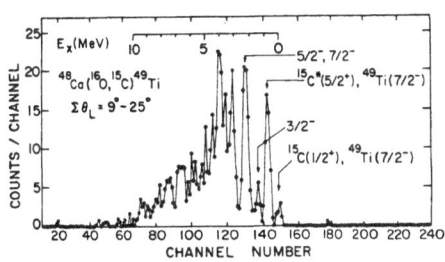

Fig. 13 Angle-integrated ^{15}C spectrum observed in the $^{16}O + ^{48}Ca$ reaction at 56 MeV

remain unchanged during the two-proton transfer process, the transition to the $2s_{1/2}$ ^{15}C g.s. proceeds via the $2s_{1/2}$ ^{17}O first excited state, and the transition to the $1d_{5/2}$ ^{15}C excited state via the $1d_{5/2}$ ^{17}O g.s., respectively.

The one-neutron pickup probability $P_{tr}|(^{16}O,^{17}O)|$ at a distance of closest approach where most of the subsequent $(^{17}O,^{15}C)$ reaction originates is determined to be $\sim 5 \times 10^{-2}$ from the experimentally known $^{48}Ca(^{16}O,^{17}O)^{47}Ca$ cross section. The total cross section for the $(^{17}O,^{15}C^*)$ reaction is taken to be 400 µb, as measured for the $^{48}Ca(^{16}O,^{14}C)^{50}Ti_{g.s.}$ reactions (DWBA calculations for both reactions indicate that this assumption is justified). Thus this estimate predicts $\sigma(^{16}O,^{15}C^*) \sim 20$ µb, which is consistent with the experimental value, $\sigma_{exp} \sim 15$ µb.

Furthermore, one expects a simple result for the ratio of the $^{15}C^*$ excited state to the ^{15}C g.s. cross section, with ^{49}Ti in its ground state:

$$\frac{\sigma(^{15}C^*(5/2^+))}{\sigma(^{15}C_{g.s.}(1/2^+))} \simeq K(Q) \frac{\sigma(^{17}O_{g.s.}(5/2^+))}{\sigma(^{17}O^*(1/2^+))} \quad ,$$

where the cross sections to the right are those observed for the $^{48}Ca(^{16}O,^{17}O)^{47}Ca_{g.s.}$ reaction, and $K(Q)$ corrects for the different Q-values. The measured ratio for ^{15}C, having ^{49}Ti in its ground state is $\sigma(^{15}C^+)/\sigma(^{15}C_{g.s.}) = 5 \pm 2$; the expected ratio is ~ 8, in rather good agreement with each other.

6. Concluding Remarks

Experimental techniques have been improved considerably over the last years: a) Much better energy resolution than previously possible was achieved; thus many more final states could be resolved. b) Cross sections were measured at forward angles, and rapid oscillations were observed in inelastic scattering and transfer reactions. c) Time-of-flight techniques allowed to accumulate an enormous amount of data, in particular to measure simultaneously most of the cross sections for all quasielastic channels in specific cases. d) More and more information about fusion cross sections is becoming available.

We have tried to demonstrate that large ambiguities in optical model parameters remain, even if the analysis includes experimental information on various transfer processes obtained at the same incident energy. We have therefore argued that it may be useful to investigate the various "absorption phenomena" in somewhat greater detail. It appears that backward-angle excitation functions are especially useful to scan the radial dependence of the surface-peaked reaction channels, like inelastic excitation and few-nucleon transfer processes, for incident energies near to the Coulomb barrier. Such data are complementary to those about the fusion process.

We have found examples for $^{16,18}O$ induced reactions, where the quasielastic processes appreciably contribute to the total reaction cross section at incident energies near to and above the Coulomb barrier. For ^{32}S induced reactions on light targets fusion appears to be the dominant reaction channel. Hopefully, this direct information on "absorption phenomena" can be combined with the indirect information about the surface part of the optical model obtained from fits to forward-rising, oscillating transfer angular distributions.

The concept of transfer probabilities was used to estimate enhancements in two nucleon transfer reactions between highly correlated ground state configurations in N = 82 and Z = 50 nuclei. It was shown that reliable estimates of the Q-value dependence can be made for two neutron transfer reactions induced by oxygen ions.

The question of additional inelastic excitations and successive transfer in multi-nucleon transfer reactions was investigated. The importance of the first of these processes appears to be established by a number of experiments, the second one is difficult to isolate experimentally. It is suggested that a comparison of $(^{16}O, ^{18}O_{g.s.})$ and $(^{16}O, ^{18}O^+(2^+))$, and the corresponding $(^{18}O, ^{20}O)$ reactions may be a way to establish the relative importance of successive transfer processes in two-neutron transfer reactions. The $(^{16}O, ^{15}C)$ reaction was discussed as an example for an "exotic reaction" in which at least two transfer processes must be involved.

References

1) Earlier investigations are summarized in the ANL Symposium on Heavy Ion Transfer Reactions, Argonne Physics Division Informal Report PHY-1973 (1973)

2) As an example, see the comparison of $(^6Li,d)$ and $(^{16}O, ^{12}C)$ reactions by J. R. Erskine, W. Henning, and L. G. Greenwood, Phys. Lett. 47 B (1973) 335

3) P. Colombani, J. C. Jacmart, N. Puffé, M. Riou, C. Stéphan, P. P. Singh, and A. Weidinger, Phys. Lett. 48 B (1974) 315

4) The following discussion is due to W. Henning, Argonne National Laboratory

5) R. M. DeVries, Phys. Rev. C 8 (1973) 951

6) W. Henning, D. G. Kovar, J. R. Erskine, and L. R. Greenwood, to be published

7) K. W. Ford and J. A. Wheeler, Ann. of Phys. 7 (1959) 259

8) R. A. Broglia, S. Landowne, R. A. Malfliet, V. Rostokin, and Aa. Winther, Physics Reports 11 C, number 1 (1974) 1

9) R. A. Malfliet, S. Landowne, and V. Rostokin, Phys. Lett. 44 B (1973) 238

10) P. R. Christensen, V. I. Manko, F. D. Becchetti, and R. J. Nickles, Nuclear Physics A 207 (1973) 33

11) Y. Eisen, R. A. Eisenstein, U. Smilansky, and Z. Vager, Nuclear Physics A 195 (1972) 513

12) K. E. Rehm, H. J. Körner, M. Richter, H. P. Rother, J. P. Schiffer, and H. Spieler, to be published

13) R. A. Broglia, Int. Conf. on Reactions Between Complex Nuclei (Nashville, 1974) Vol.2

14) R. A. Broglia, S. Landowne, and Aa. Winther, Phys. Lett. 40 B (1972) 293

15) B. Zeidmann, W. Henning, and D. G. Kovar, Nucl. Instr. and Meth. 118 (1974) 361

16) H. Spieler, Thesis, Technische Universität München, 1974

17) W. von Oertzen, C. E. Thorn, A. Z. Schwarzschild, and J. D. Garret, Int. Conf. on Reactions Between Complex Nuclei (Nashville, 1974) Vol. 1, p. 83

18) H. H. Gutbrod, W. G. Winn, and M. Blann, Nuclear Physics A 213 (1973) 267

19) M. Richter, Thesis, Technische Universität München, 1973

20) H.L.Harney, P.Braun-Munzinger, and C.K.Gelbke, Z.Physik 269 (1974) 339

21) W. von Oertzen, H. G. Bohlen, and B. Gebauer, Nuclear Physics A 207 (1973) 91

22) H. G. Bohlen, K. D. Hildenbrand, K. I. Kubo, and A. Gobbi, Int. Conf. on Reactions Between Complex Nuclei (Nashville, 1974) Vol.1, p. 75

23) R. A. Broglia, O. Hansen, and C. Riedel: Two-Neutron Transfer Reactions and the Pairing Model; in Advances in Nuclear Physics (ed. M. Baranger and M. Vogt) Vol. 6 (1973)

24) N. K. Glendenning and R. J. Ascuitto, ANL Symposium on Heavy Ion Transfer Reactions, Argonne Physics Division Informal Report PHY-1973 B (1973) 513

25) K. Yagi, D. L. Hendrie, L. Kraus, C. F. Maguire, J. A. Mahoney, D. K. Scott, and Y. Terrien, Int. Conf. on Reactions Between Complex Nuclei (Nashville, 1974) Vol. 1, p. 99; M.-C. Lemaire, M. C. Mermaz, H. Sztark, and A. Cunsolo, ibid, p. 69.

26) D. K. Scott, contribution to this conference

27) U. Götz, M. Ichimura, R. A. Broglia, and A. Winther, Int. Conf. on Reactions Between Complex Nuclei (Nashville, 1974) Vol. 1, p. 74

28) D. G. Kovar, W. Henning, B. Zeidmann, Y. Eisen, and H. T. Fortune, to be published

OSCILLATIONS IN THE ANGULAR DISTRIBUTIONS OF HEAVY-ION TRANSFER REACTIONS[*]

J. D. Garrett
Brookhaven National Laboratory
Upton, New York, USA 11973

Oscillations in the angular distributions are established as a general feature of heavy-ion induced, few-nucleon transfer reactions at incident energies sufficiently above the Coulomb barrier. For proper Q-matching with incident and exiting energies well above the Coulomb barrier, the angular position of the most forward maxima in the differential cross section is characteristic of the transferred angular momentum. The threshold for such oscillations is particularly sensitive to the absorptive potential near the nuclear surface requiring potentials that are strongly absorbing in the nuclear interior but nearly transparent at the surface. Evidence also is presented for an energy dependent surface absorption. Certain anomalous cases where the oscillating cross sections are predicted to be out of phase with the data are discussed.

1. Introduction

Recent experimental studies (c.f. refs. [1-10]) have indicated that large forward angle cross sections which oscillate as a function of angle are a general feature of heavy-ion induced, few-nucleon transfer reactions at incident energies sufficiently above the Coulomb barrier. This development has stimulated renewed experimental and theoretical activity in such measurements that previously had been considered by many as uninteresting. I should like to discuss some of the features of these angular distributions concentrating on the heavy-ion physics that can be learned from such studies. Since most of the audience is familiar with the subject, I intend to outline only briefly the general features of such angular shapes, and then to discuss some of the recent data and what can be learned from it. Finally, perhaps I will be able to speak to some questions that were raised at our Brookhaven informal study on this subject last summer. I will use Brookhaven data to illustrate my points, but I wish to stress that similar data has been measured at many other European and American laboratories.

It must be remarked that large, oscillating forward angle cross sections previously had been anticipated by several theorists (c.f. refs. [11-13]) and in fact studied in detail for light targets here at the Max-Planck-Institute[14]) as well as

[*] This work performed under the auspices of the U. S. Atomic Energy Commission.

elsewhere. As early as 1961 large cross sections were observed forward of the grazing angle in radiochemical total cross section data obtained at Yale[15]). The new element revealed by the recent detailed measurements is the low threshold for the large forward angle cross sections in certain cases. Furthermore, measurements at higher energies suggest[3]) that such structure is a general feature of few-nucleon transfer reactions sufficiently above the Coulomb barrier. Oscillating forward angle cross sections have been observed[10]) for few nucleon transfer induced by light heavy ions on intermediate mass targets having Sommerfeld parameters, $\eta = Z_1 Z_2 e^2 / \hbar v$ ~ 20.

2. General features of oscillating angular distributions

Typical angular shapes observed in such transfer reactions are shown in fig. 1

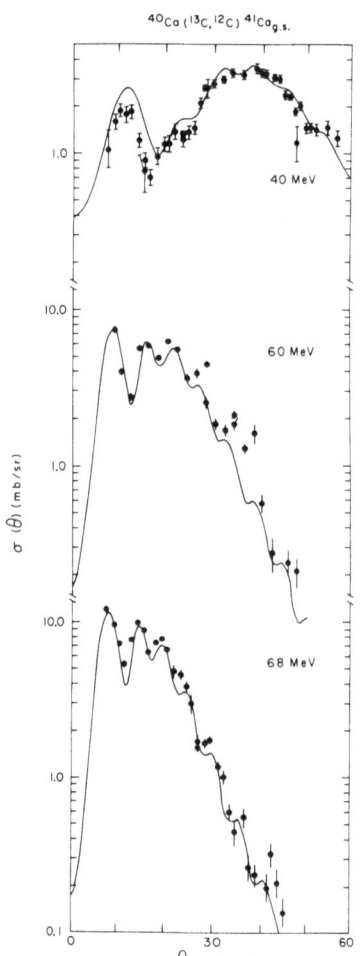

$^{40}Ca\,(^{13}C,^{12}C)\,^{41}Ca_{g.s.}$

40 MeV

60 MeV

68 MeV

$\sigma\,(\theta)$ (mb/sr)

$\theta_{c.m.}$(deg)

Fig. 1. Angular distributions of the ^{40}Ca $(^{13}C, ^{12}C)$ transition to the ^{41}Ca ground state at 40, 60, and 68 MeV incident energies (refs. 3 and 16). The DWBA predictions (curves) reproduce the shape change between 40 and 68 MeV.

for the ^{40}Ca$(^{13}C, ^{12}C)$ transition populating the ^{41}Ca ground state[3,16]) measured at incident energies of 40, 60, and 68 MeV. At 60 and 68 MeV pronounced oscillations are observed in the data. At 40 MeV incident energy, however, a well developed, broad peak is observed near the angle corresponding to a grazing trajectory and at more forward angles a smaller peak is observed. The detailed change in shape and magnitude is reproduced in terms of DWBA predictions.

The shape change from a relatively smooth grazing peak at 40 MeV to rapid oscillations at 68 MeV also has been explained[3]) in terms of a diffractive model based on a partial wave expansion of the cross section[1,11]). Such a parameterization approximates the ℓ space localization of the transfer amplitude by a Gaussian distribution and assumes a linear variation with ℓ of the transfer phase. Until recently such a semiclassical description

had been applied only to L=0 or to the M=0 magnetic substates for L≠0 transfer.
Even though such a model is qualitatively successful in describing the existence of
oscillations, it fails to describe the richness of heavy-ion induced transfer
phenomena. One improvement would be to extend such an approach to include M≠0 components and to incorporate an expansion which is valid at small angles. Such a
description has been developed by Kahana, Bond, and Chasman[17,18]. I do not wish
to discuss the details of such a description, since it has been covered explicitly
in a recent publication[17] and in the talk of Kahana at the Nashville Conference on
Reactions Between Complex Nuclei[18]. Such a description, however, shows that the
partial cross section corresponding to the $|M|$=L magnetic substate will dominate over
the $|M|$<L substates when: i) the reaction is well matched, i.e. when the localiza-
tion in ℓ space is the same for entrance and exit channels, and ii) the incident and
exiting energies are well above the Coulomb barrier. For such cases, the angular
position of the forward most peak is given by the first maximum in the Bessel function
$J_M[(\ell_f^o + 1/2)\,\theta_f]$ and satisfies the inequality

$$(M(M+2))^{1/2} < (\ell_f^o + 1/2)\,\theta_M < (2M(M+1))^{1/2}.$$

Such "L" or really M dependence is observed in experimental measurements and reproduced in the DWBA analysis[1,6]. Figure 2 demonstrates such dependence for the
$^{48}Ca(^{16}O,^{14}C)$ transitions[6] to the lowest 0^+, 2^+, 4^+, and 6^+ states of ^{50}Ti. The
solid curves in the figure are DWBA cross sections calculated using the finite range,
recoil code LOLA[19] with a two-nucleon cluster form factor as described in ref.[6].
This figure indicates the necessary precision in the extreme forward angle data to
distinguish transferred L values using such reactions. It is apparent that it is
easier to extract transferred L values from light-ion induced transfer if the equivalent transfer is possible with available projectiles.

3. DWBA analysis

Although the simple models are successful in describing the general features
of heavy-ion induced transfer reactions, and are an aid in understanding the physics
involved, the DWBA formalism still remains the main prescription in the analysis
of such data. The remainder of this talk will be based on such an analysis. It
already has been indicated that the energy dependence (see fig. 1) of the shape and
magnitude of the $^{40}Ca(^{13}C,^{12}C)^{41}Ca_{g.s.}$ transition and the observed L dependence (see
fig. 2) is reproduced in DWBA predictions.

The DWBA differential cross sections for single-nucleon transfer studies discussed were calculated using the finite-range DWBA code SRC of Baltz[20]. This code
includes the effects of recoil in a proper Taylor series expansion[21] keeping terms
to second order. For single-nucleon transfer the predicted cross section for the
reaction A(a,b)B is related to the stripping cross section through spectroscopic

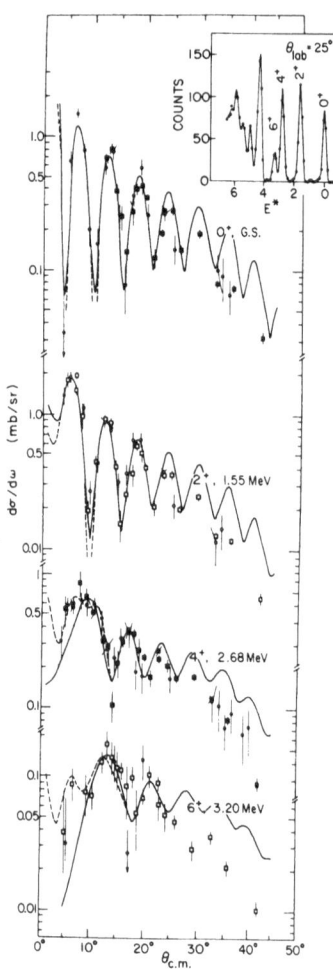

Fig. 2. Angular distributions for
$^{48}Ca(^{16}O,^{14}C)$ transitions
at an incident energy of
56 MeV populating the low-
est 0^+, 2^+, 4^+, and 6^+ states
in ^{50}Ti (ref. [6]). The solid
lines are DWBA predictions
calculated using the code LOLA.
The inset shows an energy spectrum.

factors.

$$\sigma_{exp}(\theta) = N\ c^2 S_{ab}\ c^2 S_{AB}\ \sigma_{SRC}(\theta)$$

A value of the normalization constant N
near one, therefore, indicates agreement
between measured and predicted absolute
cross sections. The two-nucleon transfer
version[22]) of the finite range code RDRC[23])
was used to predict the $^{60}Ni(^{18}O,^{16}O)^{62}Ni$
and $^{26}Mg(^{16}O,^{14}C)^{28}Si$ cross sections.
This code uses a microscopic form factor
but does not include the effects of recoil.
The parameters, spectroscopic factors and
two-nucleon spectroscopic amplitudes used
in these calculations are tabulated in the
appendix.

    4. Sensitivity to optical potentials
        The ℓ-space localization of the heavy-
ion induced transfer amplitude is a consequence of the decay of the radial form
factor affecting the fall off toward
higher ℓ and the absorptive effects of the
ion-ion optical potential governing the
decay toward lower ℓ's. The phase in the
region of the ℓ window also is quite
sensitive to the nuclear absorption. For
larger ℓ's the phases will be identical to
Coulomb phases. Near the ℓ window the
phases will deviate from the Coulomb phase
(for energies above the Coulomb barrier)
as the effects of the real nuclear potential
become important. However as the absorption increases the effects of the nuclear
potential will decrease (see e.g. figs.
19 and 21). It, therefore, is not surpris-
ing that the details of the imaginary potential near the nuclear surface are important
for transfer reactions.

    Angular distribution shapes with very large forward peaks and a well formed
"bell-shaped" peak at the "grazing" angle are particularly sensitive to the details

of the imaginary potential. Such shapes represent the onset of oscillations. A typical example of such an angular shape is the ^{40}Ca(^{13}C, ^{14}N) transition to the ^{39}K ground state, $J^{\pi} = 3/2^{+}$ (fig. 3), measured[16,24]) at an incident energy of 40 MeV.

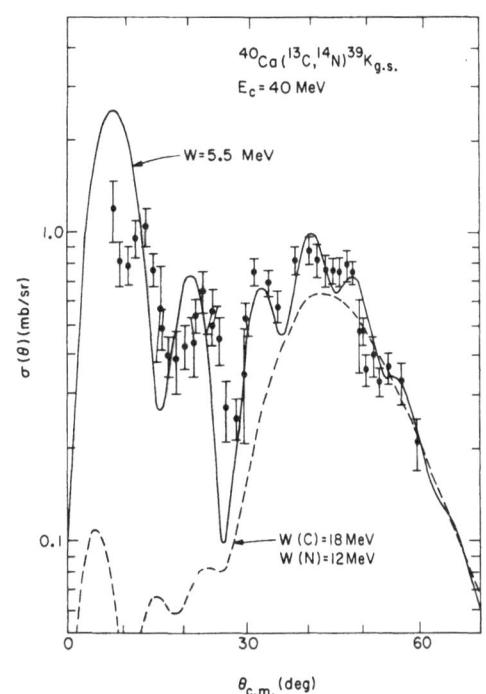

^{40}Ca(^{13}C,^{14}N)^{39}K$_{g.s.}$

$E_c = 40$ MeV

W = 5.5 MeV

W (C) = 18 MeV
W (N) = 12 MeV

$\theta_{c.m.}$ (deg)

Fig. 3. Angular distribution of the ^{40}Ca (^{13}C, ^{14}N) transition to the ground state of ^{39}K measured at an incident energy of 40 MeV (refs.[16,24]). The curves are DWBA predictions using optical-model parameter set 2, table A1. Only the magnitude of the absorption has been varied for the two different calculations shown.

A large, rapidly varying differential cross section is observed at the most forward angles while smaller amplitude oscillations, superimposed on a broad peak, are observed near angles corresponding to the classical grazing trajectory. Shown with the data are DWBA calculations based on optical-model parameters having identical geometry for both the real and imaginary potential wells. The dashed curve represents calculations based on parameters (set 2 of Table A1) which reproduce ^{13}C and ^{14}N elastic scattering on ^{40}Ca over a range of incident energies[3,24]). The calculation reproduces neither the large forward angle cross section nor the variation of the cross section as a function of angle. If the imaginary potentials in both the entrance and exit channels are reduced to W = 5.5 MeV, while leaving the other parameters unchanged, the forward angle cross sections are strongly enhanced and strong oscillations are predicted (solid curve in fig. 3). However, the oscillations of the curves based on these weakly absorbing potentials are out of phase with the most forward angle data points. The angular positions of the maxima at the forward angles also have changed relative to their predicted positions when the more strongly absorbing potentials are used.

The change in the angular positions of the forward angle maxima is the result of contributions from amplitudes corresponding to large nuclear overlaps. The predicted angular distributions for the ^{40}Ca(^{13}C, ^{14}N)^{39}K$_{g.s.}$ reaction as a function of a lower radial cutoff radius, R_{co}, are shown in fig. 4. These calculations indicate that for such weakly absorbing potentials contributions are obtained well inside the channel radii. In fact, at forward angles large contributions are observed for less than a 5 fm separation of the nuclear centers. The angle of the most forward maxima

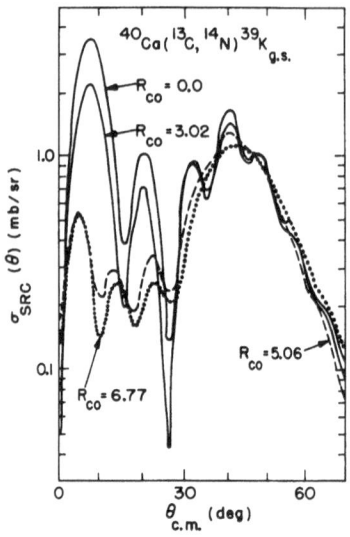

Fig. 4. Predicted angular distribution of the $^{40}Ca(^{13}C,^{14}N)^{39}K_{g.s.}$ cross section at an incident energy of 40 MeV as a function of a lower cutoff on the radial integration, R_{co}. Optical-model parameters set 2 of table A1 with an absorption of 5.5 MeV, (which correspond to the solid curve in fig. 3) were used in all the calculations.

also changes when the contributions from the nuclear interior are removed.

With this extremely weak absorptive potential the lower partial waves have a significant effect upon the cross section. In fig. 5 the transition amplitudes B_ℓ^{LM} as a function of the outgoing partial wave ℓ are shown for the M=1 substate. The change in the position of the calculated maxima in the weakly absorbing case arises from a beating of the frequencies of oscillation corresponding to the two peaks in B_ℓ^{LM} of fig. 5. When the first 15 partial waves are removed from this calculation the position of the maxima returns to that calculated using the strong absorptive potential (see figs. 3 and 6).

The contributions to calculated heavy-ion transfer cross sections from deep within the nuclear interior that arise as a result of weakly absorbing ion-ion potentials are unphysical. The absorption in the nuclear interior must be increased to reduce such contributions, and yet it is necessary to maintain a weak absorption in the nuclear surface region to obtain the large forward-angle

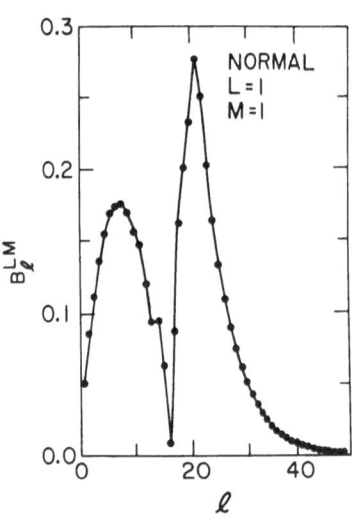

Fig. 5. Calculated M=1 transition amplitude as a function of outgoing ℓ corresponding to the angular distribution shown as a solid curve in fig. 3. Large values are predicted for low partial waves.

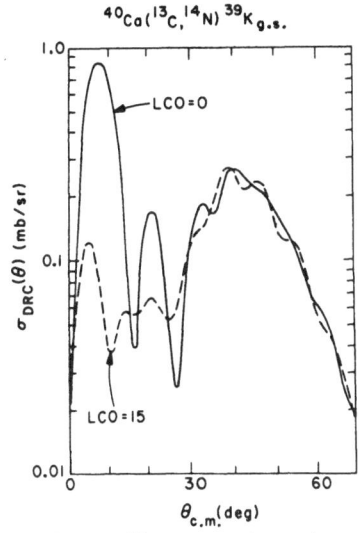

$^{40}Ca(^{13}C,^{14}N)^{39}K_{g.s.}$

Fig. 6. Comparison of predicted angular distribution shown as a solid curve in fig. 3 with a calculation having the contributions for all partial waves < 15 removed (see fig. 5).

cross sections that are observed, e.g., in the $^{40}Ca(^{13}C,^{14}N)^{39}K$ ground-state transition (see fig. 3). It is particularly important that potentials which do not have unphysical contributions from the nuclear interior should be used when spectroscopic information is to be obtained from the analysis. As seen in figs. 4 and 6 such contributions can shift the position of the most forward maxima which is sensitive to the L dependence of the transfer.

At Brookhaven we have been using an imaginary potential shape which combines a strong absorption in the nuclear interior and weak absorption at the nuclear surface with an imaginary diffusivity at the nuclear surface similar to that of nuclear matter. The sum of a Woods-Saxon shape with a small diffusivity and a surface derivative term with a larger diffusivity is used for the imaginary potential. Strong absorption from the Woods-Saxon shape reduces the contributions from the nuclear interior while the weaker imaginary potential at the nuclear surface with a "reasonable" diffusivity gives the absorption at the radii where transfer takes place. Predictions based on such surface transparent potentials are shown in fig. 7 for the 40 MeV $^{40}Ca(^{13}C,^{14}N)$ and $^{40}Ca(^{13}C,^{12}C)$ transitions to the ground states of ^{39}K and ^{41}Ca. The shape and magnitude of the $(^{13}C,^{12}C)$ angular distribution is well reproduced by these calculations. The general shape and the magnitude of the more structured $(^{13}C,^{14}N)$ angular distribution is given, but the details of the most forward angle data still are not completely reproduced. However, predictions based on these potentials give a better description of the forward angle $^{40}Ca(^{13}C,^{14}N)^{39}K$ g.s. data than calculations using potentials with the same real and imaginary Woods-Saxon geometry (see, e.g., fig. 3).

The $^{40}Ca(^{13}C,^{14}N)^{39}K$ g.s. transition amplitudes, B_ℓ^{LM}, calculated for the M=1 magnetic substate using these potentials are shown as a function of ℓ in fig. 8. As a result of the stronger absorption in the nuclear interior, the large contributions that were observed for the small partial waves using the weakly absorbing Woods-Saxon potentials (fig. 5) are greatly reduced in these calculations. In fig.9 the elastic scattering of ^{13}C from ^{40}Ca at 40 MeV is calculated with these surface transparent potentials and compared with experimental data[16]. The experimental cross sections are reproduced.

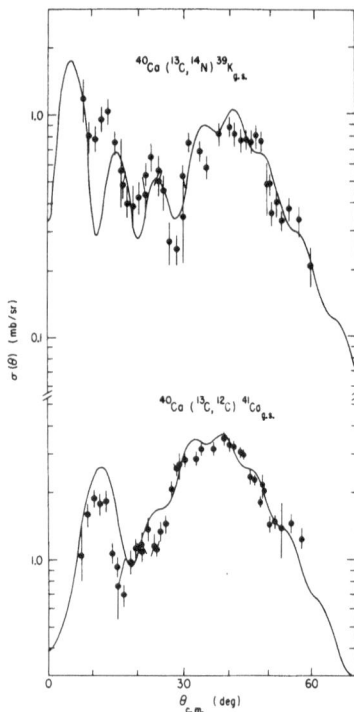

Fig. 7.   Comparison of measured and
predicted angular distributions for the 40 MeV ^{40}Ca
(^{13}C,^{14}N)^{39}K$_{g.s.}$ and ^{40}Ca
(^{13}C,^{12}C)^{41}Ca$_{g.s.}$ transitions.
The calculated angular shapes
used surface transparent
potentials as discussed in the
text and given as set 1 in
table A1.

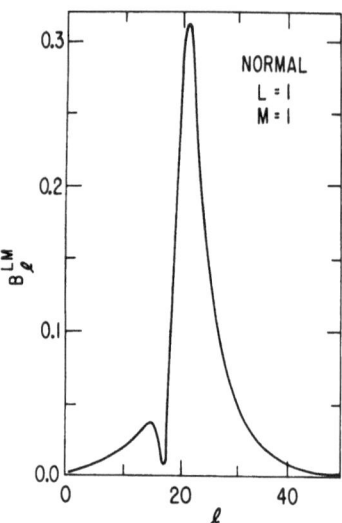

Fig. 8.   Calculated M=1 transition
amplitude as a function of
outgoing ℓ using surface
transparent potentials.
These potentials were used to
calculate the (^{13}C,^{14}N) angular distribution shown in
fig. 7.   The large values predicted for low partial waves
using weak absorptive potentials of Woods-Saxon geometry
(fig. 5) have been reduced.

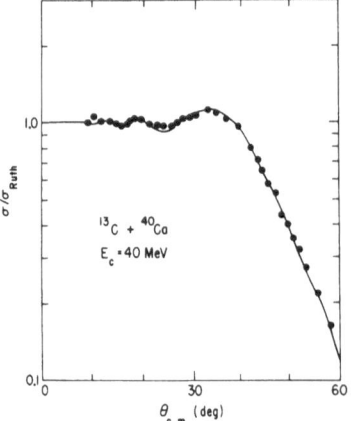

Fig. 9.   Comparison of measured and
predicted 40 MeV ^{13}C elastic
scattering from ^{40}Ca using the
surface transparent potentials
(set 1 table A1) that reproduce the transfer cross sections
(fig. 7).

In figs. 10 and 11 the calculated differential and total cross sections for the $^{40}Ca(^{13}C, ^{14}N)^{39}K$ ground-state transition at an incident energy of 40 MeV are shown

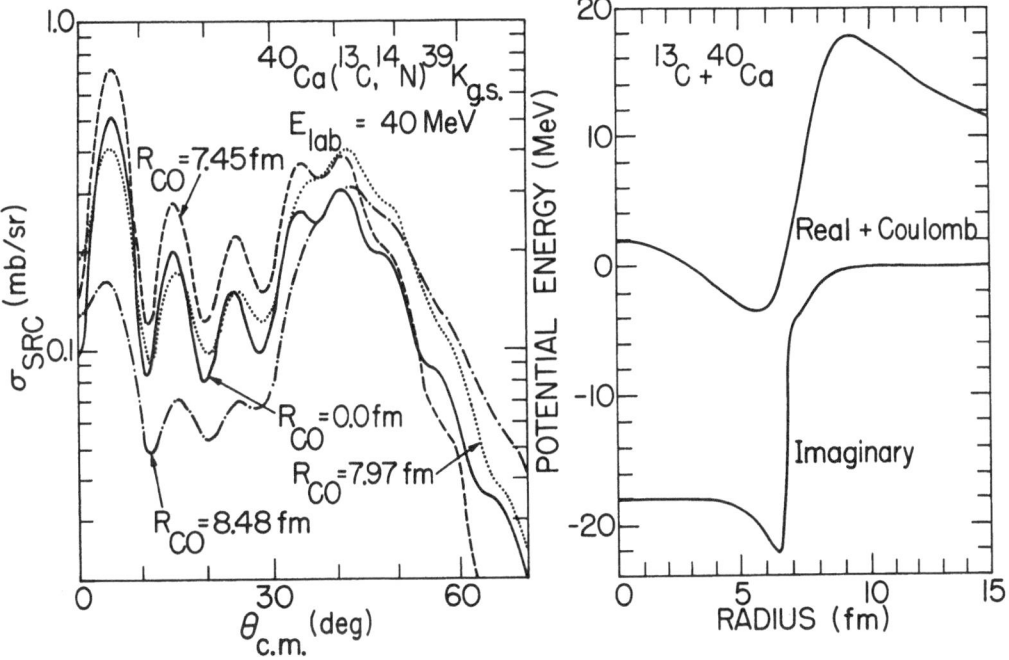

Fig. 10. Calculated 40 MeV $^{40}Ca(^{13}C, ^{14}N)^{39}K_{g.s.}$ differential cross sections as a function of a lower cutoff on the radial integration, R_{co}. These calculations used the surface transparent potentials (set 1 of table A1) discussed in the text. Predictions based on these potentials are shown with the experimental data in fig. 7. The radial shape of the Coulomb plus real nuclear and imaginary nuclear potentials for $^{13}C + ^{40}Ca$ also are shown.

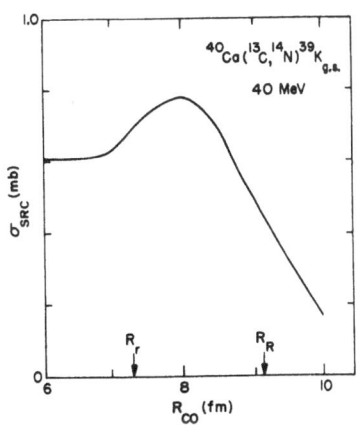

Fig. 11. Predicted total cross section for the 40 MeV $^{40}Ca(^{13}C, ^{14}N)^{39}K_{g.s.}$ transition as a function of a lower cutoff on the radial integration. See fig. 10 and text for potentials used. R_r and R_R indicate the real nuclear radius and the Rutherford radius in the entrance channel.

as a function of lower cutoffs on the radial integration. These calculations are based on the surface transparent potentials. For comparison the real plus Coulomb and imaginary nuclear potentials for the $^{13}C + ^{40}Ca$ channel are shown in fig. 10.

It must be emphasized that the use of a radial integration cutoff is a computational technique and that it does not completely remove the effects of the nuclear interior. It does indicate, however, at what radii large contribution to the cross section are obtained.

The predicted cross sections corresponding to radial cutoffs of 6-7 fm were nearly identical both in magnitude and shape to the calculation with no radial cutoff. Such was not the case for the weak Woods-Saxon imaginary potential which had significant contributions from the nuclear interior (see fig. 4). The rise in the total cross section for cutoffs between 7 and 8 fm result from removing certain contributions which cause cancellations in the cross section. The main contributions to the total cross section are still obtained near the Rutherford radius (i.e. the radius at which the Coulomb plus real nuclear potential is maximum--see fig. 10). The forward-angle cross section, however, starts to decrease relative to the "grazing" peak cross section for radial cutoffs between 7.5 and 8.0 fm and has nearly disappeared after R_{co} = 8.5 fm. Therefore the large forward peak seems to have its origin at radii inside the Rutherford radius, suggesting that this peak is connected to contributions from the opposite side of the nucleus that is either diffracted or refracted around the nucleus.

By using surface transparent potentials (set 1 of table A1) it also has been possible to reproduce the details of the oscillations observed in the $^{60}Ni(^{18}O,^{16}O)$ transition to the ^{62}Ni ground state[10]) (see fig. 12). Prediction using surface

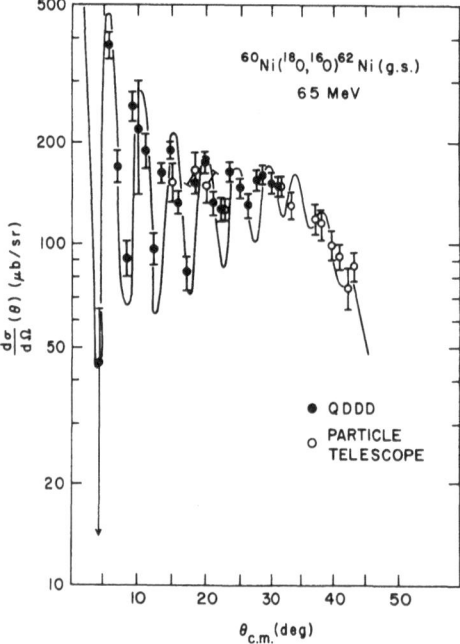

Fig. 12. Angular distribution corresponding to the $^{60}Ni(^{18}O,^{16}O)$ transition to the ground state of ^{62}Ni measured at an incident energy of 65 MeV (ref. [10]). Data indicated by solid points were recorded using the BNL QDDD spectrometer. The DWBA curves were calculated using surface transparent potentials (set 1, table A1).

transparent potentials which reproduce the experimental data[10]) are compared in fig. 13 with predictions based on weakly absorbing imaginary potentials of Woods-Saxon geometry (set 2 of table A1) which have been used in a previous analysis of this reaction[4]). The strong absorption in the nuclear interior has decreased the magnitude of the oscillation at the more backward angles. The different period of the oscillations at the forward angles in the calculations based on weakly absorbing imaginary potentials of Woods Saxon geometry probably results from a beating between contributions from the interior and the surface of the nucleus. Furthermore, nearly all the angular dependence on the single-particle configurations which is predicted using the potentials that are too weakly absorbing in the nuclear interior is removed in the calculation based on the surface transparent potentials fig. 14.

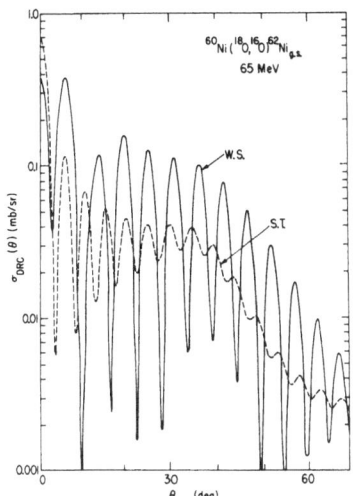

Fig. 13. Predictions based on surface transparent
potentials (set 1 table A1) which reproduce the experimental data (fig. 12) are
compared with predictions based on weakly
absorbing potentials of Woods-Saxon
geometry (set 2 of table A1) which were
used in a previous analysis of this
reaction (ref. 4). We have made progress
in a year, haven't we!

The important features of the potentials which
we have been using are: i) strong absorption in the
nuclear interior, ii) weak absorption near the nuclear surface, and iii) a diffusivity near the
nuclear surface that approximates the nuclear matter distribution. The suggested parameterization
is only one that incorporates these features. In
fact, imaginary potentials with a small diffusivity
and a radius somewhat smaller than the real nuclear radius are presently being used
by several groups, c.f. refs.[6,8]). Such an imaginary potential incorporates the first
two of the proposed features.

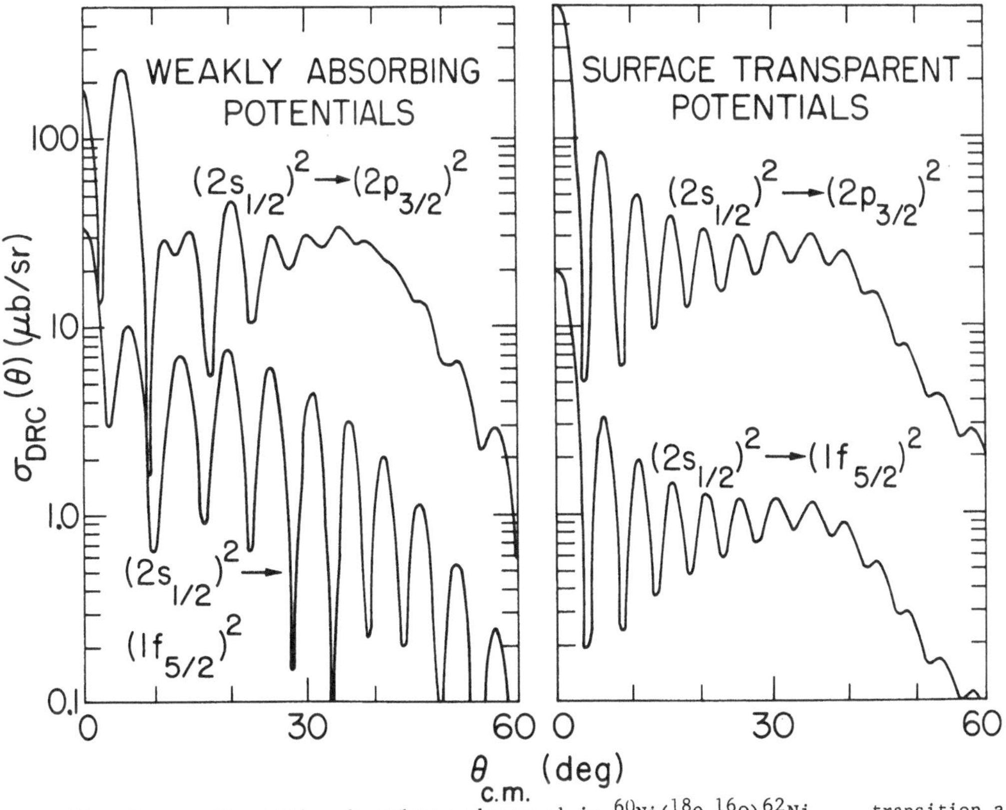

Fig. 14. The configuration dependence observed in $^{60}\text{Ni}(^{18}\text{O},^{16}\text{O})^{62}\text{Ni}_{g.s.}$ transition at
65 MeV using weakly absorbing potentials (set 2, table A1) is almost com-
pletely removed when surface transparent potentials are used (set 1, table A1).

5. Total quasielastic cross sections

Such imaginary potential shapes may have a physical basis. The interior and surface absorption almost certainly arise from different processes--compound processes dominating the interior and quasielastic processes important in the surface region. A sharp increase in the absorption might then be expected at the radius where the more violent compound processes become important. In order to determine if quasielastic processes contain sufficient total cross section to account for the partial cross sections of the surface partial waves, we have measured[25]) the total few-nucleon transfer reaction and inelastic scattering cross sections induced by ^{16}O on ^{208}Pb at bombarding energies of 82, 88, and 94 MeV. Heavy targets were chosen to allow the transfer cross sections to be measured above the Coulomb barrier without competition with reaction products from light target impurities. For the cases measured the reaction kinematics completely separate the groups of interest and products resulting from light target impurities in the angular region which contribute significantly to the total cross section. Angular distributions of the total yield of inelastically scattered ^{16}O and O, N, and C resulting from transfer is shown for 82 MeV ^{16}O incident on $^{206,208}Pb$ in fig. 15. Much smaller yields also were observed for Be and B. Total reaction cross sections were determined from the optical model using parameter giving a best fit to the

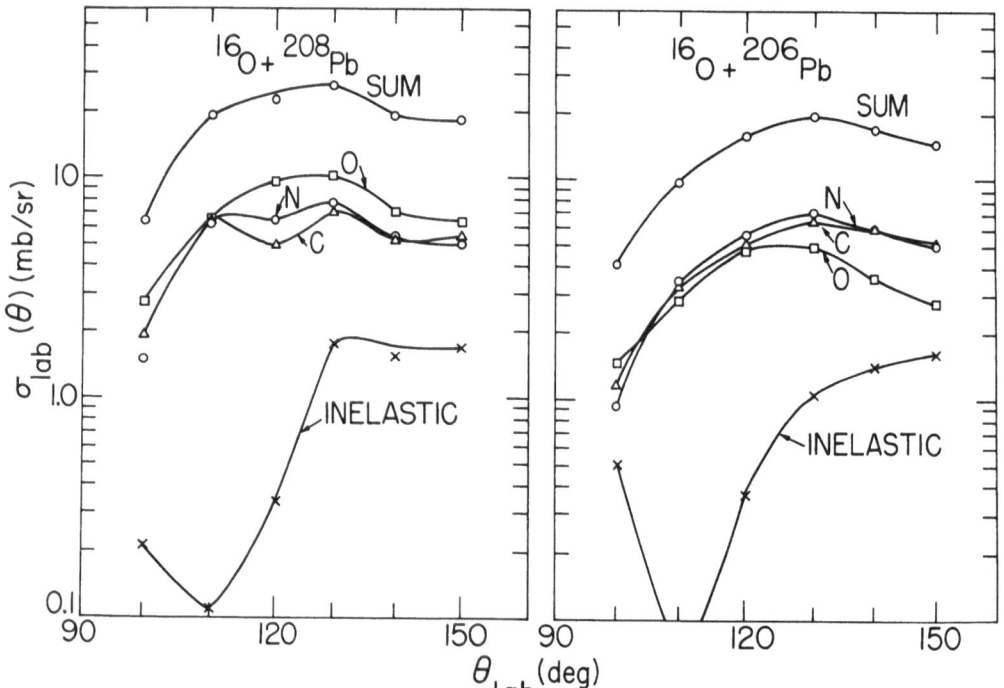

Fig. 15. Differential cross section as a function of angle for the total yields of O, N, and C resulting from transfer and inelastically scattered ^{16}O induced by 82 MeV ^{16}O incident on ^{206}Pb and ^{208}Pb (see ref.[25]). The summed cross section for all channels also is shown.

elastic scattering at each energy. The total reaction cross section is the sum over ℓ's of the partial cross section $\sigma_\ell(\theta)$ for each ℓ given by

$$\sigma_\ell = \pi \chi^2 (2\ell + 1) T_\ell$$

where T_ℓ is the transmission coefficient as calculated using the optical model code ABACUS[26]). The results of a preliminary analysis for ^{208}Pb are given in table 1.

Table 1. Summary of ^{16}O + ^{208}Pb Total Cross Section Data.

E_{lab} (MeV)	$E_{c.m.}$ (MeV)	σ_T [a] (mb)	σ_{QE} [b] (mb)	σ_{QE}/σ_T	ℓ_{crit} [c]
82	76.14	219	94	0.43	16
88	81.71	553	257	0.46	23
94	87.29	876	331	0.38	32

[a] From optical model fit of elastic scattering.

[b] Integrated experimental cross sections.

[c] See fig. 16--assumes sharp cutoff.

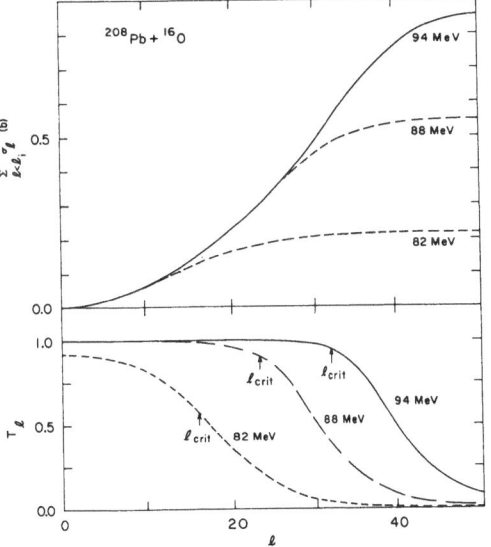

A surprisingly large fraction of the total reaction cross section ~ 40% is accounted for by the quasielastic processes even as high as 15 MeV above the Coulomb barrier. The sum of the partial cross sections σ_ℓ for $\ell < \ell_i$ is shown as a function of ℓ_i in fig. 16. The values of ℓ dividing the quasielastic cross section from other processes are shown as ℓ_{crit} on the plot of transmission coefficients in fig. 16. The ℓ_{crit} are obtained assuming that the partial cross sections for large ℓ's are all quasielastic and a sharp division in ℓ space between quasielastic and other processes. It is seen for the system studied in this energy region that the quasielastic cross sections are sufficiently large to account for the cross sections of the surface partial waves. Such a division between quasielastic and other more violent processes in ℓ-space is consistent with a difference in the radial dependence of the absorption near the nuclear surface and in the nuclear interior.

Fig. 16. Sum of partial cross section σ_ℓ for $\ell < \ell_i$ as a function of ℓ_i. For comparison transmission coefficients based on a fit to the ^{16}O + ^{208}Pb elastic scattering data at each energy is shown. The values of ℓ_{crit} separating contributions from quasielastic and other processes in a sharp cutoff approximation are indicated with the transmission coefficients. These values are determined from the measured quasielastic and the calculated total reaction cross sections given in table 1.

6. Energy dependence of the surface absorption from the $^{26}Mg(^{16}O,^{14}C)^{28}Si_{g.s.}$
    transition

I would now like to turn my attention to the $^{26}Mg(^{16}O,^{14}C)$ transition to the ^{28}Si ground state. Experimental data now exists for this transition at 33 and 40 MeV (ref.[27]), 45 MeV (refs.[2,7]), 60 MeV, (ref.[28]) and 128 MeV (ref.[29]) (see fig. 17).

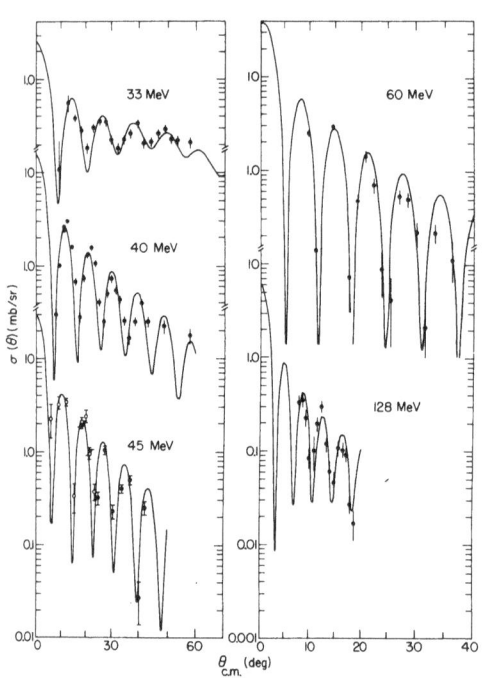

Oscillating cross sections are observed at all energies. Once again the lowest energy data is particularly sensitive to the imaginary potential shape. Calculations based on three different sets of optical model parameters are shown in fig. 18. The two dashed curves which have relatively weak absorption in the nuclear surface but are of Woods Saxon geometry with a diffusivity ~ 0.5-0.6 do not reproduce the oscillating experimental cross sections. Attempts to "fit" the data by reducing the imaginary potential keeping a fixed geometry have been unsuccessful. The solid curve in fig. 18 is based on the same real geometry as the long dashed curve, but with surface transparent imaginary potentials given as set 1 in table A1. The amplitudes, B_ℓ^{LM} and phase derivatives $\psi(\ell)$ are compared in fig. 19 for the calculations using weak Woods-Saxon imaginary potentials (long dashed curve of fig. 18) and surface transparent potential (solid curve of fig. 18). The surface transparent potential which produces oscillating cross sections is more localized in ℓ space and has a more pronounced dip in the phase derivative, $\psi(\ell)$, in the "ℓ-window".

Fig. 17. Differential cross sections for the $^{26}Mg(^{16}O,^{14}C)^{28}Si$ ground-state transition at 33 (ref.[27]), 40 (ref.[27]), 45 (refs.[2,7]), 60 (ref.[28]), and 128 (ref.[29]) MeV. DWBA predictions based on identical optical-model parameters (set 1, table A1) are shown with the 33-60 MeV data. At 128 MeV it is necessary to increase the absorption in the region of the nuclear surface to W_{SD}=10 MeV to reproduce the shape of the angular distribution. Note the change in angular scale between the left and right portions of the figure.

The weaker absorption for the surface region in these potentials presumably allows the transfer amplitude to build up faster, but then the rapid increase in absorption at smaller radii causes a sharp drop in the amplitude on the low ℓ side of the ℓ window. Similarly for the phase derivative a weaker absorption at large radii permits the real nuclear potential to have a greater effect and $\psi(\ell)$ deviates more from the

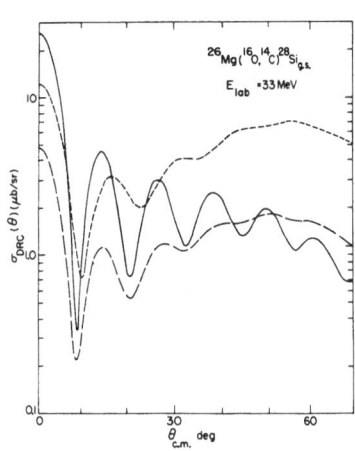

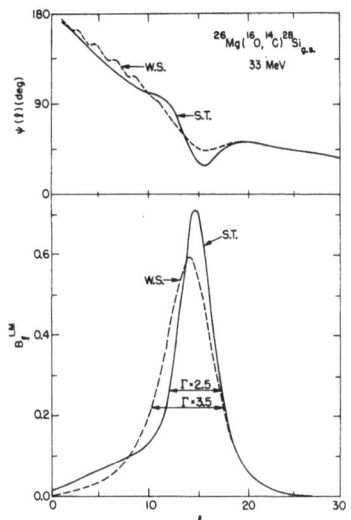

Fig. 18. Comparison of DWBA predictions for the $^{26}Mg(^{16}O,^{14}C)^{28}Si_{g.s.}$ transition at an incident energy of 33 MeV using three different optical potentials. The solid curve corresponds to surface transparent potentials (set 1, table A1) and is shown with data in fig. 17. The long and short dashed curves correspond to sets 2 and 3, respectively in table A1.

Fig. 19. Amplitudes, B_ℓ^{LM}, and phase derivatives, $\psi(\ell)$, as a function of ℓ for the solid and long dashed calculations shown in fig. 18. The 1/e half widths, Γ, are indicated for the B_ℓ^{LM}'s.

Coulomb phase change for surface transparent potentials. The added localization and a smaller $\psi(\ell)$ both enhance the probability of oscillations[1,3]).

The 33 MeV $^{26}Mg(^{16}O,^{14}C)^{28}Si_{g.s.}$ data are sensitive to the magnitude of the surface absorption in the surface transparent potentials (see fig. 20). An adequate

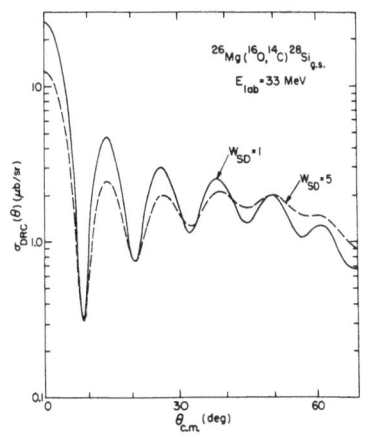

Fig. 20. Effect of increasing the surface absorption from W_{SD} = 1 to 5 for 33 MeV $^{26}Mg(^{16}O,^{14}C)^{28}Si_{g.s.}$ angular distribution.

description of the experimental data can be obtained for $W_{SD} \leqslant 2$ MeV. Using identical parameters with a W_{SD} = 1 MeV absorption in the surface region, the shapes of the oscillating cross sections can be reproduced for 33, 40, 45, and 60 MeV. The ratio of the experimental to calculated cross section, however, decreases a factor of 2 between 33 and 60 MeV incident energy (see table 2). The

Table 2. Surface Absorption and Normalizations for $^{26}\text{Mg}(^{16}\text{O},^{14}\text{C})^{28}\text{Si}_{\text{g.s.}}$ Transition analysis.

E_{lab} (MeV)	$E_{\text{c.m.}}$ (MeV)	W_{SD}(MeV) to fit shape[a)	$\sigma(\theta)_{\text{exp}}/\sigma(\theta)_{\text{DRC}}$	
			$W_{SD} = 1$ MeV	Energy Dep W_{SD}[b)
33	20.43	$\leqslant 2$	130	130
40	24.76	$\leqslant 4$	100	130
45	27.86	$\leqslant 8$	90	130
60	37.14	$\leqslant 10$	65	130
128	79.24	$\geqslant 5$[d)	c)	37

a)Range of surface absorption with parameter set 1, table A1 which fits angular shape.

b)$W_{SD} = 0.54$ ($E_{\text{c.m.}}$ - 18.58) MeV.

c)Shape of 128 MeV angular distribution not reproduced for $W_{SD} = 1.0$. $\sigma_{\text{exp}}(\theta)/\sigma_{\text{DRC}}(\theta)$ =20 for $W_{SD} = 10$ MeV.

d)Values as large as 100 MeV still reproduce the angular shape.

calculated cross section for the 128 MeV data using parameters identical to those used at the lower incident energies do not reproduce the shape of the experimental angular distribution. This problem is remedied by increasing the surface absorption to $W_{SD} = 10$ MeV. A calculation based on such parameters is shown with the experimental data in fig. 17. Therefore, an energy dependent surface absorption is necessary to repro- duce simultaneously the shape of the angular distribution at 33 and 128 MeV incident energy. An energy dependent surface absorption decreases the variation with incident energy of the normalization between calculated and experimental cross sections. In fact, for a surface absorption of

$$W_{SD} = 0.54 \ (E_{\text{c.m.}} - 18.58) \text{ MeV}$$

the variation of the normalization with energy is removed between 33 and 60 MeV. Such an energy dependence also is consistent with the surface absorption needed to reproduce the shape of the single-nucleon transfer angular distributions for 60 MeV ^{16}O incident on ^{26}Mg. The normalization between experimental and calculated cross sections for the 128-MeV data is increased by the energy dependent absorption; how- ever, it still is a factor of ~ 3.5 less than that of the lower energy data. This is not surprising, since the recoil corrections, which are not included in the present two-nucleon transfer calculations, are expected to vary considerably between 60 and 128 MeV.

Using such potentials the absolute experimental $^{26}\text{Mg}(^{16}\text{O},^{14}\text{C})^{28}\text{Si}$ ground state cross section is under-predicted by a factor of ~ 130 (see table 2). It is known[30] that a proper treatment of recoil increases the calculated cross section for this transition at 45 MeV incident energy. This is in contrast to a predicted[31] decrease in the $^{62}\text{Ni}(^{18}\text{O},^{16}\text{O})^{64}\text{Ni}$ ground state cross section at an incident energy of 65 MeV. Recent calculations[32] for $(^{16}\text{O},^{14}\text{C})$ reactions on ^{208}Pb and ^{48}Ca targets suggest that

increasing the basis of single particle states over what are used in the present calculations might increase the predicted cross sections by as much as an order of magnitude. It seems imperative that large-basis full-recoil calculations be applied to
this and similar cases where data exists over a large range of energies.

The transfer amplitudes, and change of phases, are shown in fig. 21 for the DWBA

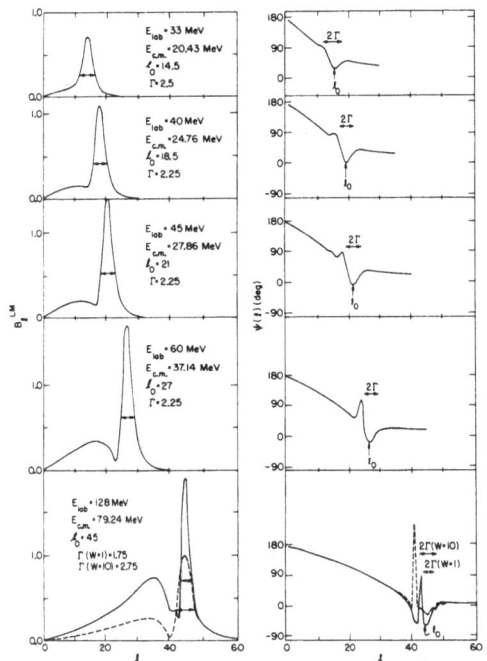

Fig. 21. Transfer amplitudes, B_ℓ^{LM}, and
phase derivatives, $\psi(\ell)^\ell$, for the
DWBA calculations shown with the
$^{26}Mg(^{16}O,^{14}C)^{28}Si_{g.s.}$ data in
fig. 17. Values both for $W_{SD}=1$.
(solid curve) and 10 (dashed
curve) are shown for 128 MeV.

calculations shown with the data in fig.
17. Several interesting features are
noted. Between 33 and 60 MeV the ℓ space
localization remains nearly constant. The
increased oscillations for higher incident
energies probably are the result of a decreased ψ. The amplitudes for small ℓ's
increase in magnitude with increasing
incident energy. Contributions from such
interior partial waves probably causes the
anomalous angular shape predicted for 128
MeV incident energy and too weak of a surface absorption.

An anomalous behavior is observed in the phase derivative, $\psi(\ell)$, for ℓ values
just below the ℓ window (see fig. 21). A shoulder is observed in the ℓ dependence of
ψ for the lower energy data, but a sharp peak develops at higher energies. The peak
is correlated in ℓ with a minimum observed in the amplitudes B_ℓ^{LM}; therefore it has been
associated[33,34]) with a zero in the transfer amplitude at a nonphysical value of the
angular momentum not far from a physical angular momentum. Similar behavior also has
been observed in calculations based on weakly absorbing potentials for several other
reactions, e.g., $^{40}Ca(^{13}C,^{14}N)^{39}K_{g.s.}$ and $^{48}Ca(^{14}N,^{13}C)^{49}Sc^1)$. The anomalous phase
changes are reduced by increasing the diffusivity in the volume absorption a'_{ws}.
Lower cutoffs in the radial integration indicate that these anomalous phase changes
are associated with a radius near that of the rapid increase in absorption. They,
therefore, may be the result of some sort of a reflection phenomena associated with
the sharp increase in the absorption. Calculations with the more standard Woods-Saxon
imaginary geometry having a relatively large diffusivity, e.g., the $^{48}Ca(^{14}N,^{13}C)^{49}Sc$
$_{g.s.}$ transition[1]) at 50 MeV incident energy, do in some cases show similar rapid changes
in phase. The effect of such an anomaly on the cross section is not large. In fig.22
the calculated angular distribution that was shown with the 60 MeV $^{26}Mg(^{16}O,^{14}C)^{28}Si_{g.s.}$

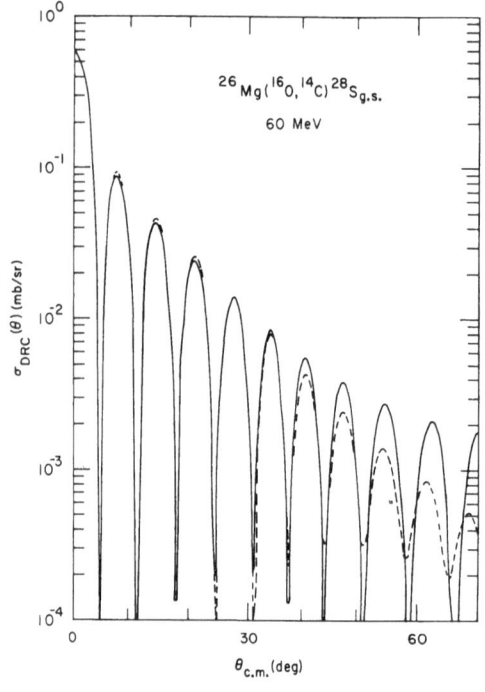

$^{26}\text{Mg}(^{16}\text{O},^{14}\text{C})^{28}\text{S}_{\text{g.s.}}$

60 MeV

Fig. 22. Comparison of calculated angular distribution for the 60 MeV ^{26}Mg $(^{16}\text{O},^{14}\text{C})^{28}\text{Si}_{\text{g.s.}}$ transition shown with the data in fig. 17 with a calculated distribution having the contributions of the partial waves below $\ell = 25$ removed.

data in fig. 17 is compared with a calculation based on identical parameters in which the contributions from the partial waves below $\ell = 25$ were removed. Such a comparison indicates that the anomalous phase change near partial waves 23 and 24 and the other contributions from lower partial waves do not significantly change the character of the forward angle cross sections. Such contributions do, however, alter the pattern of amplitude modulations which Friedman, McVoy, and Shuy have shown[34,35]) can result from the dip in $\psi(\ell)$ in the ℓ window. The phases in the ℓ window are identical for the two curves shown in fig. 22, and indeed amplitude modulations are observed in both. Contributions from the lower partial waves (see fig. 21), which are removed in the case of the dashed curve, do change the modulation. Such amplitude modulations may be observed in our 68 MeV $^{40}\text{Ca}(^{13}\text{C},^{14}\text{N})$ data in fig. 1.

7. Difference in angular distribution shapes for single- and two-nucleon transfer

Significant differences between the angular shapes of single-nucleon and two-nucleon transfer are well known (c.f. ref.[2,36,37]). Smoother single-nucleon transfer angular distributions have been explained[37]) as less ℓ-space localization resulting from a less rapidly decaying form factor in single-nucleon transfer and the incoherent contributions from several magnetic substates $|\Delta M| \leq L$. A dramatic example of the difference is shown in fig. 23 where experimental angular distributions for the $^{26}\text{Mg}(^{16}\text{O},^{14}\text{C})$ and $^{26}\text{Mg}(^{16}\text{O},^{15}\text{N})$ transitions to the ground states of ^{28}Si and ^{27}Al are compared. These two transitions were measured[28]) simultaneously at an incident energy of 60 MeV using the MIT-Brookhaven time-of-flight system. In fig. 24 calculated angular shapes are shown for a progression of changes between correct two-proton and single-proton transfers. The amplitudes and phase derivatives are shown in fig. 25 as a function of ℓ corresponding to the various curves shown in fig. 24. The change in the angular shape produced by changing the form factor in two-nucleon transfer to that of a single proton having the same binding as one of the protons in the two

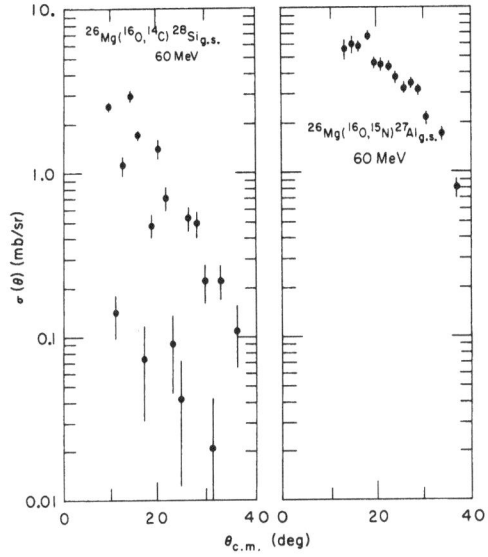

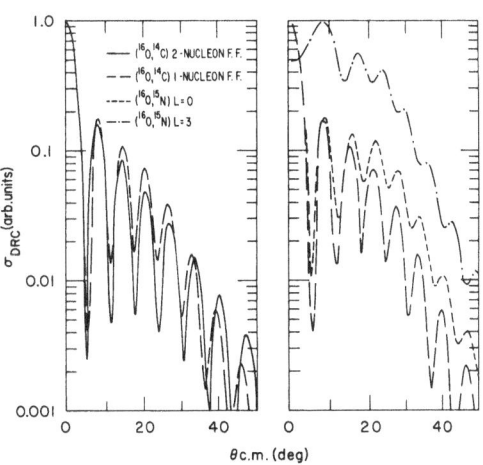

Fig. 23. Comparison of the experimental angular distributions (ref.[28]) for the 60 MeV $^{26}Mg(^{16}O,^{14}C)$ $^{28}Si_{g.s.}$ and $^{26}Mg(^{16}O,^{15}N)^{27}Al_{g.s.}$ transitions.

Fig. 24. Comparison of predicted angular shapes showing a progression of changes between the correct two-proton transfer $^{26}Mg(^{16}O,^{14}C)$ $^{28}Si_{g.s.}$ and one-proton transfer $^{26}Mg(^{16}O,^{15}N)^{27}Al_{g.s.}$ The calculations were for 60 MeV incident energy and used optical-model parameter set 1, table Al with $W_{SD}=10$ MeV. The solid and long dashed curves are for identical $^{26}Mg(^{16}O,^{14}C)^{28}Si_{g.s.}$ calculations except the former has a correct microscopic two-proton form factor, whereas the latter has a single-proton form factor. The short dashed and dot-dash curves are for $^{26}Mg(^{16}O,^{15}N)^{27}Al$ transitions with spins changed to give L=0 and the correct L=3 transfer respectively.

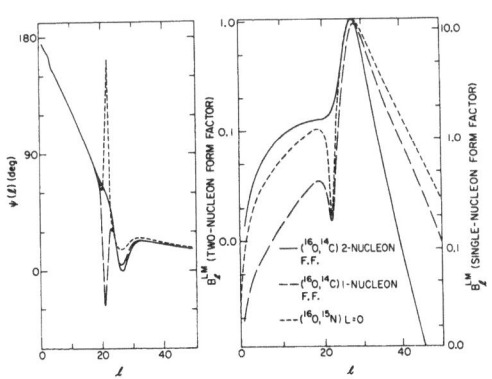

Fig. 25. Transfer amplitudes, B_ℓ^{LM}, and phase derivatives, $\psi(\ell)$, corresponding to the calculated angular shapes shown in fig.24. The various curves correspond to calculations as described in the caption of fig. 24.

proton form factor is not as large as might be expected[37]), even though the ℓ window is now about two units wider at the 1/e width. Changing the kinematic from that of $^{26}Mg(^{16}O, ^{14}C)^{28}Si_{g.s.}$ to that of $^{26}Mg(^{16}O, ^{15}N)^{27}Al$ makes a more dramatic change in the angular shapes (compare the long dashed and short dashed curves in fig. 24) than a change between two-nucleon and single-nucleon form factors. The correct ($^{16}O, ^{15}N$) kinematics gives an even wider ℓ-window and makes a large difference for the phase derivative, $\psi(\ell)$, in the ℓ window. Finally, additional smoothing is obtained from contributions of the various magnetic substates and the angles of the peaks are moved when the correct angular momentum is included in the L = 3 $^{26}Mg(^{16}O, ^{15}N)^{27}Al$ calculation. Thus for the case studied the kinematics and the different angular momentum are important effects in the observed shape change between single-proton and two-proton transfer. The role of the phase in producing more or less oscillatory angular distributions also must be stressed. Similar conclusions have been made[38]) for comparison of two- and single-nucleon transfer using other targets and projectiles.

8. Difficulties in reproducing the correct phase of oscillations for Ca($^{13}C, ^{14}N$)K transitions

Finally, I would like to discuss an anomalous behavior of the forward angle oscillations. The DWBA cross sections predicted for the $^{40,42,44}Ca(^{13}C, ^{14}N)$ transitions[9,24,39]) to the $^{39,41,43}K$ ground states shown in figs. 26 and 27 correctly reproduce the magnitude and the general shape of the angular distributions as well as

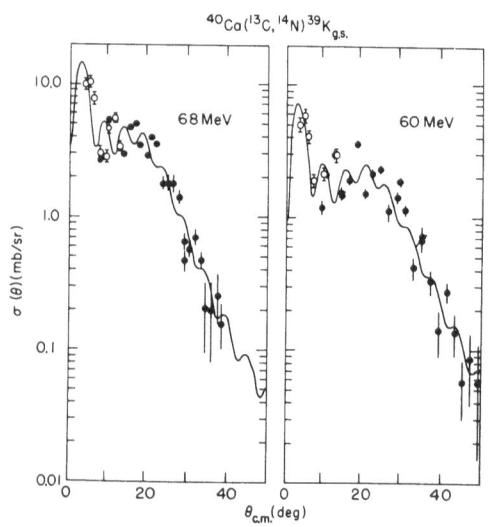

Fig. 26. Angular distributions of the $^{40}Ca(^{13}C, ^{14}N)^{39}K_{g.s.}$ transition at incident energies of 60 and 68 MeV (refs. [9,24]). The data represented by open circles were obtained using the QDDD spectrometer. The curves correspond to DWBA calculations as described in the text.

the period of the oscillations; however, the oscillations predicted are out of phase with the 68 MeV data for all targets and with both 60 and 68 MeV data for the ^{40}Ca target. This is in contrast to the case of the $^{40,42,44}Ca(^{13}C, ^{12}C)^{41,43,45}Ca_{g.s.}$ transitions at the same incident energies as the Ca($^{13}C, ^{14}N$)K transitions

figs. 1 and 28 where the details of the oscillations are reproduced. The shape of the $^{40}Ca(^{13}C, ^{14}N)^{39}K_{g.s.}$ transition at an incident energy of 40 MeV also is reproduced using surface transparent potentials (see fig. 3). A wide variety of optical parameters have been tried; however, we have not been able to reproduce this angular shape. It

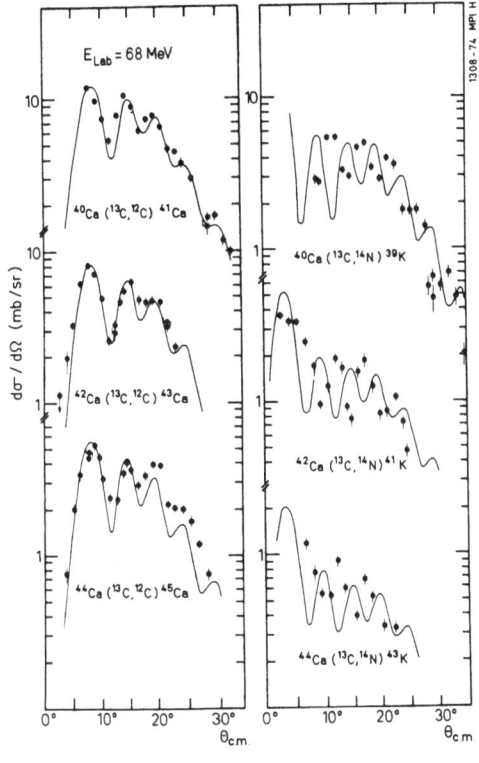

Fig. 27. Angular distributions of the $40,42,44Ca(^{13}C, ^{12}C)41,43,45Ca$ and $^{40,42,44}Ca(^{13}C, ^{14}N)39,41,43_{g.s.}$ $^{43}K_{g.s.}$ transitions at an incident energy of 68 MeV (refs.[9,39]). The data on $^{42,44}Ca$ were obtained using the QDDD spectrometer. The DWBA calculations shown as curves reproduce the shape of the $(^{13}C, ^{12}C)$ angular distributions, but the predicted oscillations are out of phase with the $(^{13}C, ^{14}N)$ data for all three targets. The optical model parameters used were set 1, table A1. For the $^{42,44}Ca$ targets the radii were scaled to account for the increased A.

is easy to change the period of the oscillations, but that does not allow a "fit" over a large angular range. We originally thought that the explanation might be a large non-normal contribution from recoil. The L = 2 non-normal cross section is indeed out of phase with the L = 1 normal predicted cross section (fig. 28); however, the normal term is predicted to dominate. Furthermore, the oscillations also are predicted to be out of phase for the $^{40}Ca(^{13}C, ^{14}N)$ transition to the $J^{\pi} = 1/2^{+}$, 2.53 MeV excited state of ^{39}K (fig. 29). Since a $s_{1/2}$ proton is picked up from ^{40}Ca, there are no non-normal contributions for this case. Therefore, recoil effects apparently are not the source of the discrepancy.

Similar difficulties in reproducing the phase of the oscillations were experienced by deVries, et al.[40] for the $^{12}C(^{14}N, ^{13}N)$ transition to the $J^{\pi} = 1/2^{+}$ state at an excitation of 3.09 MeV in ^{13}C. Therefore, similar problems exist in fitting the detailed oscillation of five different transitions--all L = 1! In the case of ^{40}Ca $(^{13}C, ^{14}N)^{39}K_{g.s.}$ the Q matching is reasonably good (see fig. 30) and the differential cross sections are several mb/sr. It, therefore, would require an unusually large second order process to explain this anomaly. Some type of a resonance phenomena seems equally unlikely, since similar effects are observed at different incident energies and at different Q values. It is imperative that such discrepancies be understood before heavy-ion transfer reactions are used as a spectroscopic tool.

9. Acknowledgements

I must acknowledge the contributions of several collaborators (listed in table 3) among the Brookhaven theory and experimental groups as well as several summer visitors and a collaboration with several experimentalist from MIT. I also wish to thank D. Sinclair for allowing the use of his 128 MeV $^{26}Mg(^{16}O,^{14}C)^{28}Si_{g.s.}$ data before it appeared in publication.

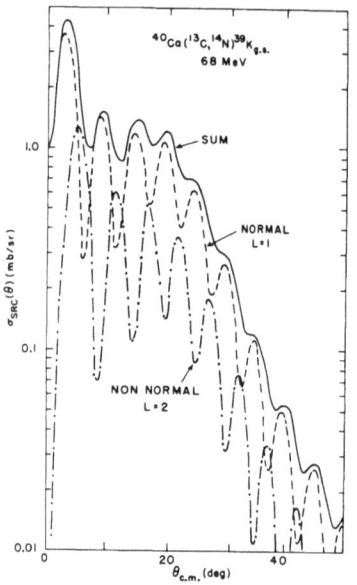

Fig. 28. DWBA predicted angular distribution for the $^{40}Ca(^{13}C,^{14}N)^{39}K_{g.s.}$ transition at 68 MeV incident energy showing the normal L = 1 and non-normal L = 2 contributions. This predicted angular shape is shown with the data in figs. 26 and 27.

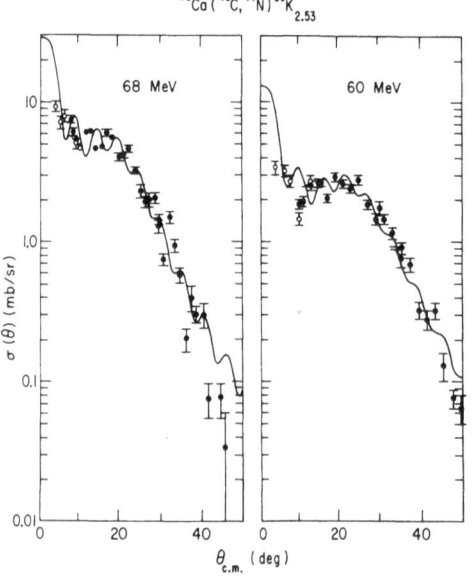

Fig. 29. Angular distributions of the $^{40}Ca(^{13}C,^{14}N)$ transition to $1/2^+$ state at 2.53 MeV excitation in ^{39}K. The open circles show data measured using the QDDD spectrometer. The curves correspond to DWBA calculations described in the text.

81

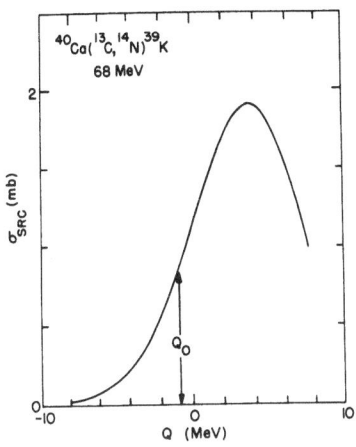

Fig. 30. Plot of the total cross section as a function of varying Q-value for the reaction $^{40}Ca(^{13}C, ^{14}N)^{39}K$ at an incident energy of 68 MeV and assuming the con-figuration of the ^{39}K ground state. The proper Q-value for the transition to the ground state of ^{39}K is indicated by Q_o.

Table 3 Collaborators

	BNL		MIT
Experimentalists	Theorists	Visitors	
P. D. Bond	E. H. Auerbach	C. K. Gelbke	T. M. Cormier
C. Chasman	A. J. Baltz	O. Hansen	E. R. Cosman
M. J. LeVine	R. Fuller	W. von Oertzen	A. Sperduto
A. Z. Schwarzschild	S. Kahana		K. van Bibber
C. E. Thorn			A. J. Lazzarini
H. E. Wegner			A. Graue

Appendix

Table A1. Optical-model parameters used in calculation.

$$U(r) = V_c - V \frac{1}{1 + e^x} - i W_{WS} \frac{1}{1 + e^{x'}} + i W_{SD} 4a'_{SD} \frac{d}{dr} \left(\frac{1}{1 + e^{x''}} \right)$$

where $x = \frac{r - R}{a}$ $x' = \frac{r - R'_{WS}}{a'_{WS}}$ $x'' = \frac{r - R'_{SD}}{a'_{SD}}$.

Reaction/Channel	V (MeV)	R (fm)	$a=a'_{SD}$ (fm)	W_{WS} (MeV)	W_{SD} (MeV)	R'_{WS} (fm)	R'_{SD} (fm)	a'_{WS} (fm)
$^{40}Ca(^{13}C,^{14}N)^{39}K$ and $^{40}Ca(^{13}C,^{12}C)^{41}Ca$								
$^{13}C + ^{40}Ca$								
40 MeV Set 1	33.4	7.33	0.55	18.	4.5	6.75	6.75	0.05
Set 2	33.4	7.33	0.55	18.	-	7.33	-	0.55
60 MeV	33.4	7.33	0.55	18.	9.0	6.90	6.90	0.05
68 MeV	33.4	7.33	0.55	18.	9.0	7.00	7.00	0.05
$^{14}N + ^{39}K$								
40 MeV Set 1	34.2	7.37	0.55	18.	0.3	7.10	7.10	0.05
Set 2	34.2	7.37	0.55	12.	-	7.37	-	0.55
60 and 68 MeV	34.2	7.37	0.55	18.	4.5	7.00	7.00	0.05
$^{12}C + ^{41}Ca$								
40 MeV	33.4	7.29	0.55	18.	4.5	6.75	6.75	0.05
60 MeV	33.4	7.29	0.55	18.	9.0	6.90	6.90	0.05
68 MeV	33.4	7.29	0.55	18.	9.0	7.00	7.00	0.05
$^{60}Ni(^{18}O,^{16}O)^{62}Ni$								
$^{18}O + ^{60}Ni$	70.0	8.68	0.40	18.	1.	8.38	8.68	0.05
$^{16}O + ^{62}Ni$	70.0	8.31	0.40	18.	1.	8.01	8.31	0.05
$^{26}Mg(^{16}O,^{14}C)^{28}Si$								
$^{16}O + ^{26}Mg$								
Set 1	100.	6.69	0.50	65	1.[a]	6.40	6.40	0.05
Set 2	100.	6.69	0.50	20	-	6.69	-	0.50
Set 3	35	6.25	0.69	35	-	5.70	-	0.61
$^{14}C + ^{28}Si$								
Set 1	100	6.64	0.50	65	1.[a]	6.35	6.35	0.05
Set 2	100	6.64	0.50	20	-	6.64	-	0.50
Set 3	35	6.21	0.69	35.	-	5.66	-	0.61
Bound State								
$^{40}Ca(^{13}C,^{14}N)$	b	$1.2A^{1/3}$	0.65					
$^{40}Ca(^{13}C,^{12}C)$	b)	$1.2A^{1/3}$	0.65					
Other Reactions	b)	$1.25A^{1/3}$	0.65					

[a] For 128 MeV it was necessary to increase this value. The calculation in fig. 17 used $W_{SD} = 10$ for 128 MeV. See also text for discussion of energy dependence of W_{SD} for this reaction. The comparison of one-proton and two-proton transfer in figs. 24 and 25 used $W_{SD} = 10$ as single-proton data seemed to prefer such a value.

[b] The bound state well depths were adjusted to give the transferred nucleon the proper binding energy.

Table A2. Summary of transitions analyzed.

Reaction	Incident Energy (MeV)	J^{π}	Transferred L Nor.	NN	C^2S_{ab} [a]	C^2S_{AB} [a]	N [a]	Figure [b]
$^{40}Ca(^{13}C,^{14}N)^{39}K_{g.s.}$								
	40	$3/2^+$	1	2	0.69	4.0	1.25	7
	60	$3/2^+$	1	2	0.69	4.0	1.00	26
	68	$3/2^+$	1	2	0.69	4.0	1.25	26
$^{40}Ca(^{13}C,^{14}N)^{39}K_{2.53}$								
	60	$1/2^+$	1	-	0.69	1.53	1.60	29
	68	$1/2^+$	1	-	0.69	1.53	2.00	29
$^{40}Ca(^{13}C,^{12}C)^{41}Ca$								
	40	$7/2^-$	4	3	0.77	0.80	1.25	1&7
	60	$7/2^-$	4	3	0.77	0.80	1.25	1
	68	$7/2^-$	4	3	0.77	0.80	1.50	1

[a] Normalization and spectroscopic factors as described in text. For sources of spectroscopic factors see ref. 41.

[b] Fig. in which calculated cross section shown with data.

Table A3. Two-nucleon spectroscopic amplitudes used in calculations.

$^{18}O \rightarrow ^{16}O$ [a]		$^{60}Ni \rightarrow ^{62}Ni$ [b]		$^{16}O \rightarrow ^{14}C$ [c]		$^{26}Mg \rightarrow ^{28}Si$ [c]	
$2s_{1/2}$	0.450	$2p_{3/2}$	0.795	$1p_{1/2}$	0.88	$1d_{5/2}$	1.03
$1d_{5/2}$	0.893	$1f_{5/2}$	0.991	$1d_{5/2}$	0.08	$2s_{1/2}$	0.51
		$2p_{1/2}$	0.395	$2s_{1/2}$	0.11	$1d_{3/2}$	0.25
				$1d_{3/2}$	0.06		

[a] Ref. 22).

[b] Determined from pairing wave functions of ref. 42.

[c] Ref. 37).

References

Proceedings of the International Conferences on Reactions between Complex Nuclei, Nashville (North-Holland, Amsterdam, 1974) is abbreviated as Proceedings of the Nashville Conference.

1) C. Chasman, S. Kahana, and M. J. Schneider, Phys. Rev. Lett. 31 (1973) 1074; M. J. Schneider, C. Chasman, E. H. Auerbach, A. J. Baltz, and S. Kahana, ibid. 31 (1973) 320.

2) P. R. Christensen, O. Hansen, J. S. Larsen, D. Sinclair, and F. Videbaek, Phys. Letts. 45B (1973) 107.

3) P. D. Bond, J. D. Garrett, O. Hansen, S. Kahana, M. J. LeVine, and A. Z. Schwarzschild, Phys. Letts. 47B (1973)231.

4) E. H. Auerbach, A. J. Baltz, P. D. Bond, C. Chasman, J. D. Garrett, K. W. Jones, S. Kahana, M. J. LeVine, M. Schneider, A. Z. Schwarzschild, and C. E. Thorn, Phys. Rev. Letts. 30 (1973) 1078.

5) P. Braun-Munzinger, W. Bohne, C. K. Gelbke, W. Grochulski, H. L. Harney, and H. Oeschler, Phys. Rev. Letts. 31 (1973) 1423.

6) W. Henning, D. G. Kovar, B. Zeidman, and J. R. Erskine, Phys. Rev. Letts. 32 (1974) 1015.

7) J. B. Ball, D. Sinclair, J. S. Larsen, O. Hansen, and F. Videbaek, Phys. Letts. 49B (1974) 348.

8) M.-C. Lemaire, M. C. Mermaz, H. Sztark, and A. Cunsolo, Phys. Rev. C10 (1974) 1103.

9) P. D. Bond, J. D. Garrett, O. Hansen, S. Kahana, M. J. LeVine, and A. Z. Schwarzschild, Proceedings of the Nashville Conference, Vol. I, p. 55 and to be published.

10) M. J. LeVine, A. J. Baltz, P. D. Bond, J. D. Garrett, S. Kahana, and C. E. Thorn, Phys. Rev. C10, No. 4 (1974).

11) V. M. Strutinskii, Sov. Phys. JETP 46 (1964) 2078 (Eng. trans. 19 (1964) 1401).

12) W. E. Frahn and R. H. Venter, Nucl. Phys. A59 (1964) 651.

13) A. Dar, Phys. Rev. 139 (1965) 1193.

14) U. C. Schlotthauer-Voos, R. Bock, H. G. Bohlen, H. H. Gutbrod, and W. vonOertzen, Nucl. Phys. A186 (1972) 225 and references therein.

15) R. Kaufmann and R. Wolfgang, Phys. Rev. 121 (1961) 206.

16) P. D. Bond, J. D. Garrett, S. Kahana, M. J. LeVine, and A. Z. Schwarzschild, Proceedings of the Nashville Conference, Vol. I, p. 54.

17) S. Kahana, P. D. Bond, and C. Chasman, Phys. Letts. 50B (1974) 199.

18) S. Kahana, Proceedings of the Nashville Conference, Vol. II.

19) R. M. DeVries and K. I. Kubo, Phys. Rev. Letts. 30 (1973) 325.

20) A. J. Baltz, Proceedings of the Nashville Conference, Vol. I, p. 60.

21) A. J. Baltz and S. Kahana, Phys. Rev. C9 (1974) 2243.

22) A. J. Baltz and S. Kahana, Phys. Rev. Letts. 29 (1972) 1267.

23) F. Schmittroth, W. Tobocman, and A. A. Golestaneb, Phys. Rev. C1 (1970) 377.

24) P. D. Bond, C. Chasman, J. D. Garrett, O.Hansen, M.J.LeVine, A.Z. Schwarzschild, and C. E. Thorn (to be published).

25) W. vonOertzen, C. E. Thorn, A. Z. Schwarzschild, and J. D. Garrett, Proceedings of Nashville Conference, Vol. I, p. 83; J. D. Garrett, A. Z. Schwarzschild, C. E. Thorn, and W. vonOertzen, Bull. Am. Phys. Soc. Oct. 1974; and to be published.

26) E. H. Auerbach (unpublished).

27) C. K. Gelbke, J. D. Garrett, M. J. LeVine, and C. E. Thorn (to be published).

28) T. M. Cormier, E. R. Cosman, A. Sperduto, K. vanBibber, A. J. Lazzarini, A. Graue, H. E. Wegner, and J. D. Garrett, Bull. Am. Phys. Soc. $\underline{19}$ (1974) 46 and to be published.

29) D. Sinclair, Oxford preprint, 1974.

30) B. F. Bayman, private communication.

31) B. F. Bayman, Phys. Rev. Letts. $\underline{32}$ (1974) 71.

32) J. Bang, C. H. Dasso, F. A. Gareev, and B. S. Nilsson, Copenhagen preprint, 1974.

33) R. Fuller, private communication.

34) K. W. McVoy, in these proceedings.

35) W. A. Friedman, K. W. McVoy, and G. W. T. Shuy, Phys. Rev. Letts. $\underline{33}$ (1974) 308; K. W. McVoy, in these proceedings.

36) E. H. Auerbach, A. J. Baltz, P. D. Bond, C. Chasman, J. D. Garrett, K. W. Jones, S. Kahana, M. J. LeVine, M. Schneider, A. Z. Schwarzschild, and C. E. Thorn, in Proceedings of the Symposium on Heavy-Ion Transfer Reactions, Argonne National Laboratory, 1973, ANL Report PHY-1973B, vol. II, p. 419.

37) B. Nilsson, R. A. Broglia, S. Landowne, R. Liotta, and A. Winther, Phys. Letts. $\underline{47B}$ (1973) 189.

38) A. J. Baltz (private communication).

39) C. K. Gelbke, J. D. Garrett, M. J. LeVine, A. Z. Schwarzschild, C. E. Thorn (to be published).

40) R. M. deVries, M. S. Zisman, J. G. Cramer, K.-L. Liu, F. D. Becchetti, B. G. Harvey, H. Homeyer, D. G. Kovar, J. Mahoney, and W. vonOertzen, Phys. Rev. Letts. $\underline{32}$ (1974) 680.

41) A. J. Baltz, P. D. Bond, J. D. Garrett, and S. Kahana (to be published).

42) L. S. Kisslinger and R. A. Sorensen, Kgl. Dan. Vidensk. Selsk., Mat.-Fys. Medd. $\underline{32}$, No. 9 (1960).

SEMI-CLASSICAL DESCRIPTION OF ELASTIC

AND INELASTIC HEAVY ION SCATTERING

R.A.Malfliet
Kernfysisch Versneller Instituut
University of Groningen
The Netherlands

A semi-classical approximation derived from Feynman's path-integral representation of the S-matrix is applied for the case of elastic and inelastic heavy ion scattering. The method has been modified to allow for complex potentials and can be extended readily for more complicated reactions. Comparison of the calculated elastic and inelastic cross-sections as well as the individual phase shifts with the same quantities obtained quantum-mechanically, demonstrates the quantitative power of the method.

1. Introduction.

In the theory of special relativity there is a fundamental constant, the velocity of light c, which determines whether a physical system behaves non-relativistic or relativistic. There is also a fundamental constant to "decide" whether a physical system behaves classical or quantum-mechanical, i.e. Planck's constant h. The physical dimensions of this constant are [time] × [energy] or [length] × [momentum] or [angular momentum] and such a quantity is known as action, $1h$ being the quantum of action. If, for a physical system any dynamical variable with the dimension of action assumes a value much larger than $1h$, then the system is said to behave classically, otherwise it is quantum-like. In other words: the classical limit is obtained by letting $h \to 0$. The purpose of these introductory remarks is to indicate roughly what is meant by a classical limit and should not be taken as a clear-cut criterion. The question whether such an approximation is valid has to be considered for each individual situation and ultimately should be tested in practice. In order to prepare the way, we will discuss some simple illustrations which already pin down certain important features.

Consider the partial wave expansion for the elastic scattering amplitude:

$$f(\theta) = \frac{1}{2ik} \sum_{\ell=0}^{\infty} (2\ell + 1)(e^{2i\delta_\ell} - 1)P_\ell(\cos \theta) \qquad (1.1)$$

In quantum mechanics only integer values for ℓ are allowed, but in classical physics ℓ can have any value. If one considers scattering of protons, then the phase shift δ_ℓ is only sparsely sampled as a function of the integer ℓ-values, while for scattering of heavy ions the density of angular momentum states per unit impact parameter $(\frac{\ell}{b} = k)$ is large[1], and the ℓ-values involved are also appreciably larger than in the proton case. For heavy ions this situation occurs because the wavelength in the relative motion is very short compared to the nuclear dimensions (the region over which the interaction changes most rapidly). We can then interpret the phase shift as the classical action function which is the same as using the WKB-approximation for δ_ℓ. Furthermore we can replace the ℓ-summation by an integral and the Legendre polynomials by their asymptotic expression:

$$P_\ell(\cos\theta) \simeq \left[\frac{2}{\pi(\ell+\frac{1}{2})\sin\theta}\right]^{\frac{1}{2}} \cos\left[(\ell+\frac{1}{2})\theta - \frac{\pi}{4}\right] \qquad (1.2)$$

As a result, the elastic scattering amplitude (1.1) contains terms of the form:

$$I(\theta) = \int_0^\infty d\ell \, \ell^{\frac{1}{2}} \, e^{2i\delta(\ell)-i\ell\theta} \qquad (1.3)$$

$$\delta(\ell) = \delta_{WKB}(\ell)$$

which can be evaluated by the method of stationary phase to give:

$$I(\theta) \simeq \sum_i \left[\frac{\mathcal{L}_i}{2\delta''(\mathcal{L}_i)}\right]^{\frac{1}{2}} e^{2i\delta(\mathcal{L}_i)-i\mathcal{L}_i\theta} \qquad (1.4)$$

and where the contributing ℓ-values $\mathcal{L}_i$ are the solutions of the equation:

$$2\delta'(\mathcal{L}_i) = \theta \qquad (1.5)$$

For a pure real interaction, the left-hand side of (1.5) can be related to the classical deflection function and we have a simple correspondance between the real $\mathcal{L}_i$-values and the scattering angle θ[2]. The scattering amplitude is expressed in terms of a coherent superposition of all orbit contributions starting at different impact parameters but converging in the same "detection" angle θ. As is demonstrated for instance in ref.3, this semi-classical prescription works out very well. Furthermore it offers a nice conceptual picture which can be useful in finding a physical interpretation of the observed experimental patterns. We have introduced here the term semi-classical because it reflects the essentials of the method i.e. the dynamics are governed by classical mechanics to which we have added the superposition principle, responsible for the main quantum-type effects.

The extension of the semi-classical method for complex potentials has been concentrated mainly on the absorptive effects, i.e. a damping factor is introduced which simply multiplies the different orbit amplitudes without affecting the phase (still given by the WKB-approximation using only the real part of the interaction), or the particle trajectories. This procedure excludes in principle the refractive and diffractive effects originating from the imaginary part of the interaction, although diffractive effects can be treated approximately by not performing the stationary phase approximation but instead summing all partial waves as in eq.(1.1).[3]

In order to account fully for the complex interaction we have to extend our prescription for calculating the trajectories. As an illustration of the direction one has to follow we consider again the elastic scattering amplitude eq.(1.1) or alternatively the term $I(\theta)$. Suppose now that $\delta(\ell)$ is purely imaginary:

$$2\delta(\ell) = i \frac{(\ell-\ell_0)^2}{\Gamma^2} \tag{1.6}$$

which means that we have introduced a Gaussian-type slit function. This gives rise to a diffraction pattern in $I(\theta)$ again of Gaussian form but with its width inversely proportional to the ℓ-space width. Another way of obtaining the same result is by the saddle point method which is nothing else than the extension of the stationary phase method leading to eqs.(1.4) and (1.5) but now allowing for complex $\mathcal{L}_i$-solutions. The "dominant" $\mathcal{L}_i$-value in this example is then:

$$\mathcal{L} = \ell_0 - i \frac{\theta \Gamma^2}{2} \tag{1.7}$$

which means that there is no longer a simple correspondence between a real $\mathcal{L}_i$-value and θ, but instead there is a whole band of ℓ-values contributing, centered around ℓ_0 with a range $\Delta \ell = \frac{\theta \Gamma^2}{2}$.

The hint of this example is that we have to continue classical mechanics into the complex plane (by analytic continuation) and as a consequence the classical limit is no longer restricted to the simple classical particle dynamics in real coordinate- and momentum space. This goal will be achieved by starting from the Feynman-path integral representation of quantum mechanics[4] and formulating the "classical" approximation. As we will show in the next section this method will also be suitable for treating more complex reactions, such as inelastic excitation, in a semi-classical approximation. The use of saddle point (stationnary phase) methods will be crucial in these discussions and will provide us with the classical limit of quantum mechanics.

2. The semi-classical approximation.

In order to obtain the "classical" limit, we start from Feynman's path integral approach to quantum mechanics[4] which in a sense is based on classical concepts. The quantum-mechanical propagator

$$K(q_2 t_2 | q_1 t_1) = \langle q_2 | \exp[-\frac{i}{\hbar} H(t_2-t_1)] | q_1 \rangle \tag{2.1}$$

can be written, according to Feynman, as a path integral over <u>all</u> possible paths $q_\alpha(t)$ in time and coordinate space satisfying the boundary conditions: $q_\alpha(t_1) = q_1$, $q_\alpha(t_2) = q_2$. The propagator (2.1) is then written formally as:

$$K(q_2 t_2 | q_1 t_1) = \int_\alpha D[q_\alpha(t)] \exp[\frac{i}{\hbar} \phi_\alpha(q_2, q_1)] \tag{2.2}$$

where $\phi_\alpha(q_2, q_1)$ is the "classical" Hamilton principal function calculated along the

the path $q_\alpha(t)$ and restricted in the endpoints by the boundary conditions. If we denote the "classical" Lagrangian function by L, the principal function takes the following form:

$$\phi_\alpha(q_2, q_1) = \int_{t_1}^{t_2} dt\ L(q_\alpha(t), \dot{q}_\alpha(t)) \tag{2.3}$$

The expression (2.2) is still exact and contains two important features which will be retained further on: a) the superposition principle, and b) the uncertainty principle (because we specify the boundary conditions in one representation and not the initial conditions q_1, $\dot{q}_1$ for instance).

We now introduce the only approximation in the derivation of the semi-classical expression for the propagator. It is based on the observation that the integrand in (2.2) is generally a very rapidly oscillating functional of the fictitious path $q_\alpha(t)$ in situations where the action integral $\phi_\alpha(q_2, q_1)$ is large (measured in units of $\hbar$) so that the contribution from a path $q_\alpha(t)$ is cancelled by the nearby path $q_\alpha(t) + \delta q_\alpha(t)$. The only exception occurs for a path $q_0(t)$, which is determined by the condition:

$$\delta \int_{t_1}^{t_2} L(q(t), \dot{q}(t))\ dt = 0 \tag{2.4}$$

This equation results in the well-known Lagrange equations and establishes the connection with classical mechanics. The value of the integral (2.3) along this classical path $q_0(t)$ is the classical action function, and by expanding the exponent in (2.2) to second order in the departure $\delta q(t)$ from the path $q_0(t)$ one can evaluate the resulting gaussian integral to give:

$$K(q_2 t_2 | q_1 t_1) \simeq \left[\frac{2\pi i\hbar}{\dfrac{\partial^2 \phi_0}{\partial q_2 \partial q_1}} \right]^{-\frac{1}{2}} \exp \left[\frac{i\phi_0(q_2, q_1)}{\hbar} \right] \tag{2.5}$$

It is interesting to note that in the limit $\hbar \to 0$ the approximation is exact and classical mechanics can thus be considered as the stationary phase approximation to quantum mechanics. The pre-exponential factor in (2.5) has a classical interpretation: the square of it is the classical probability to find at time t_2 the system at position q_2, provided it was in position q_1 at time t_1, with q_1 fixed.

At this stage it is convenient to change towards the hamiltonian formulation. The momentum $p(t)$, conjugate to $q(t)$, is defined as:

$$p = \frac{\partial L(q, \dot{q})}{\partial \dot{q}} \tag{2.6}$$

and the hamiltonian H:

$$H(p, q) = p\dot{q} - L(q, \dot{q}) \tag{2.7}$$

while the equations of motion are now Hamilton's equations. In order to find the semi-classical approximation for the quantum-mechanical propagator in momentum representation we first transform from coordinate - to momentum space by a unitary transformation:

$$K(p_2 t_2 | p_1 t_1) = \int dq_2' \int dq_1' \langle p_2 | q_2' \rangle K(q_2' t_2 | q_1' t_1) \langle q_1' | p_1 \rangle \tag{2.8}$$

As is shown by Miller[5] in detail, this transformation reduces to a canonical one by stationary phase requirements: the particular coordinates $q_2' = q_2$ and $q_1' = q_1$ contributing mostly to the integral are the ones related to p_2 and p_1 by the classical equations of motion. The phase factor (action integral) $\phi_0(p_2, p_1)$ in momentum representation is then given by:

$$\phi_0(p_2, p_1) = -(p_2 q_2 - p_1 q_1) + \int_{t_1}^{t_2} dt \, [\, p\dot{q} - H(p,q)] \tag{2.9}$$

with q_1, q_2 determined by p_1, p_2. In this way we obtain the semi-classical approximation for the propagator (2.8) i.e.

$$K(p_2 t_2 | p_1 t_1) \simeq \left[\frac{2\pi i \hbar}{\dfrac{\partial^2 \phi_0}{\partial p_2 \partial p_1}} \right]^{-\frac{1}{2}} \exp \left[\frac{i\phi_0(p_2, p_1)}{\hbar} \right] \tag{2.10}$$

with $\phi(p_2, p_1)$ given by eq.(2.9).

Since the S-matrix is also a propagator-type quantity we can now easily write down the corresponding semi-classical expression. The S-matrix is defined as:

$$\langle p_2 | S | p_1 \rangle = \lim_{\substack{t_1 \to -\infty \\ t_2 \to +\infty}} \langle p_2 | e^{+iH_0 t_2/\hbar} e^{-iH(t_2 - t_1)/\hbar} e^{-iH_0 t_1/\hbar} | p_1 \rangle \tag{2.11}$$

where $|p_1\rangle$, $|p_2\rangle$ denote eigenstates of the unperturbed hamiltonian H_0. They appear as "indices" (or quantum numbers) in the S-matrix and correspond classically to the constants of motion of H_0. The semi-classical phase factor (action integral) associated with the propagator or S-matrix element (2.11) is given by ($t_2 \to +\infty$, $t_1 \to -\infty$):

$$\phi_0(p_2, p_1) = E(t_2 - t_1) + \int_{t_1}^{t_2} dt \, [\, p\dot{q} - H] - (p_2 q_2 - p_1 q_1) \tag{2.12}$$

and by using energy conservation (H is time-independent) the final semi-classical expression reads[5]:

$$\langle p_2|S|p_1\rangle \approx \left[\frac{2\pi i\hbar}{\dfrac{\partial^2\phi_0}{\partial p_2\,\partial p_1}}\right]^{-\frac{1}{2}} \exp\left[\frac{i\phi(p_2,p_1)}{\hbar}\right]$$

(2.13)

$$\phi(p_2,p_1) = -\int_{t_1}^{t_2} dt\, q(t)\dot{p}(t)$$

Implicitly we have assumed that if there is more than one path for which the action integral (2.3) is stationary (eq.(2.4)) the corresponding semi-classical expression for the S-matrix (2.13) contains a summation over all these paths which of course must satisfy the boundary conditions (i.e. $p(t_1)=p_1$, $p(t_2)=p_2$). Furthermore it is a trivial matter to generalise the derivation to a multi-dimensional system.

As an especially important generalisation we note that, apart from the translational motion, it is possible to include intrinsic degrees of freedom describing transitions between specific quantum states, as long as there exists a classical analogue for the internal hamiltonian. An example is the excitation of collective states in an inelastic collission for which the internal hamiltonian can be thought to describe a simple classical vibrational or rotational system. The internal degrees of freedom (deformation parameters) are expressed as action-angle variables and the quantisation of these is achieved by fixing the boundary conditions in the appropriate way. By solving the equations of motion for all the degrees of freedom one effectively solves a coupled channel problem for which there is a feed back mechanism by which the trajectory is affected because of internal excitations[11]. If, however, these processes can be treated in first-order perturbation it should be correct to decouple the relative motion and the internal "motion". A more detailed discussion of the material presented in this section can be found in ref.5.

3. Elastic and Inelastic Scattering.

In this section we will discuss the actual form of the semi-classical approximation in case of elastic and inelastic heavy-ion scattering. Introducing a complex potential V+iW, Hamilton's principal function reads (in coordinate representation):

$$\phi(q_2,q_1) = \int_{t_1}^{t_2} [\tfrac{1}{2}m\dot{q}^2 - V(q) - iW(q)]\,dt$$

(3.1)

Restricting the q-variable to be real, the phase function contains a real and imaginary part, ϕ_R and ϕ_I respectively:

$$\phi_R(q_2,q_1) = \int_{t_1}^{t_2} [\tfrac{1}{2}m\dot{q}^2 - V(q)]\,dt$$

(3.2)

$$\phi_I(q_2,q_1) = \int_{t_1}^{t_2} W(q)dt$$

If we consider ϕ_I to vary smoothly as a function of the different paths $q(t)$ we can, in principle, apply the stationary phase prescription towards ϕ_R only. In this way we obtain the well-known classical trajectory $q_0(t)$ and the corresponding real phase shift $\phi_0(q_2,q_1)$, governed by the real part of the interaction. The form taken by the real phase shift is equivalent to the well-known WKB-expression for it. The imaginary potential W appears only in the magnitude of the S-matrix (2.13) which acts as a mean free path type of absorption i.e.:

$$\exp\left[\frac{1}{\hbar}\int_{t_1}^{t_2} W(q_0(t))dt\right] \tag{3.3}$$

One recognises in this prescription the usual way (denoted s.c. for later purposes) in which the semi-classical approximation is presented[6]. The range of applicability depends obviously on the "smallness" of W such that a first-order treatment is valid.

In order to correct for this deficiency we have to take into account the total complex phase function in the extremal condition for the least action trajectory. This, of course, is in accordance with applying the saddle point method to the full Feynman path integral expression for the propagator. The "stationary" trajectory is then the solution of the equations

$$\delta\phi_R(q_2,q_1) = 0$$
$$\tag{3.4}$$
$$\delta\phi_I(q_2,q_1) = 0$$

which have to become complex in order to guarantee a solution. This means that we have to analytically continue the Lagrangian and the equations (3.4) in order to find the "minimal" trajectory. The variables q and $\dot{q}$ become complex but the boundary conditions q_1,q_2 remain real (and thus observable). In the same way as before we can define the conjugate complex momentum p, the hamiltonian and the corresponding Hamiltonian equations by analytic continuation. The appropriate semi-classical expression for the S-matrix is still given by (2.13) but with complex variables p, q which are found by solving the Hamilton equations with the real boundary conditions $p(t_1) = p_1$, $p(t_2) = p_2$. Thus, in order to find the main contribution in the integral expression for the propagator, the saddle point method forces us to use complex trajectories which are obtained by analytic continuation of all quantities and equations involved. In this way we have enlarged the framework to deal with complex potentials, while the dynamics are still determined by "classical"-type equations of motion.

If we transform to polar coordinates (r,θ) the corresponding momenta are (p_r,p_θ) and the boundary conditions are formulated accordingly. Since θ is still a cyclic coordinate the conjugate momentum p_θ is conserved:

$$\dot{p}_\theta = -\frac{\partial H}{\partial \theta} = 0 \qquad (3.5)$$

Since the real part of p_θ can be interpreted as the orbital angular momentum ℓ and the imaginary part is simply zero (these are the values given by the boundary condition at $t_1 = t \to -\infty$) one sees that because of the conservation law (3.5) the boundary condition for p_θ at $t = t_2 \to +\infty$ is automatically fullfilled. The other conserved quantity which is of importance is the total energy. By trivial algebra one can show that

$$\frac{dH}{dt} = \frac{\partial H}{\partial t} = 0 \qquad (3.6)$$

where H is a complex-valued quantity. This implies that H is a complex constant, the real part of which is equal to the energy E, while the imaginary part is zero. This is so, because the limits $t_1 = t \to -\infty$ and $t_2 = t \to +\infty$ are defined such that the interaction vanishes. As a consequence the boundary conditions take the following form:

$t \to -\infty$	$t \to +\infty$
$\text{Re}\, p_r(t) = -[\,2mE\,]^{\frac{1}{2}}$	$\text{Re}\, p_r(t) = [\,2mE\,]^{\frac{1}{2}}$
$\text{Im}\, p_r(t) = 0$	$\text{Im}\, p_r(t) = 0$
$\text{Re}\, p_\theta(t) = \ell$	$\text{Re}\, p_\theta(t) = \ell$
$\text{Im}\, p_\theta(t) = 0$	$\text{Im}\, p_\theta(t) = 0$

$$(3.7)$$

and it is clear that once they are fixed for $t \to -\infty$ they automatically have the correct value for $t \to +\infty$ because of the conservation laws (3.4) and (3.5). Furthermore, since the semi-classical expression for the S-matrix i.e. eq.(2.13) is only dependent on p_1 and p_2 the initial value used for q_1 is irrelevant and can even be choosen complex, as long as $\text{Re}\, r(t \to -\infty)$ is large enough so that in the asymptotic region the interaction vanishes. Complex-valued "classical" trajectories may sound unphysical but all physically meaningful quantities are real in the asymptotic region (i.e. the observables p_1, p_2) while the others (i.e. q_1, q_2) are not observable so that their actual value is of no importance except for the fact that their magnitude should be large.

The semi-classical expression for the elastic scattering amplitude

$$f(\theta) = \frac{1}{2ik} \sum_\ell (2\ell+1) P_\ell(\cos\theta)\, [\, <p_2|S|p_1> - 1\,] \qquad (3.8)$$

is obtained by identifying p with (p_r, p_θ) and using the approximation (2.13) for the S-matrix. The phase function $\phi(p_2, p_1)$ is then a function of ℓ and E, and will be calculated by solving the corresponding complex-valued hamilton equations. It can be shown that the phase shifts one finds in this way are related to the ones derived in WKB-approximation allowing for complex turning points. This latter method has been discussed and illustrated in ref.7 starting from a different viewpoint.

As a further simplifying step one can apply the saddle point method for evaluating the partial wave sum in (3.8). This procedure has been followed by Knoll and
Schaeffer[8] and it goes as follows: one tries to find those ℓ-values for which the
complex deflection angle (i.e. $\theta(t)$ for $t \to +\infty$) is equal to the real scattering angle
θ. This results in complex ℓ-values which give the dominant contributions in the sum
(3.8) in a similar manner as discussed in the introduction for the Gaussian slit-
system. This means that one has to solve the equations of motion with a complex impact
parameter in order to get trajectories which end at a real angle. Although this method
is as good as performing the partial wave summation, it doesn't give an indication
on the correctness of the more detailed input like the phase shifts and furthermore
one has to restart the whole procedure for each angle θ, which means that effectively
one deals with a large number of complex ℓ-values.

The excitation process in inelastic scattering is described in terms of the
collective motion of, in the present case, the target nucleus. Considering only the
$\lambda=2$ vibrational modes, the surfaces of constant density are given by:

$$R = R_0[1 + \sum_\mu \alpha^*_{\lambda\mu}(t) Y_{\lambda\mu}(\theta,\psi)] \tag{3.9}$$

with θ and ψ polar angles with respect to some arbitrarily choosen space-fixed axis.
The collective motion is then initiated by the variations of the coefficients $\alpha_{\lambda\mu}$
with time. The corresponding intrinsic hamiltonian can be written in the form:

$$H_{intr} = \tfrac{1}{2} B_\lambda \sum_\mu |\dot{\alpha}_{\lambda\mu}(t)|^2 + \tfrac{1}{2} C_\lambda \sum_\mu |\alpha_{\lambda\mu}(t)|^2 \tag{3.10}$$

and the coupling between the relative motion and the collective motion appears through
the "deformation" of the interaction, which in first-order perturbation reduces to:

$$V_{coupl} = \sum_\mu \alpha^*_{\lambda\mu}(t) Y_{\lambda\mu}(\theta(t),0) f_\lambda(R(t)) \tag{3.11}$$

where $f_\lambda(R(t))$ is the usual radial form factor[9] for Coulomb and nuclear (both real
and imaginary) excitation. In accordance with the first-order treatment we neglect
now the coupling between relative and collective motion. This means that the total
Lagrangian appearing in the action integral contains a term for the relative motion
(which essentially is the same as the one for elastic scattering), a term for the
intrinsic motion (associated with the hamiltonian (3.10)) and a perturbation given
by eq.(3.11). The stationary phase condition i.e. eq.(3.4) is then applied only for
the first two terms in this Lagrangian, the perturbation being evaluated over the
resulting trajectory. Thus, the relative motion corresponds to a complex trajectory
encountered already for elastic scattering while the remaining intrinsic coordinates
$\alpha_{\lambda\mu}(t)$ are given by

$$\alpha_{\lambda\mu}(t) = \alpha_{\lambda\mu}(0) \ e^{i\omega_\lambda t} \qquad (3.12)$$

corresponding to the solutions of the harmonic oscillator hamiltonian (3.10). The quantity ω_λ is the oscillator frequency for the vibrational excitation λ of energy $\hbar\omega_\lambda$ corresponding to the Q-value involved. The amplitude $\alpha_{\lambda\mu}(0)$ is related to the deformation parameter β_λ by the equation

$$\alpha_{\lambda\mu}(0) = \frac{\beta_\lambda}{(2\lambda+1)^{\frac{1}{2}}} \qquad (3.13)$$

which is obvious from the definition of β_λ.

The semi-classical approximation for the transition amplitude, exciting a state of spin λ in substate μ is then given under the simplest considerations by:

$$T_{\lambda\mu}(\theta) = \frac{\sqrt{\pi}}{ik} \sum_{\ell=0}^{\infty} (2\ell+1)^{\frac{1}{2}} \ C_{\lambda\mu}(\ell) \ e^{2i\delta_\ell^{S.C.}} \ Y_{\ell,-\mu}(\theta,0) \qquad (3.14)$$

where $2\delta_\ell^{S.C.}$ represents the semi-classical phase shift associated with the complex trajectories discussed in connection with the elastic scattering. The excitation amplitudes $C_{\lambda\mu}(\ell)$ are given by the integrals ($t_1 \to -\infty$, $t_2 \to +\infty$):

$$C_{\lambda\mu}(\ell) = \frac{i}{\hbar} \int_{t_1}^{t_2} dt \ \frac{\beta_\lambda}{(2\lambda+1)^{\frac{1}{2}}} \ Y_{\lambda\mu}(\theta(t),0) \ f_\lambda(R(t))e^{i\omega_\lambda t} \qquad (3.15)$$

where the points $R(t)$, $\theta(t)$ locate the complex trajectory . The expression (3.14) corresponds to the quantum-mechanical Distorted Wave Born Approximation for inelastic scattering and will be compared to it in the next section. The derivation sketched above for the inelastic transition amplitude gives the same result as presented in ref.3 but is virtually simpler and more consistent in the sense that it treats the quantized degree of freedom (vibration) also classically. More details of the topics discussed in this section will be published elsewhere[10].

4. Illustrations.

In this section we present numerical results for elastic and inelastic scattering of ^{16}O on ^{58}Ni at 60 MeV bombarding energy. The calculations were done using three methods i.e. a full quantum-mechanical treatment (Q.M.), the semi-classical approximation (S.C.) which takes into account complex trajectories (section 3), and what we will call the simple semi-classical prescription (s.c.) mentionned in the beginning of section 3. The last procedure, based on real trajectories governed by the real part of the interaction only and a mean free path type of absorption (i.e. eq.(3.3)), has been extensively studied in ref.3 for three different potential parameter sets (I, II, III) which will also be used here. Their actual values are displayed in table 1.

Table 1.

Parameters of the Saxon-Woods potentials

	V_0 (MeV)	r_V (fm)	a_V (fm)	W_0 (MeV)	r_W (fm)	a_W (fm)
I	2.	1.65	0.6	20.	1.25	0.54
II	7.	1.51	0.6	20.	1.25	0.54
III	20.	1.39	0.6	20.	1.25	0.54

Since the three cases (I, II and III) differ by their real potentials, they give rise
to different types of classical deflection functions[3] i.e.: almost Coulomb-like (I),
rainbow scattering (II) and orbiting (III). This behaviour plays an important role
in the simple semi-classical approximation (s.c.) since it offers a simple picture
of interfering real orbits. If, however, one allows for complex trajectories its significance becomes disputable, as we will show later on.

The elastic cross-sections are obtained by summing the full partial wave expansion eq.(1.1) with the complex phase shifts calculated by means of the three methods
outlined before. The results are compared in fig.1, which speaks for itself.

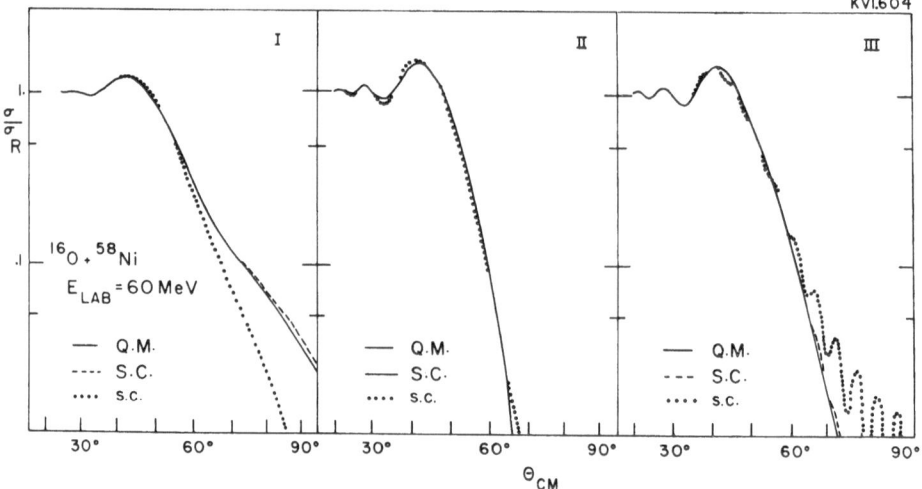

Fig.1. Quantal (Q.M.), semi-classical (S.C.) and simple semi-classical (s.c.)
calculations of the elastic scattering of 60 MeV ^{16}O ions on ^{58}Ni for the three
different types of potentials given in table 1. All calculations were done with
100 partial waves.

The agreement between the Q.M.curves and the S.C.curves is impressive and can be better understood by a closer inspection of the phase shifts and the reflection coefficients as a function of orbital momentum ℓ. These are displayed in fig.2 from which it is clear that the S.C.method reproduces the exact behaviour in quite some detail. The s.c.results are markedly different, especially for the smaller ℓ-values and the region around the so-called critical ℓ-value ℓ_c ($\eta(\ell_c) = \frac{1}{2}$). The fact that

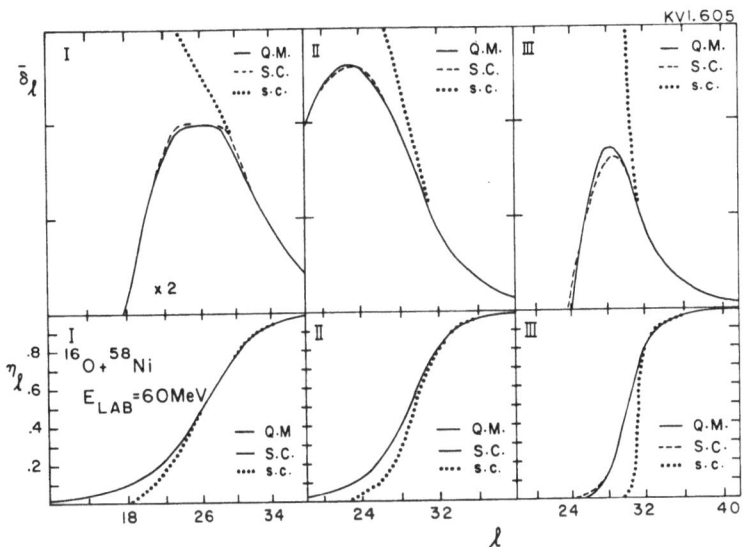

Fig.2. The nuclear phase shifts $\bar{\delta}_\ell$ and the reflection coefficients η_ℓ displayed as functions of the angular momentum ℓ for the scattering of ^{16}O on ^{58}Ni at 60 MeV bombarding energy. For reasons of clearness only part of the phase shift behaviour is shown. The curves were obtained using the three methods under study and are labeled as such.

for case II the s.c. cross section compares favourably with the exact calculation must be regarded as fortuitous in view of the poor agreement obtained for the phase shifts and the reflection coefficients. If we define the deflection function $\theta(\ell)$ as:

$$\theta(\ell) = 2 \frac{d\delta_\ell}{d\ell} \tag{4.1}$$

which becomes equal to the classical deflection angle if δ_ℓ is calculated by the s.c. prescription (real trajectories), the quantal deflection function will agree with the semi-classical one but not with the classical deflection angle. Especially the deflection to negative angles is suppressed by the repulsive influence of the imaginary potential.

This effect is also illustrated in fig.3, where for the three potential sets I, II and III, the values for the turning points $r_0(\ell)$ as a function of ℓ are compared in case of real trajectories (s.c.) and complex trajectories (S.C.). The latter situation will give rise to complex turning points for which only the real part is depicted in fig.3. The imaginary part of this complex turning point is also a smoothly behaving function of ℓ. At this point, the question may arise, which complex turning point has been selected since there are several ones, each giving rise to a genuine complex trajectory and a corresponding phase factor (see eq.(2.13) and the discussion following it). The trajectory which has been picked out is the one which gives the smallest imaginary phase contribution and corresponds to the weakest damping. For a more extensive discussion, see ref.10.

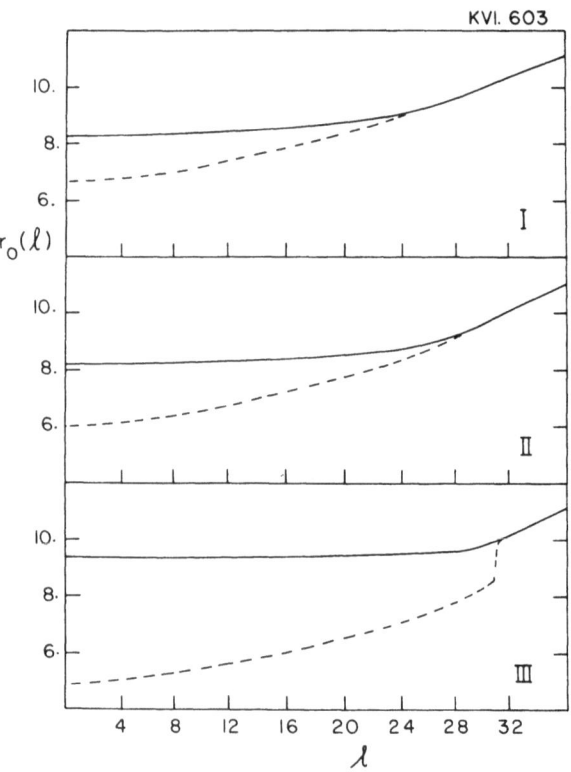

KVI. 603

Fig.3. Behaviour of the turning points $r_0(\ell)$ as a function of angular momentum ℓ for the three potential parameter sets (I, II and III). The dashed curves correspond to the s.c. approximation (real orbits) while the full curves represent the real part of the complex turning points associated with the complex trajectories in the S.C. approximation.

As can be seen from fig.3, the imaginary potential produces a significant repulsion, especially striking for case III where the sudden jump due to orbiting has disappeared. This effect can be important in view of applications of the semi-classical approximation to reactions where exchange of mass, energy, angular momentum etc. takes place. Here the simple semi-classical approximation would induce an "orbit mismatch" around the critical ℓ-value which, apart from not being correct, would result in a funny behaving form factor (which is evaluated along the orbit).

The procedure adapted by Knoll and Schaeffer is demonstrated in fig.4, taken from their work, for ^{16}O on ^{16}O scattering. It shows a beautiful agreement between the S.C. calculation and the exact result which again testifies in favour of the semiclassical approximation.

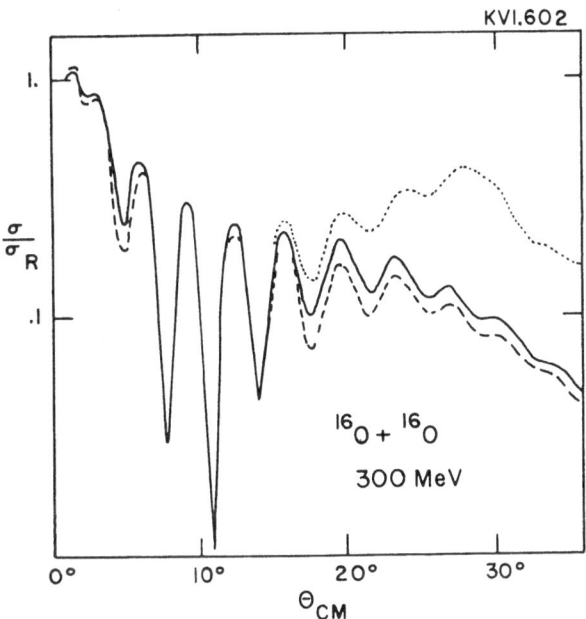

Fig.4. Elastic cross sections for scattering of ^{16}O ions on ^{16}O at 300 MeV[8]. The dashed curve corresponds to the S.C.approximation and the dotted one to the s.c.approximation (paths given by the real part of the potential). Both were obtained using the saddle point method for the partial wave summation. The exact result is denoted by the full curve.

Finally, we present in fig.5 cross sections obtained using the Q.M., S.C. and s.c. method for inelastic scattering. The last two prescriptions have been used in eqs.(3.14 and (3.15) based on complex and real trajectories respectively. We have choosen case III since of all three cases this is the most instructive one. The agreement between Q.M. and S.C. results for the other two cases (I, II) is of a similar quality[10] while the s.c. calculations[3] are quite different at large angles as could be anticipated from the elastic scattering results (fig.1).

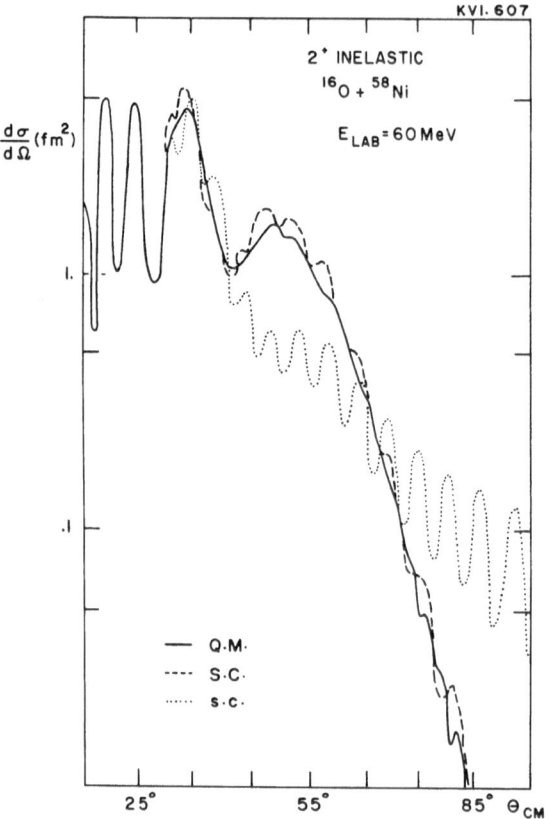

Fig.5. Quantal (Q.M.), semi-classical
(S.C.) and simple semi-classical (s.c.)
calculations for the inelastic scattering of 60 MeV ^{16}O ions on ^{58}Ni. The
quantal calculations were carried out
with the code DWUCK, utilizing an integration step of 0.1 fm and a maximum
radius of 40 fm. The semi-classical
calculations (both S.C. and s.c.) based
on eq.(3.14), were also done by integrating up to 40 fm. The number of
partial waves was in all three cases
set equal to 150.

5. Conclusion.

The semi-classical approximation has been discussed as the stationary phase
approximation of quantum mechanics. The primary objects are probability amplitudes
which are added according to the principle of superposition, and the dynamical relations between coordinates and momenta are those of classical mechanics which, for
complex interactions, are analytically continued. The problem of constructing the
semi-classical S-matrix is then reduced to finding complex-valued solutions of the
classical equations of motion which do satisfy the correct boundary conditions.

Intrinsic degrees of freedom can be treated in the same way, as long as there exists a classical analogue for the hamiltonian.

We have demonstrated the quantitative properties of the semi-classical approximation for elastic and inelastic heavy ion scattering. Refractive as well as diffractive effects are correctly taken into account. However, the physical insight one could gain from this procedure is somehow obscured as compared to the simple semi-classical prescription which is based on real particle orbits.

Altogether, the method has appeared to be very reliable and makes one believe that in cases where a macroscopic understanding is needed and many degrees of freedom are excited, it may turn out to be a powerful method and to be preferred over a more complicated quantal calculation.

The author is greatly indebted to T.Koeling for his many contributions to this work and W.Ockels for stimulating discussions.

References.

1) N.K.Glendenning, International Conference on Reactions between Complex Nuclei, Nashville (1974).

2) K.W.Ford and J.A.Wheeler, Annals of Phys.7(1959)259

3) R.A.Broglia, S.Landowne, R.A.Malfliet, V.Rostokin and Aa.Winther, Phys.Reports 11C(1974)1.

4) R.P.Feynman and A.R.Hibbs, Quantum Mechanics and Path Integrals (McGraw-Hill, New York, 1965)

5) W.H.Miller, J.Chem.Phys.53(1970)1949
   W.H.Miller, Advan.Chem.Phys.25(1974)

6) R.A.Broglia and Aa.Winther, Nucl.Phys.A182(1972)112
   R.A.Broglia and Aa.Winther, Phys.Reports 4C(1972)153

7) R.A.Malfliet, Extended Seminar on Nuclear Physics, I.C.T.P. Trieste (1973)

8) J.Knoll and R.Schaeffer, Extended Seminar on Nuclear Physics, I.C.T.P. Trieste (1973)
   J.Knoll and R.Schaeffer, to be published

9) R.A.Broglia, S.Landowne and Aa.Winther, Phys.Lett.40B(1972)293

10) R.A.Malfliet and T.Koeling, to be published.

11) See ref.5 and S.Levit, U.Smilansky and D.Pelte (Weizmann Institute preprint) for an application of this procedure to Coulomb excitation.

WAVE MECHANICAL ASPECTS OF HEAVY ION COLLISIONS

W.E. Frahn

Physics Department, University of Cape Town
Rondebosch (Cape) 7700, South Africa

A unified quantal description of the wave mechanical and semiclassical aspects of heavy ion collisions is given by means of closed expressions derived from partial-wave expansion of the transition amplitudes.
The results depend entirely on the asymptotic properties of the scattering wave functions (the partial-wave S matrix). It is shown, in particular, how the "refractive" features of semiclassical potential scattering
are changed, in the presence of absorption, into characteristic quantal
interference patterns of "diffractive" type. Only the simplest idealized
limits are considered as examples of how insight is gained into the physical mechanisms that operate under various scattering conditions. Elastic
scattering, inelastic scattering, and direct transfer reactions are described and the close physical connexion between these processes is made
explicit.

1. Introduction

The most successful description of heavy ion collisions so far is
given by the optical model (OM) for elastic scattering and the distorted
wave (DW) theory for direct reactions. These are of course fully quantal
formalisms, based on the one-body Schrödinger equation for the relative
motion of the colliding nuclei. The many-body aspects of the interaction
are globally represented by average complex potentials depending mainly
on the relative coordinate, and by simple interaction operators depending
in addition on only a few degrees of freedom involved in a direct reaction.

In spite of these simplifications the evaluation of the OM and DW
formulae to calculate angular distributions and excitation functions
requires extensive numerical computation, particularly for heavy ion
collisions which involve many partial waves. Another difficulty is the
well-known ambiguity of optical potentials for heavy particles. Even
though OM and DW practitioners acquire considerable intuitive experience
with the effects of the many adjustable parameters involved, the connexion
between input and output is by no means simple, and the physics often
remain hidden in the computer.

To elucidate the physical mechanisms, to understand "what happens",

it is necessary to make further approximations which enable us to derive
closed expressions. It so happens that the very features that complicate
numerical computations for heavy ion interactions, high orbital angular
momenta ℓ and large values of the Sommerfeld parameter n, may be turned
to advantage in an analytic formulation. Large values of ℓ and n charact-
erize the regime near the classical limit of quantal description. This
has led to the recent development of semiclassical theories for heavy ion
collisions[1]. Earlier formulations, the strong absorption or diffraction
models[2], emphasized the quantal aspects associated with the absorptive
properties of heavy ion interaction due to the presence of many nonelastic
channels; these manifest themselves in quantum diffraction effects and
characteristic patterns resulting from quantal interference between
diffractive scattering and strong Coulomb scattering. Both these descriptions, in their original form, were incomplete in complementary ways:
the diffraction theories, with their main emphasis on absorptive properties, gave a too simplified account of the real part of the nuclear
interaction, while the semiclassical theories, with their main emphasis
on the classical features of potential scattering and the trajectory
concept, encountered difficulties in incorporating the effects of strong
absorption.

As I have been given the task of discussing the wave mechanical
aspects of heavy ion collisions, I shall endeavour to present a unified
description, based on quantum theory, of the effects of both the real and
absorptive parts of the nuclear interaction, as well as those of the
Coulomb interaction, in the elastic scattering, inelastic scattering and
direct transfer reactions of heavy ions. I hope to show that it is possible to give such a description in closed form which, while being a reasonably accurate approximation to the OM and DW theory, explicitly identifies
the physical mechanisms that operate in various heavy ion scattering
situations.

2. Elastic scattering

We start with the partial wave expansion of the elastic scattering
amplitude for spinless particles

$$f(k,\theta) = \frac{i}{k} \sum_{\ell=0}^{\infty} (\ell+\tfrac{1}{2})[1-S_\ell(k)]P_\ell(\cos\theta). \qquad (1)$$

(Spin effects, which are still difficult to observe experimentally at
present, are ignored for simplicity.) The basic ingredients of our formalism are the complex quantities $S_\ell(k)$ which, regarded as a function of

orbital angular momentum ℓ and wave number k, define the <u>elastic scatter-ing function</u>. They are the diagonal elements of the partial-wave S-matrix $S_{fi,\ell}$, whose unitarity implies $|S_\ell(k)| \leq 1$. Thus we may write

$$S_\ell(k) = \eta_\ell(k) e^{i2\delta_\ell(k)} \quad , \tag{2}$$

where the reflection coefficients $\eta_\ell(k)$ and phase shifts $\delta_\ell(k)$ are real quantities. Elastic unitarity gives

$$[\eta_\ell(k)]^2 = 1 - \sum_r |S_{ri,\ell}|^2 = 1 - \frac{k^2}{2\pi(\ell+\frac{1}{2})} \sigma_\ell^{(r)}(k) \tag{3}$$

in terms of the partial-wave reaction cross section $\sigma_\ell^{(r)}$. The presence of open reaction channels r causes a depletion of flux in the elastic channel which we call "absorption". A given partial wave ℓ is called "strongly absorbed" or "weakly absorbed" according to whether

$$\eta_\ell(k) \ll 1 \quad \text{or} \quad 1 - \eta_\ell(k) \ll 1. \tag{4}$$

Under conditions of strong absorption, elastic scattering may be regarded as the "shadow" of all nonelastic processes and is then essentially a quantal diffraction process. When absorption is weak, the scattering is dominated by the real part of the interaction, essentially embodied in the phase shifts $\delta_\ell(k)$. For heavy ions, the Coulomb potential forms a substantial part of the real interaction, and it is convenient to separate the Rutherford scattering phase shifts σ_ℓ from the remaining "nuclear" phase shifts $\delta_\ell^{(N)}$,

$$S_\ell = S_\ell^{(N)} e^{i2\sigma_\ell} , \qquad S_\ell^{(N)} = \eta_\ell e^{i2\delta_\ell^{(N)}} , \tag{5}$$

where

$$\sigma_\ell = \arg \Gamma(\ell+1+in), \qquad n = Z_1 Z_2 e^2/\hbar v . \tag{6}$$

Now we introduce some basic assumptions (for details see ref.[3]). Since many partial waves are involved in heavy ion collisions, we may replace the summation over ℓ in eq.(1) by an integration over $\lambda=\ell+\frac{1}{2}$, using the appropriate asymptotic form of the Legendre polynomials,

$$f(\theta) = \frac{i}{k}\left(\frac{\theta}{\sin\theta}\right)^{\frac{1}{2}}\int_0^\infty d\lambda\,\lambda[1 - S(\lambda)]J_0(\lambda\theta),\tag{7}$$

where we have assumed that η_ℓ and δ_ℓ can be interpolated by smooth (diff-erentiable) functions $\eta(\lambda)$ and $\delta(\lambda)$. (In the presence of effects associated with individual partial waves, e.g. resonances, the corresponding terms have to be taken out of the ℓ-sum in (1) before the remainder is replaced by an integral.)

By a partial integration, eq.(7) can be rewritten as

$$f(\theta) = \frac{i}{k}\frac{1}{(\theta\sin\theta)^{\frac{1}{2}}}\int_0^\infty d\lambda\,\lambda\,a(\lambda)e^{i2\delta(\lambda)}J_1(\lambda\theta),\tag{8}$$

where

$$a(\lambda) = D(\lambda) + i\Theta(\lambda)\eta(\lambda).\tag{9}$$

Here we have introduced the <u>absorptive shape function</u> $D(\lambda)$ and the <u>quantal deflection function</u> $\Theta(\lambda)$ defined by

$$D(\lambda) = \frac{d\eta(\lambda)}{d\lambda},\qquad\qquad \Theta(\lambda) = \frac{d2\delta(\lambda)}{d\lambda}.\tag{10}$$

We need not specify any explicit form of the functions $\eta(\lambda)$ and $\delta(\lambda)$, or $D(\lambda)$ and $\Theta(\lambda)$; we assume these functions to be determined as uniquely as possible by analysis of elastic scattering data in terms of a dynamical model. They may be generated for instance from an optical potential (of given phenomenological form or derived from microscopic theory) by numerical solution of the Schrödinger equation, or by using a semiclassical (WKB) or high-energy (Glauber) approximation if appropriate.

Explicit evaluation of the integral in (8) is achieved by different methods depending on the <u>general forms</u> of the functions $D(\lambda)$ and $\Theta(\lambda)$. These in turn depend on the various scattering situations of which we consider, for brevity, only the simplest limiting cases according to the following scheme:

		real part of nuclear interaction	
		weak	strong
absorptive part of nuclear interaction	strong	(A_S,R_W)	(A_S,R_S)
	weak	(A_W,R_W)	(A_W,R_S)

In this rough classification, "strong absorptive interaction" refers to a rapid variation of $\eta(\lambda)$, i.e. to an absorptive shape function $D(\lambda)$ which is localised in λ-space within a narrow range of width Δ about a critical angular momentum Λ defined by $\eta(\Lambda)=\frac{1}{2}$. By contrast, "weak absorption" refers to slowly varying $\eta(\lambda)$,i.e. to a broad distribution of $D(\lambda)$. As to the real part of the interaction, heavy ion collisions are characterized by a strong Coulomb contribution; for large values of n we have

$$2\sigma(\lambda) = 2\sigma(o) - 2n\ell n(\sin\tfrac{1}{2}\phi) + n\phi\cot\tfrac{1}{2}\phi - \tfrac{1}{2}\pi + o(n^{-1}) \qquad (11)$$

and

$$\Theta^{(R)}(\lambda) = \frac{d2\sigma(\lambda)}{d\lambda} = \phi = 2\arctan(n/\lambda). \qquad (12)$$

The real part of the nuclear interaction is called "strong" if $\Theta^{(N)}=\Theta-\Theta^{(R)}$ varies rapidly,giving rise to rainbow scattering, orbiting etc.[4]; it is called "weak" if $\Theta^{(N)}(\lambda)$ varies slowly enough so that the inverse of the total deflection function $\Theta(\lambda)$ is single-valued.

In each of these cases a different method of evaluating the integral in eq.(8) or (7) is appropriate. Let us briefly consider the procedure and results.

$\underline{A_S,R_W}$. For strong absorption we use eq.(8), giving a sum of two amplitudes, $\overline{f_1}$ and f_2 , from the terms $D(\lambda)$ and $i\Theta(\lambda)\eta(\lambda)$ of the partial wave amplitude function $a(\lambda)$, respectively. The integral in $f_1(\theta)$ receives its main contribution from the vicinity of $\lambda=\Lambda$; linear expansion of $\delta(\lambda)$ about this value yields[3]

$$f_1(\theta) = \frac{i}{2k} \frac{\Lambda}{(\theta\sin\theta)^{\frac{1}{2}}} e^{i2\delta(\Lambda)} \left[(J_1-iJ_2)F_- + (J_1+iJ_2)F_+ \right] , \qquad (13)$$

with $J_{1,2} \equiv J_{1,2}(\Lambda\theta)$ and $F_\pm \equiv F[\Lambda(\theta_\Lambda\pm\theta)]$, where

$$\theta_\Lambda = \Theta(\Lambda) \qquad (14)$$

is the critical angle associated with Λ, and the function

$$F(\Delta x) = \int_{-\infty}^{\infty} D(\lambda)e^{i(\lambda-\Lambda)x}d\lambda = \int_{-\infty}^{\infty} \tilde{D}(\mu)e^{ix\mu}d\mu \qquad (15)$$

is the Fourier transform of the absorptive shape function $D(x)\equiv\tilde{D}(\lambda-\Lambda)$. In the "sharp cutoff" limit $\Delta\to o$, eq.(13) reduces to

$$f_1(\theta) \to f_{FRA}(\theta) = \frac{i}{2k} \left[\frac{\theta}{\sin\theta}\right]^{\frac{1}{2}} \Lambda^2 \left[\frac{2J_1(\Lambda\theta)}{\Lambda\theta}\right] e^{i2\delta(\Lambda)} \quad , \tag{16}$$

the amplitude for Fraunhofer diffraction scattering.

In evaluating the amplitude $f_2(\theta)$ we note that there are now several domains of λ-space yielding main contributions to the integral; in addition to the vicinity of $\lambda=\Lambda$ there are contributions from possible points of stationary phase. For weak nuclear attraction there is only one such point, λ_s , from the "negative frequency" branch of $J_1(\lambda\theta)$, defined by

$$\Theta(\lambda_s) = \theta . \tag{17}$$

By expanding the total phase of the integrand about $\lambda=\lambda_s$ up to second order, $f_2(\theta)$ can be evaluated in closed form. I shall not give the general result (which is derived in ref.[3]) but consider only the sharp cutoff limit $\Delta\to o$, for which

$$f_2(\theta) \to f_{FRE}(\theta) = f_{sc}(\theta) Fr^{\frac{1}{2}}\left[(\tfrac{1}{2}|\Theta'(\lambda_s)|)^{\frac{1}{2}} (\Lambda-\lambda_s)\right] . \tag{18}$$

Here

$$f_{sc}(\theta) = -\frac{i}{k}\left[\frac{\lambda_s}{|\Theta'(\lambda_s)|\sin\theta}\right]^{\frac{1}{2}} e^{i[2\delta(\lambda_s)-\lambda_s\theta]} \tag{19}$$

is the semiclassical potential scattering amplitude, and $Fr^{\frac{1}{2}}$ denotes the function $Fr^{\frac{1}{2}}(x) \equiv \tfrac{1}{2}erfc(e^{i\frac{1}{4}\pi}x)$ whose square modulus is the "Fresnel function"

$$Fr(x) \equiv |F^{\frac{1}{2}}(x)|^2 = \tfrac{1}{4}\left|erfc(e^{i\frac{1}{4}\pi}x)\right|^2 . \tag{20}$$

The angular distribution described by eq.(18) has the same form as that of Fresnel diffraction in classical optics. Fresnel patterns in heavy ion scattering (first identified in ref.[5]) are quantal interference phenomena resulting from the modification of potential scattering by strong absorption[6]. In addition there is the quantal analogue of Fraunhofer diffraction as a result of shadow scattering alone. Thus the total amplitude in the sharp cutoff limit is a superposition of diffractive components of Fraunhofer and Fresnel type,

$$f(\theta) = f_{FRA}(\theta) + f_{FRE}(\theta) . \tag{21}$$

Their interference results in characteristic oscillatory structures in elastic angular distributions, showing a gradual change from the Fresnel to the Fraunhofer limit with increasing energy[6].

A_S, R_S. For strong nuclear attraction there will be more than one point of stationary phase, and the component $f_2(\theta)$ consists of several "Fresnel-type" contributions of the form (18) (in the limit $\Delta \to o$). In addition there are certain points λ_r at which $\Theta'(\lambda_r) = o$, and the stationary phase approximation breaks down. Such points give rise to rainbow scattering[4]. Accurate approximation methods (such as the "uniform approximation"[7]) have been developed for dealing with rainbows and other caustics in non-absorptive potential scattering, and applied in analyses of heavy ion data[8]. A generalization to strong absorption situations has been given in ref.[3]. For simplicity I confine myself to the Airy approximation, valid only in the neighbourhood of the rainbow angle $\theta_r = \Theta(\lambda_r)$. Absorption of partial waves $\lambda < \Lambda$ modifies the semiclassical Airy pattern involving the (real-valued) function Ai(x) to

$$f_2(\theta) \simeq \frac{1}{k} \left(\frac{2\pi\lambda_r}{\sin\theta}\right)^{\frac{1}{2}} \frac{\text{Ai}(x_r, y_r)}{\left[\frac{1}{2}\Theta''(\lambda_r)\right]^{\frac{1}{3}}} e^{i[2\delta(\lambda_r) - \lambda_r\theta - \frac{1}{4}\pi]} , \qquad \theta \approx \theta_r \tag{22}$$

where I have defined the (complex-valued) "incomplete Airy function"

$$\text{Ai}(x,y) = \frac{1}{2\pi} \int_y^\infty e^{i(x\mu + \frac{1}{3}\mu^3)} d\mu, \qquad \text{Ai}(x,-\infty) = \text{Ai}(x) , \tag{23}$$

with arguments

$$x_r = \frac{\theta_r - \theta}{\left[\frac{1}{2}\Theta''(\lambda_r)\right]^{\frac{1}{3}}} , \qquad y_r = \left[\frac{1}{2}\Theta''(\lambda_r)\right]^{\frac{1}{3}} (\Lambda - \lambda_r) \tag{24}$$

A_W, R_W. When both the absorption and the real part of the nuclear inter-action are "weak", the main contribution to the amplitude comes only from the vicinity of a single point of stationary phase. For weak absorption in general, the form (7) of $f(\theta)$ is more appropriate than eq.(8), and the stationary point λ_s defined by $\Theta(\lambda_s)=\theta$ arises from the "negative frequency" branch of $J_o(\lambda\theta)$. Expansion of the total phase about $\lambda=\lambda_s$ to second order yields

$$f(\theta) = f_{sc}(\theta)\eta(\lambda_s) \quad , \tag{25}$$

where $f_{sc}(\theta)$ is the semiclassical potential scattering amplitude given by eq.(19).

$\underline{A_W, R_S}$. For strong nuclear attraction there will again be more than one stationary point λ_s, and the total amplitude consists of several contributions of the form (25) which interfere through the phase factors $\exp i[2\delta(\lambda_s)-\lambda_s\theta]$; each term is modulated in amplitude by an attenuation coefficient $\eta(\lambda_s)\leq 1$. In addition, rainbow scattering occurs whenever $\Theta'(\lambda_r)=o$. Depending on the strength and shape of the real potential, there may be several rainbow angles $\theta_r=\Theta(\lambda_r)$. The largest positive and the largest negative value of $\Theta(\lambda_r)$ give rise to what has been termed "Coulomb rainbow scattering" and "nuclear rainbow scattering", respectively[9]. In either case the scattering amplitude may be accurately calculated in the uniform approximation, weak absorption being accounted for by the attenuation coefficients $\eta(\lambda_r)$. Confining ourselves again to the Airy approximation, we have

$$f(\theta) \approx \frac{1}{k}\left(\frac{2\pi\lambda_r}{\sin\theta}\right)^{\frac{1}{2}}\frac{Ai(x_r)\eta(\lambda_r)}{\left[\frac{1}{2}\Theta''(\lambda_r)\right]^{\frac{1}{3}}}\,e^{i[2\delta(\lambda_r)-\lambda_r\theta-\frac{1}{4}\pi]} \quad , \quad \theta\approx\theta_r \quad . \tag{26}$$

Finally, there may be contributions from "glory scattering" to extreme forward and backward angles, arising from points λ_g where $\Theta(\lambda_g)=o$ and $\Theta(\lambda_g)=\pm\pi$, respectively. These too can be calculated by accurate "uniform approximation" methods[10].

Comparing the results for weak and strong absorption we note the following differences. It is characteristic of the weak absorption situation that the amplitude for each stationary point λ_s or rainbow point λ_r etc., _factorizes_ into a semiclassical amplitude $f_{sc}(\theta)$ for potential scattering or non-absorptive rainbow scattering etc., and a real attenuation coefficient $\eta(\lambda_s)$ or $\eta(\lambda_r)$. When averaged over the quantal interference terms arising from the phase factors $\exp i[2\delta(\lambda_s)-\lambda_s\theta]$

in case of more than one stationary contribution, the differential cross section is a (sum of) product(s) of the classical cross section and an attenuation factor $P_A(\theta)$ representing the probability for absorption,

$$\frac{d\sigma_{e\ell}}{d\Omega} = \left(\frac{d\sigma}{d\Omega}\right)_{class} \cdot P_A(\theta) \ , \qquad P_A(\theta) = [\eta(\lambda_s)]^2 \ . \qquad (27)$$

The dependence of P_A on θ is mediated by the dependence of η^2 on $\lambda_s = \theta^{-1}(\theta)$. A similar factorization property will be found for inelastic scattering and direct transfer reactions under conditions of weak absorption. In our formalism this is a consequence of the slow variation of $\eta(\lambda)$ under such conditions, coupled with the stationary phase approximation. The semiclassical description based on time-dependent perturbation theory[1,11] leads to expressions of the same form, the absorption probability being given in terms of the imaginary part $W(r)$ of the optical potential, integrated along the classical trajectory pertaining to $\lambda_s = \theta^{-1}(\theta)$,

$$P_A(\theta) = \exp\left\{ -\frac{2}{\hbar} \int_{-\infty}^{\infty} W[r(t)]dt \right\} \ . \qquad (28)$$

Since the trajectory concept plays a central role in semiclassical theories, let us briefly consider its significance in the context of a quantal description. For heavy ions with energies near to and above the Coulomb barrier, that part of the incoming wave packet which is affected by the interaction potential $V(r)$ contains a large number of partial waves, each undergoing a phase shift $\delta(\lambda)$. Their superposition results in large-scale destructive interference which is survived only by those with λ in the vicinity of certain values λ_s (the stationary points) to contribute to the scattering in a given direction θ. We may then describe the scattering as if classical particles moved along trajectories with impact parameters $b_s = \lambda_s/k$ related to θ via the classical equations of motion in the potential $V(r)$. It must, however, be emphasized that this interpretation is purely formal and does not imply that the particles can be represented by microscopically localized wave packets moving along the trajectories. The wave packets that represent particles issuing from an accelerator are always of macroscopic size, only their centres are determined within microscopic distances of order $\lambdabar$ and move, under semiclassical conditions, according to the classical laws of motion.

In the presence of strong absorption, even this formal interpretation of scattering in terms of trajectories is insufficient. The absorptive interaction affects partial waves that contribute to scattering at all angles θ. In other words, in addition to the contributions from variable (θ-dependent) points of stationary phase λ_s there are contributions from fixed (θ-independent) domains of λ-space where $\eta(\lambda)$ changes rapidly (i.e., around the point Λ where $D(\lambda)$ is sharply peaked). These give rise to quantal diffraction in contrast to the "refractive" effects associated with the points $\lambda_s, \lambda_r, \lambda_g$, etc.

The results are twofold. Firstly there is Fraunhofer-type diffraction due to shadow scattering only. Secondly there is Fresnel-type diffraction resulting from the interference between absorptive and refractive scattering; this causes the factors $\eta(\lambda_s)$ multiplying each "trajectory" contribution to be replaced by functions such as $Fr^{\frac{1}{2}}(x)$ and $Ai(x,y)$ depending on the relative values $\Lambda-\lambda_s$, and $\Lambda-\lambda_r$, respectively. These specific patterns of course result only in the limit of sharp cutoff absorption; in general the diffractive structures depend on the shape of $\eta(\lambda)$ or $D(\lambda)$. The angular dependence is then determined also by the Fourier transforms $F(\Delta x)$ of the absorptive shape function $D(\lambda)$ for $x=\theta_\Lambda-\theta$ and $x=\theta_\Lambda+\theta$ (see eq. (13)), which in general cause a damping of the strong oscillations in the idealized patterns.

Without going into further detail it will be clear that the interplay between the absorptive and refractive properties of heavy ion interactions is capable of producing a great variety of quantal interference phenomena.

3. Inelastic scattering

Turning now to inelastic scattering of heavy ions via nuclear and Coulomb excitation of low-lying collective states, we start with the transition amplitude as given by DW theory,

$$T_{fi}^{DW} = \int d^3r \; \chi_f^{(-)^*}(\underline{k}_f, \underline{r}) <\Phi_f|u|\Phi_i> \chi_i^{(+)}(\underline{k}_i, \underline{r}) \; , \qquad (29)$$

where χ_i, χ_f are the distorted waves and Φ_i, Φ_f the internal wave functions in the initial and final channels, respectively, and u denotes the interaction operator. For simplicity we confine ourselves to single excitation of deformed, axially-symmetric even nuclei. Then the differential cross section for multipolarity L becomes (see for instance ref.[12])

$$\frac{d\sigma}{d\Omega} \ (o \rightarrow L) \ = \ \left[\frac{\mu}{2\pi\hbar^2}\right]^2 \ \frac{k_f}{k_i} \ \delta_L^2 \ \sum_{M=-L}^{L} |\tilde{B}_{LM}|^2 \, , \tag{30}$$

where

$$\tilde{B}_{LM} \ = \ \sum_{\ell_i \ell_f} \ i^{\ell_i - \ell_f - L} \ <\ell_f o, Lo|\ell_i o><\ell_f M, L-M|\ell_i o>$$

$$\cdot (2\ell_f + 1)^{\frac{1}{2}} \ R_{\ell_f \ell_i}^{L} \ e^{i[\sigma_{\ell_f}^{(f)} + \sigma_{\ell_i}^{(i)}]} \ Y_{\ell_f, -M}(\theta, o) \ . \tag{31}$$

The radial integrals

$$R_{\ell_f \ell_i}^{L} \ = \ i^L \ \frac{4\pi}{k_f k_i} \ \int_0^{\infty} dr \ f_{\ell_f}^{(f)}(k_f, r) F_L(r) f_{\ell_i}^{(i)}(k_i, r) \tag{32}$$

are determined by the overlap between the radial distorted wave functions f_ℓ and the form factor

$$F_L(r) \ = \ - \ \frac{dU_N}{dr} \ + \ Z_1 Z_2 e^2 \ \frac{3}{2L+1} \cdot \begin{cases} r^L/R_C^{L+2} \qquad & r \leq R_C \ , \\ \\ R_C^{L-1}/r^{L+1} & r \geq R_C \ , \end{cases} \tag{33}$$

where R_C is the Coulomb radius and $\delta_L = \beta_L^{(N)} R_o = \beta_L^{(C)} R_C$ is the deformation length for multipolarity L.

Now it is well known (for a thorough discussion see ref.[12]) that for heavy particles the radial integrals have appreciable values only for ℓ_i and ℓ_f in the neighbourhood of a critical angular momentum ℓ_o, semi-classically associated with the nuclear radius R_o. As a result the amplitude (31) may be simplified to yield

$$\frac{d\sigma}{d\Omega} \ (o \rightarrow L) \ = \ \sum_{M=-L}^{L} |B_{LM}|^2 \ , \tag{34}$$

where

$$B_{LM} = i^{-(M+1)} \pi^{\frac{1}{2}} \delta_L \frac{Y_{LM}(\frac{1}{2}\pi, 0)}{(2L+1)^{\frac{1}{2}}} \sum_{\ell=0}^{\infty} (2\ell+1)^{\frac{1}{2}} b_{\ell} e^{i2\delta_{\ell}} Y_{\ell,-M}(\theta, 0), \quad (35)$$

and the "inelastic partial-wave amplitudes" b_{ℓ} are given by

$$b_{\ell} = \frac{dn_{\ell}}{d\ell} + i \left[\Theta_{\ell}^{(N)} + \gamma_L I_{LM}(\Theta_{\ell}^{(R)}, \zeta) \right] n_{\ell} . \quad (36)$$

Details of the derivation are given in ref.[13]; here I can only indicate the main ingredients. Firstly, for $L \ll \ell_o$ the summations over ℓ_i and ℓ_f in (31) may be reduced to a sum over the mean angular momentum $\ell = \frac{1}{2}(\ell_i + \ell_f)$, and another over the Clebsches yielding $Y_{LM}(\frac{1}{2}\pi, 0)$. For excitation of low-lying states we may replace k_i and k_f by the mean wave number $k = \frac{1}{2}(k_i + k_f)$, and the Sommerfeld parameters by their average value $n = \frac{1}{2}(n_i + n_f)$. Secondly, the contribution to the radial integrals from the nuclear part $-dU_N/dr$ of the form factor $F_L(r)$ is expressed in terms of the derivative of the "nuclear" part $S_{\ell}^{(N)}$ of the average elastic scattering function S_{ℓ} by virtue of the Austern-Blair relation[14]. The latter accounts for the first two ("nuclear") terms of b_{ℓ} in eq. (36). Thirdly, the Coulomb contribution to the radial integrals is evaluated by employing the Sopkovich prescription. The result is a product of $S_{\ell}^{(N)}$ and the standard radial integrals of Coulomb excitation theory; the latter may be expressed in terms of the quantities $I_{LM}(\Theta_{\ell}^{(R)}, \zeta)$, defined in ref.[15], as functions of the quantal deflection function for Rutherford scattering $\Theta_{\ell}^{(R)}$ (given by eq. (12)) and $\zeta = n_f - n_i$. Finally, the coefficients γ_L are defined by

$$\gamma_L = \frac{3}{2L+1} \left(\frac{kR_c}{n} \right)^{L-1} . \quad (37)$$

Next we proceed as in elastic scattering by replacing the summation over ℓ in eq. (35) by an integration over $\lambda = \ell + \frac{1}{2}$ and using the asymptotic form of the spherical harmonics,

$$Y_{\ell,-M}(\theta, 0) \simeq i^{M-|M|} \left(\frac{\lambda\theta}{2\pi\sin\theta} \right)^{\frac{1}{2}} J_{|M|}(\lambda\theta), \quad \lambda = \ell + \frac{1}{2} . \quad (38)$$

The result is

$$B_{LM} = i^{-(|M|+1)} \delta_L \frac{Y_{LM}(\frac{1}{2}\pi,0)}{(2L+1)^{\frac{1}{2}}} \left[\frac{\theta}{\sin\theta}\right]^{\frac{1}{2}} \mathcal{F}_M(\theta) \quad , \tag{39}$$

where

$$\mathcal{F}_M(\theta) = \int_0^\infty d\lambda\lambda \; b(\lambda) e^{i2\delta(\lambda)} J_{|M|}(\lambda\theta) , \tag{40}$$

with the "inelastic partial-wave amplitude" $b(\lambda)$ given by

$$b(\lambda) = D(\lambda) + i\left[\Theta^{(N)}(\lambda) + \gamma_L \tilde{I}_{LM}(\lambda)\right] \eta(\lambda) \quad , \tag{41}$$

where $\tilde{I}_{LM}(\lambda) \equiv I_{LM}(\Theta^{(R)}(\lambda), \zeta)$.

Comparison of eqs. (40) and (41) with eqs. (8) and (9) shows how closely inelastic scattering is related to elastic scattering. If in eq. (9) we formally split the quantal deflection function into a "nuclear" and a Rutherford part, $\Theta(\lambda) = \Theta^{(N)}(\lambda) + \Theta^{(R)}(\lambda)$, we have a one-to-one correspondence between the three contributions to the elastic and inelastic partial-wave amplitudes: (i) from the absorptive part of the nuclear interaction, given by the absorptive shape function $D(\lambda)$, (ii) from the real part of the nuclear interaction, determined by $\Theta^{(N)}(\lambda)$, and (iii) from the Coulomb interaction, described by $\Theta^{(R)}(\lambda)$ and $\gamma_L \tilde{I}_{LM}(\lambda)$, respectively. Because the nuclear force is attractive and the Coulomb force repulsive, the contributions (ii) and (iii) in eq. (41) are mostly of opposite sign and give rise to the characteristic destructive interference effects between the amplitudes for nuclear and Coulomb excitation.

The integrals (40) may be evaluated explicitly by similar methods as for elastic scattering. This has been described in detail in ref.[13]. Here I can only point out some of the results for the simplest idealized cases. In general, the method of evaluation again depends on whether $\eta(\lambda)$ varies rapidly or slowly with λ, particularly in the vicinity of the critical value $\Lambda = \ell_0 + \frac{1}{2}$. If $D(\lambda)$ is sharply peaked at Λ, its contribution to the inelastic amplitude is of Fraunhofer diffraction type; in the sharp cutoff limit this contribution becomes

$$B_{LM}^{(FRA)} = i^{-(|M|+1)} \delta_L \frac{Y_{LM}(\frac{1}{2}\pi,o)}{(2L+1)^{\frac{1}{2}}} \left(\frac{\theta}{\sin\theta}\right)^{\frac{1}{2}} \Lambda J_{|M|}(\Lambda\theta) e^{i2\delta(\Lambda)}, \tag{42}$$

corresponding to Blair's diffraction model for inelastic scattering[16]. Generalizations for smooth cutoff absorption have been given earlier[17] and applied to heavy ion inelastic scattering[18]; the results again involve the Fourier transforms $F[\Delta(\theta_\Lambda \pm \theta)]$ of the absorptive shape function $D(\lambda)$.

The real nuclear and Coulomb excitation contributions may be evaluated in stationary phase approximation; modification by sharp cutoff absorption gives rise to inelastic scattering of Fresnel type. The result is[13]

$$\mathcal{F}_M(\theta) = \Lambda J_{|M|}(\Lambda\theta) e^{i2\delta(\Lambda)}$$

$$+ i^{|M|+1} \left[\frac{\lambda_s}{\theta|\Theta'(\lambda_s)|}\right]^{\frac{1}{2}} \left(-|\Theta^{(N)}(\lambda_s)| + \gamma_L \tilde{I}_{LM}(\lambda_s)\right) Fr^{\frac{1}{2}}\left((\frac{1}{2}|\Theta'(\lambda_s)|)^{\frac{1}{2}}(\Lambda-\lambda_s)\right)$$

$$\cdot e^{i[2\delta(\lambda_s)-\lambda_s\theta]}, \tag{43}$$

where $\lambda_s = \Theta^{-1}(\theta)$. Again the generalization to smooth cutoff absorption leads to a damping of both the Fraunhofer and Fresnel diffractive oscillations by the Fourier transforms $F[\Delta(\theta_\Lambda \pm \theta)]$. Note that our result implies a specific physical interpretation of the nature of the oscillations at small angles in heavy ion inelastic scattering: they are identified as a quantal diffraction pattern of Fresnel type, analogous to that in elastic scattering, arising from the interference between inelastic excitation by the combined nuclear and Coulomb fields and absorption in the nuclear surface. Similar conclusions have recently been reached by Landowne and Takigawa[19] from numerical calculations using WKB phase shifts. At larger angles $(\theta > \theta_\Lambda)$ the Blair-type Fraunhofer oscillations, represented by the first term in eq. (43), dominate unless damped by smooth cutoff absorption.

For sufficiently strong nuclear attraction, inelastic rainbow scattering may occur; in our formalism this is described by an absorption-modified Airy pattern involving the function $Ai(x_r, y_r)$, or more accurately by the generalized uniform approximation. Numerical analyses of inelastic heavy ion scattering based on this approximation have recently been carried out by da Silveira and Leclercq-Willain[20].

Under conditions of weak absorption, the quantal diffraction effects are negligible, and the only contributions to the integral in eq. (40) come from the points of stationary phase. In that case the inelastic amplitude is of a form given by the semiclassical theory[11,21-23], each contribution being multiplied by an attenuation coefficient $\eta(\lambda_s)$. The cross section can then be written as a product of the classical elastic scattering cross section times a probability $P_L(\theta)$ for inelastic excitation,

$$\frac{d\sigma}{d\Omega}(o \to L) = \left(\frac{d\sigma}{d\Omega}\right)_{class} \dot{P}_L(\theta) \, ,$$

(44)

$$P_L(\theta) = \frac{(\delta_L k)^2}{2L+1} \sum_{M=-L}^{L} |Y_{LM}(\tfrac{1}{2}\pi, o)|^2 \left[E_{LM}(\lambda_s)\eta(\lambda_s)\right]^2 .$$

I have already emphasized the very close connection between inelastic and elastic scattering displayed by our formalism. This relation may be expressed by a remarkably simple formal recipe: To obtain from $\mathcal{F}_M(\theta)$ the corresponding expression for elastic scattering, replace $|M| \to 1$ and $E_{LM}(\lambda_s) \to \theta$ or $E_{LM}(\Lambda) \to \theta_\Lambda$, where $E_{LM}(\lambda) \equiv -|\theta^{(N)}(\lambda)| + \gamma_L \tilde{I}_{LM}(\lambda)$. Conversely, once the average elastic scattering function $S(\lambda)$ has been determined, either by calculation from a dynamical model or by phase shift analysis of elastic scattering data, the amplitude for inelastic scattering of given multipolarity L is completely determined except for an overall spectroscopic factor (the deformation length δ_L).

4. Direct transfer reactions

Next we consider transfer reactions of the form

$$A_i + B_i \equiv (A_f + x) + B_i \longrightarrow A_f + (B_i + x) \equiv A_f + B_f , \qquad (45)$$

where the A, B represent heavy ions and x a single nucleon or light cluster. We assume that this process can be described by DW theory with a transition amplitude of the form

$$T_{fi} = \int d^3r_{xB_i} \int d^3r_{A_fx} \ \chi_f^{(-)*} (\underset{\sim}{k}_f, \underset{\sim}{r}_f) \ <A_fB_f|V_{A_fx}|B_iA_i> \ \chi_i^{(+)} (\underset{\sim}{k}_i, \underset{\sim}{r}_i) . \qquad (46)$$

To derive simple closed expressions we make further approximations which are really too drastic for heavy ion reactions: the "zero-range" and the "no recoil" approximation; however, the main quantal features of transfer angular distributions we wish to describe will not be too badly spoilt by finite range and recoil effects.

With these simplifications, the differential cross section for transfer with orbital angular momentum L may be written

$$\left(\frac{d\sigma_T}{d\Omega}\right)_L = A_L \sum_{M=-L}^{L} |C_{LM}|^2 , \qquad (47)$$

where A_L contains kinematical and spectroscopic factors and where

$$C_{LM} = \sum_{\ell_i \ell_f} i^{\ell_i - \ell_f - L} \ <\ell_f 0, L0 | \ell_i 0><\ell_f M, L-M | \ell_i 0>$$

$$(48)$$

$$\cdot (2\ell_f + 1)^{\frac{1}{2}} \ T^L_{\ell_f \ell_i} \ e^{i[\sigma_{\ell_f}^{(f)} + \sigma_{\ell_i}^{(i)}]} \ Y_{\ell_f, -M} (\theta, 0) .$$

The amplitudes C_{LM} have a similar structure as B_{LM} for inelastic scattering, except that the radial integrals are now given by

$$T^L_{\ell_f \ell_i} = i^L \frac{4\pi}{k_i k_f'} \int_0^{\infty} dr \ f_{\ell_f}^{(f)} (k_f, r') w_L (r) f_{\ell_i}^{(i)} (k_i, r) , \qquad (49)$$

where $r'=(B_i/B_f)$, $k_f'=(B_i/B_f)k_f$, and $w_L(r)$ is the radial bound state wave function of the captured particle x. Following the methods of refs.[24,25] we employ the Sopkovich prescription for approximating $T^L_{\ell_f \ell_i}$ and reduce the summations in (48) in a similar fashion as for in-elastic scattering. The result can be written as

$$C_{LM} = i^{-M} \frac{(4\pi)^{\frac{3}{2}}}{k_i k_f'} \frac{Y_{LM}(\frac{1}{2}\pi,0)}{(2L+1)^{\frac{1}{2}}} \sum_{\ell=0}^{\infty} (2\ell+1)^{\frac{1}{2}} I_\ell^{(T)} \bar{S}_\ell Y_{\ell,-M}(\theta,0) \qquad (50)$$

where

$$I_\ell^{(T)} = \int_0^{\infty} dr \, \mathcal{f}_\ell(k'_f r) \frac{e^{-\kappa r}}{\kappa r} \mathcal{f}_\ell(k_i r) \qquad (51)$$

is a radial integral involving Coulomb wave functions $\mathcal{f}_\ell(kr)$ and the asymptotic form of the bound state wave function $w_L(r)$ determined by the binding energy $\varepsilon_B = (\hbar^2/2m)\kappa^2$ of the captured particle x, and where we have defined the mean elastic scattering function

$$\bar{S}_\ell = \bar{\eta}_\ell e^{i2\bar{\delta}_\ell} \equiv \left[\eta_\ell^{(f)}(k'_f) \eta_\ell^{(i)}(k_i) \right]^{\frac{1}{2}} \cdot e^{i[\delta_\ell^{(f)}(k'_f) + \delta_\ell^{(i)}(k_i)]}. \qquad (52)$$

To proceed, we once again replace the ℓ-summation in (50) by an integration over $\lambda = \ell + \frac{1}{2}$, using the asymptotic expression (28) for $Y_{\ell,-M}(\theta,0)$.
The result may be written as

$$C_{LM} = i^{-(|M|+1)} \frac{8\pi}{k_i k_f'} \frac{Y_{LM}(\frac{1}{2}\pi,0)}{(2L+1)^{\frac{1}{2}}} \left(\frac{\theta}{\sin\theta}\right)^{\frac{1}{2}} \mathcal{T}_M(\theta), \qquad (53)$$

where

$$\mathcal{T}_M(\theta) = \int_0^{\infty} d\lambda\,\lambda\; c(\lambda)\; e^{i2\bar{\delta}(\lambda)}\, J_{|M|}(\lambda\theta) \qquad (54)$$

with the "transfer partial-wave amplitude"

$$c(\lambda) = i I^{(T)}(\lambda) \bar{\eta}(\lambda) . \qquad (55)$$

In this form the analogy with the corresponding expressions (40), (41) for inelastic scattering becomes apparent: in $c(\lambda)$ the function $I^{(T)}(\lambda)$ corresponds to $E(\lambda) = \Theta^{(N)}(\lambda) + \gamma_L \tilde{I}_{LM}(\lambda)$ in $b(\lambda)$; note, however, that there is no contribution that corresponds to the term $D(\lambda)$ in eq.(41). The function $I^{(T)}(\lambda)$ is also known explicitly from the theory of Coulomb excitation[15]; however, for most purposes it suffices to use its asymptotic form for large λ,

$$I^{(T)}(\lambda) \cong I_T \, \lambda^{-\frac{1}{2}} \, e^{-\gamma\lambda}, \qquad \gamma = \frac{1}{n}\left[(n\,\frac{\kappa}{k})^2 + \zeta^2\right], \tag{56}$$

where the constant I_T involves κ, $k = \frac{1}{2}(k_i + k_f')$, $n = \frac{1}{2}(n_i + n_f)$ and $\zeta = n_f - n_i$.

The integration in (54) can again be carried out explicitly by different methods depending on whether $c(\lambda)$ varies rapidly or slowly compared to the phase factor. This in turn depends on the strength of the absorption which tends to quench the low partial waves, and on the strength of binding of the captured particle which suppresses the high partial waves because of the exponential in (56).

In the idealized case of sharp cutoff absorption at $\lambda = \Lambda$ we obtain

$$\tau_M(\theta) = i^{|M|+1} \, I^{(T)}(\lambda_s) \left[\frac{\lambda_s}{\theta|\overline{\Theta}'(\lambda_s)|}\right]^{\frac{1}{2}} Fr^{\frac{1}{2}}\left[(\tfrac{1}{2}|\overline{\Theta}'(\lambda_s)|)^{\frac{1}{2}}(\Lambda-\lambda_s)\right] e^{i[2\overline{\delta}(\lambda_s) - \lambda_s\theta]}. \tag{57}$$

Comparison with eq.(18) shows that this can be written in the form

$$C_{LM} = i\,\frac{8\pi k}{k_i k_f'}\,\frac{Y_{LM}(\tfrac{1}{2}\pi, o)}{(2L+1)^{\frac{1}{2}}}\, I^{(T)}(\lambda_s)\,\overline{f}_{FRE}(\theta), \tag{58}$$

where $\overline{f}_{FRE}(\theta)$ is the (average) elastic Fresnel scattering amplitude; therefore in this case the transfer angular distribution is a Fresnel diffraction pattern modified by $|I^{(T)}(\lambda_s)|^2$, which causes an essentially exponential decrease toward small angles.

Actually $c(\lambda)$ will vary smoothly, with an asymmetrical distribution over λ, peaked at a critical value Λ_T determined approximately by

$\overline{D}(\Lambda_T) \approx \gamma\overline{\eta}(\Lambda_T)$. If the peak is sharp enough (strong absorption, not too weak binding), a linear expansion of the phase $\overline{\delta}(\lambda)$ about Λ_T yields[3]

$$\mathcal{T}_M(\theta) = \tfrac{1}{2}\Lambda_T e^{-\gamma\Lambda_T} e^{i2\overline{\delta}(\Lambda_T)} \left[(J_{|M|} - iJ_{|M|+1}) H(\theta_T - \theta + i\gamma) + (J_{|M|} + iJ_{|M|+1}) H(\theta_T + \theta + i\gamma) \right],$$

(59)

where $J_{|M|} \equiv J_{|M|}(\Lambda_T\theta)$ etc., where we have defined the Fourier transform of $\overline{\eta}(\lambda)$,

$$H(z) = \int_{-\infty}^{\infty} d\lambda \ \overline{\eta}(\lambda) e^{i(\lambda-\Lambda_T)z} ,$$

(60)

and where θ_T is the critical angle for transfer defined by $\theta_T = \overline{\Theta}(\Lambda_T)$.

To simplify this result further we replace the Bessel functions by their asymptotic forms, valid for $\theta \gg |M|/\Lambda_T$. We then obtain for the transfer angular distribution

$$\left(\frac{d\sigma_T}{d\theta}\right)_L = 2\pi\sin\theta \left(\frac{d\sigma_T}{d\Omega}\right)_L \sim \alpha(\theta) \left[1 + (-)^L[\mathrm{Re}\beta(\theta)\sin(2\Lambda_T\theta) + \mathrm{Im}\beta(\theta)\cos(2\Lambda_T\theta)]\right],$$

(61)

where

$$\alpha(\theta) = |H_-|^2 + |H_+|^2 , \qquad \beta(\theta) = \frac{2H_+H_-^*}{|H_-|^2 + |H_+|^2} ,$$

(62)

with the abbreviations $H_\pm \equiv H(\theta_T \pm \theta + i\gamma)$. Thus the angular distribution consists of a smooth part $\alpha(\theta)$ and an oscillatory part representing (Fraunhofer-type) diffraction oscillations of angular period π/Λ_T. In the smooth part the contribution $|H_-|^2 = |H(\theta_T - \theta + i\gamma)|^2$ is dominant; it is "bell-shaped" with a maximum at $\theta = \theta_T$ and a width that is <u>inversely</u> proportional to the width of the function $\overline{\eta}(\lambda)\exp(-\gamma\lambda)$ in λ-space, $H(x)$ being the Fourier transform of this function. The oscillatory, diffractive part arises from interference between the two amplitudes in (59) determined respectively by $H(\theta_T - \theta + i\gamma)$ and $H(\theta_T + \theta + i\gamma)$.

Since the presence of oscillatory structure in angular distributions of heavy ion transfer reactions has gained some prominence recently, it will be worthwhile discussing in more detail their properties, their physical nature, and the conditions under which they appear. To simplify this discussion we replace the partial-wave amplitude $c(\lambda)$ by a simple parameterized form. This was done in the earliest diffraction models of heavy ion transfer reactions by assuming $c(\lambda)=i\bar{D}(\lambda)$ (Frahn-Venter[26]) or $c(\lambda) = i\exp\{-[(\lambda-\Lambda_T)/\Delta_T]^2\}$ (Strutinsky[26]). If we adopt the latter form, as was also done recently by Chasman et al.[28], eq.(61) is replaced by

$$\left(\frac{d\sigma_T}{d\theta}\right)_L \sim \alpha(\theta)\left[1 + (-)^L\beta(\theta)\sin(2\Lambda_T\theta)\right] , \qquad (63)$$

where

$$\alpha(\theta) = e^{-\frac{1}{4}\Delta_T^2(\theta-\theta_T)^2} + e^{-\frac{1}{4}\Delta_T^2(\theta+\theta_T)^2} , \qquad \beta(\theta) = \frac{2e^{-\frac{1}{2}\Delta_T^2(\theta^2+\theta_T^2)}}{\alpha(\theta)} . \quad (64)$$

This shows that the amplitude of the diffraction oscillations depends sensitively on the magnitude of $\Delta_T\theta_T$. Since θ_T will be close to the grazing angle for Rutherford scattering, $\Delta_T\theta_T$ is approximately proportional to the Sommerfeld parameter n, varying with energy as $E^{-\frac{1}{2}}$.

Thus at low energies the diffraction oscillations are strongly damped ("Coulomb damping") and the angular distribution has a smooth "bell-shaped" form with a maximum at θ_T, described by $|H_-|^2$. (It should be emphasized that this is not a "semiclassical" feature, as is often stated in the literature; on the contrary it is due to the uncertainty relation between angular momentum and scattering angle[26], the width of the angular distribution varying inversely to that of $c(\lambda)$.) As the energy increases there first appears a modified Fresnel-type pattern as represented schematically by eq.(57). At still higher energy the Fraunhofer, type oscillations emerge with increasing amplitude and decreasing period. At the same time the whole pattern shifts to smaller angles ($\theta_T \sim E^{-1}$) and decreases in width ($\Delta_T^{-1} \sim E^{-\frac{1}{2}}$). This behaviour completely parallels the corresponding changes in the elastic angular distribution. Experimental evidence for the presence of diffractive oscillations in heavy ion transfer angular distributions, first predicted ten years ago[26,27], has only recently been found at higher energies by the Brookhaven[29], Copenhagen[30], Heidelberg[31], Argonne[32] and Saclay[33] groups. For further discussion of oscillatory structure in heavy ion transfer data I refer to Garrett's talk[34].

Finally, let us return to eq. (54) to consider the case that $c(\lambda)$ varies slowly (i.e., weak absorption and/or weak binding). Then the integral in (54) may be evaluated in stationary phase approximation, with the result

$$\mathcal{T}_M(\theta) = i^{|M|+1} \left(\frac{\lambda_s}{\theta |\overline{\theta}'(\lambda_s)|} \right)^{\frac{1}{2}} I^{(T)}(\lambda_s)\overline{\eta}(\lambda_s) e^{i[2\overline{\delta}(\lambda_s)-\lambda_s\theta]} \tag{65}$$

or

$$C_{LM} = i \frac{8\pi k}{k_i k_f'} \frac{Y_{LM}(\frac{1}{2}\pi,0)}{(2L+1)^{\frac{1}{2}}} I^{(T)}(\lambda_s)\overline{\eta}(\lambda_s)\overline{f}_{sc}(\theta) , \tag{66}$$

where $\overline{f}_{sc}(\theta)$ is the average semiclassical elastic scattering amplitude given by eq. (19). In this case the differential cross section for transfer is a product of the mean classical elastic scattering cross section and a transfer probability $P_T(\theta)$ whose angular distribution is given by $|c(\lambda_s)|^2$,

$$\left(\frac{d\sigma_T}{d\Omega} \right)_L = \overline{\left(\frac{d\sigma}{d\Omega} \right)}_{class} \cdot P_T(\theta) , \qquad P_T(\theta) \propto |I^{(T)}(\lambda_s)\overline{\eta}(\lambda_s)|^2 . \tag{67}$$

The angular distribution is then always smooth (i.e., non-oscillatory); its width is _directly_ proportional to that of the λ-space distribution $c(\lambda)$ and will therefore _increase_ with increasing energy. This difference in the behaviour of apparently similar shapes of transfer angular distributions under semiclassical and quantal conditions has been emphasized by Siemens and Becchetti[35]. A simple model that incorporates both types of behaviour as limiting cases was recently given by Strutinsky[36]. Note that if there is only a single point of stationary phase, the semiclassical approximation implies a factorizable cross section (67) and the absence of diffraction oscillations. Oscillatory structure of non-diffractive nature may, however, occur by interference from several stationary points or in the presence of rainbow scattering.

5. Further aspects

Within the limits of this talk I had to restrict the subject in several respects. Firstly, it was not possible to cover all the wave mechanical aspects of heavy ion collisions. Let me mention only a few that had to be left out. In elastic collisions of two heavy ions that differ in mass by only one or a few nucleons there are important contributions from elastic transfer[37], the amplitude of which is coherent with the elastic scattering amplitude. Their interference results in quantal symmetry oscillations in the angular distributions, similar to the Mott oscillations in the scattering of identical nuclei. Well above the Coulomb barrier the interference between symmetry and diffractive effects may give rise to quite complex oscillatory structure for such collisions. In inelastic scattering as well as in transfer reactions there are significant contributions from two- and multi-step processes[38]. Then there are the effects of finite range[39] and recoil[40] in transfer reactions which were ignored in our simplified treatment.

Secondly, with regard to the mathematical methods of evaluating partial-wave expansions, it would have been instructive to compare our formalism with alternative approaches. One of these, based on analytic continuation of the scattering function $S(\lambda)$ into the complex λ-plane[14], employs the Watson transformation to represent the scattering amplitude as a sum of Regge-pole contributions plus a background integral. In cases where only one or a few Regge poles dominate this method is specially adapted to describe surface resonances[42]. It has also been used recently to explain oscillatory transfer angular distributions[43]. Another approach at present being developed[44] is an extension of semiclassical theory for strongly absorbing potentials: by introducing partly complex trajectories, the main contributions to the partial-wave expansion with complex WKB phases come from saddle points in complex impact parameter space. Quantal interference of contributions from two or more saddle points may result in oscillatory structure of diffractive type[45].

Finally, the methods described in this talk lead to closed expressions that are much more general and accurate than the simple formulae given here. The formalism may also readily be extended to account for other wave mechanical effects as mentioned above. Naturally the results become more involved with each step of generalization and it is a matter of taste to decide how far one wants to go into detail. As I see it, the aim is not so much to develop an alternative to *ab initio* numerical computation than to provide physical insight and, hopefully, clues to predicting new phenomena.

References

1) R.A. Broglia and A. Winther, Phys. Reports 4C (1972) 153, and references therein.

2) W.E. Frahn, in Fundamentals in Nuclear Theory, IAEA, Vienna 1967, p.3, and references therein.

3) W.E. Frahn, Wave mechanics of heavy ion collisions, Lectures at the Extended Seminar on Nuclear Physics, ICTP, Miramare-Trieste 1973, to be published by IAEA, Vienna.

4) K.W. Ford and J.A. Wheeler, Ann. Phys. (N.Y.) 7 (1959) 259.

5) W.E. Frahn, Nucl. Phys. 75 (1966) 577.

6) W.E. Frahn, Phys. Rev. Lett. 26 (1971) 568; Ann. Phys. (N.Y.) 72 (1972) 524.

7) M.V. Berry, Proc. Phys. Soc. 89 (1966) 479.

8) R. da Silveira, Phys. Lett. 45B (1973) 211.

9) D.A. Goldberg and S.M. Smith, Univ. of Maryland report 74-084, 1974.

10) M.V. Berry, J. Phys. B (Atom. Molec. Phys.) 2 (1969) 381.

11) R.A. Broglia, S. Landowne and A. Winther, Phys. Lett. 40B (1972) 293.

12) N. Austern, Direct nuclear reaction theories, Wiley, New York 1970.

13) W.E. Frahn, Closed-form quantal description of inelastic heavy ion scattering, Informal Report, Technical University Munich, July 1974, to be published.

14) N. Austern and J.S. Blair, Ann. Phys. (N.Y.) 33 (1965) 15.

15) K. Alder, A. Bohr, T. Huus, B. Mottelson and A. Winther, Rev. Mod. Phys. 28 (1956) 432.

16) J.S. Blair, Phys. Rev. 115 (1959) 928.

17) J.M. Potgieter and W.E. Frahn, Nucl. Phys. 80 (1966) 434; Phys.Lett. 21 (1966) 211.

18) J.M. Potgieter and W.E. Frahn, Nucl. Phys. A92 (1967) 84.

19) S. Landowne and N. Takigawa, Phys. Lett. 50B (1974) 414.

20) R. da Silveira and Ch. Leclercq-Willain, Report IPNO/TH 73-52, Orsay, December 1973.

21) R.A. Malfliet, S. Landowne and V. Rostokin, Phys. Lett. <u>44B</u> (1973) 238.

22) R.A. Broglia, S. Landowne, R.A. Malfliet, V. Rostokin, and Aa. Winther, Physics Reports, to be published.

23) R.A. Malfliet, Lectures at Extended Seminar on Nuclear Physics,ICTP, Miramare-Trieste 1973, to be published by IAEA, Vienna; Proc. Argonne Symposium on Heavy Ion Transfer Reactions 1973, Vol.II, p.605; and this Symposium.

24) A. Dar, Phys. Rev. <u>139</u> (1965) B1193; Nucl. Phys. <u>82</u> (1966) 354.

25) W.E. Frahn and M.A. Sharaf, Nucl. Phys. <u>A133</u> (1969) 593.

26) W.E. Frahn and R.H. Venter, Nucl. Phys. <u>59</u> (1964) 651.

27) V.M. Strutinsky, Zh. Eksp. Teor. Fiz. <u>46</u> (1964) 2078 [Sov. Phys. JETP <u>19</u> (1964) 1401].

28) C. Chasman, S. Kahana and M. Schneider, Phys. Rev. Lett. <u>31</u> (1973) 1074.

29) M.J. Schneider et al., Phys. Rev. Lett. <u>31</u> (1973) 231; P.D.Bond et al., Phys. Lett. <u>47B</u> (1973) 231; M.J. LeVine et al., Phys. Rev. Comments, to be published.

30) P.R. Christensen et al., Phys. Lett. <u>45B</u> (1973) 107; J.B.Ball et al., Phys. Lett. <u>49B</u> (1974) 348.

31) P. Braun-Munzinger et al., Phys. Rev. Lett. <u>31</u> (1973) 1423.

32) W. Henning et al., Phys. Rev. Lett. <u>32</u> (1974) 1015.

33) M.-C. Lemaire et al., Proc. Nashville Conf. 1974, to be published.

34) J.D. Garrett, this Symposium.

35) P.J. Siemens and F.D. Becchetti, Phys. Lett. <u>42B</u> (1972) 389.

36) V.M. Strutinsky, Phys. Lett. <u>44B</u> (1973) 245.

37) W. von Oertzen, Nucl. Phys. <u>A148</u> (1970) 529; C.A. McMahan and W. Tobocman, Nucl. Phys. <u>A202</u> (1973); G. Baur and C.K. Gelbke, Nucl. Phys. <u>A204</u> (1973) 138; W. von Oertzen and W. Nörenberg, Nucl. Phys. <u>A207</u> (1973) 113.

38) S. Landowne, R.A. Broglia and R. Liotta, Phys. Lett <u>43B</u> (1973) 160; C.K. Gelbke et al., Nucl. Phys <u>A219</u> (1974) 253; G. Baur and H.H. Wolter, Proc. Nashville Conf. 1974; B.Kohlmeyer et al., *ibid.*; S. Landowne, R.A. Broglia and B. Nilsson, *ibid.*, to be published; H.H. Wolter, this Symposium.

39) N. Austern et al., Phys. Rev. <u>133</u> (1964) B3; F.D. Santos, Nucl.Phys. <u>A212</u> (1973) 341.

40) L.R. Dodd and K.R. Greider, Phys. Rev. Lett. 14 (1965) 959; Phys. Rev. 180 (1969) 1187, P.J.A. Buttle and L.J.B. Goldfarb, Nucl. Phys. A176 (1971) 299; M.A. Nagarajan, Nucl. Phys. A196 (1972) 34; R.M. DeVries and K.I. Kubo, Phys. Rev. Lett. 30 (1973) 325; R.M.DeVries, Phys. Rev. C8 (1973) 951; P. Braun-Munzinger and H.L. Harney, Nucl. Phys. A223 (1974)381; A.J. Baltz and S. Kahana, Phys. Rev. C9 (1974) 2243.

41) T. Tamura and H.H. Wolter, Phys. Rev. C6 (1972) 1976.

42) K.W. McVoy, Phys. Rev. C3 (1971) 1104; J.T. Londergan and K.W.McVoy, Nucl. Phys. A201 (1973) 390; R.C. Fuller, Nucl. Phys. A216 (1973) 199; R.C. Fuller and Y. Avishai, Nucl. Phys. A222 (1974) 365; K.W. McVoy, this Symposium.

43) R.C. Fuller and O.Dragun, Phys. Rev. Lett. 32 (1974) 617.

44) J. Knoll and R. Schaeffer, Lectures at the Extended Seminar on Nuclear Physics, ICTP, Miramare-Trieste 1973, to be published by IAEA, Vienna; Saclay Report DPh. T/74/31, 1974, to be published.

45) R. Schaeffer, private communication.

SURFACE WAVES AND SURFACE RESONANCES IN HEAVY-ION REACTIONS*

K. W. McVoy
Department of Physics
University of Wisconsin
Madison, Wisconsin 53706

The relations between surface waves, rainbows and refractive shadows are examined. These all appear to be Regge-pole phenomena, which seem to play a major dynamical role in heavy-ion transfer reactions and backward-hemisphere α-scattering.

1.   INTRODUCTION

Within the past year or so, experimental evidence strongly suggesting the appearance of surface or "creep" waves in nuclear physics has shown up in two types of angular measurements, both associated with transparent-surface optical potentials. One is the exponential angular falloff on the small-angle side of the grazing-angle-peak in heavy-ion transfer reactions, and the other is the exponential decrease, at angles beyond 60°, of the angular distribution of alpha particles elastically scattered by targets ranging from Mg to Zr.[1] The data of Put and Paans[1b] for ^{90}Zr, reproduced in Fig. 1, shows this non-oscillatory section setting in at about 80 MeV; that of Goldberg et al.[1c] at 142 MeV shows an accurately exponential falloff of over 3 orders of magnitude in $\sin \theta (d\sigma/d\Omega)$ (the relevant quantity) between 60° and 100°. As the latter authors have cogently pointed out, this is what might be called a "rainbow shadow". That is, at 118 MeV the maximum classical deflection angle θ_r through which the attractive α-Zr interaction can scatter α-particles is about 75°. This occurs at an impact parameter corresponding to $\ell \stackrel{\sim}{\sim} 28$, and the pile-up of flux near this ℓ produces the broad maximum seen at 60°. Classically the cross section at angles beyond this maximum would be zero, but wave-mechanically the transition into this dark or classically-forbidden region (which we shall simply call a "shadow" region) is characterized by a cross section which decreases approximately exponentially with increasing angle.

In this context, we would like to suggest a distinction between what might be called "diffractive shadows" and "refractive shadows". By diffraction we mean a spreading of waves into classically-inaccessible regions caused by absorption, e.g., flux-removal such as that described by the imaginary part of an optical potential. By a refractive shadow we mean the dark region of an angular distribution beyond a rainbow angle, where the _real_ part of the optical potential is too weak to channel

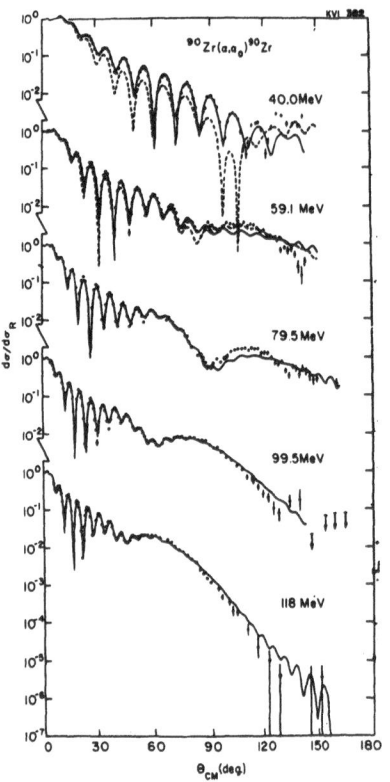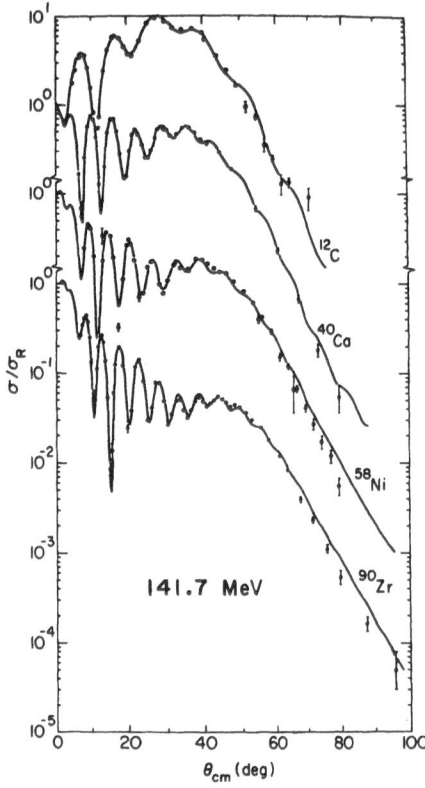

FIG. 1. Elastic $\alpha + {}^{90}Zr$ angular distributions from Groningen (left, Ref. 1b) and Maryland (right, Ref. 1c), showing a non-oscillatory rainbow shadow for $E_\alpha > 75$ MeV.

flux. Both are exponential shadows, but in a partial-wave representation, a diffractive shadow is characterized by a decrease in $|S_\ell|$, the __magnitude__ of certain S-matrix elements, while a rainbow shadow is described by a vanishing second derivative of its __phase__ $\delta(\ell)$.

Recent work by Fuller[2] strongly suggests that it is primarily this second, refractive, effect which illuminates shadow regions when waves are scattered by absorptive 3-dimensional objects __provided__ they have a transparent refractive surface. In this case, the waves are "steered" into the shadow regions by creeping along the surface of the interaction region, and give rise to a rainbow phenomenon in the phase of S_ℓ. Evidently it is not well known whether diffractive shadowing is a surface-wave phenomenon as well, but at least it seems safe to employ the occurrence of "exponential shadows" in asymptotic angular distributions as an indicator of surface creep waves whenever there is evidence that an attractive surface-interaction region is present, and it is this interpretation which we will employ here.

As will be readily evident, almost all the ideas on surface waves and Regge poles presented below have grown from the seminal work of Fuller[2a] on the subject, which in turn was in part inspired by previous work of Nussenzveig's.[2b]

2. DIFFRACTIVE BLURRING OF REFRACTIVE TRAJECTORIES

In the short-wavelength or large-ℓ limit, the partial-wave expansion of a scattering or reaction amplitude takes the form (in the spin-zero case)

$$f(\theta) = \frac{1}{2ik} \Sigma \ (2\ell+1) \ f(\ell) \ P_\ell(\cos \theta)$$

$$\underset{\sim}{} (ik)^{-1} \ (2\pi \sin \theta)^{-\frac{1}{2}} \ \{\Sigma \ \ell^{\frac{1}{2}} \ f(\ell) \ e^{-i[(\ell+\frac{1}{2})\theta-\pi/4]}$$

$$+ \ \Sigma \ \ell^{\frac{1}{2}} \ f(\ell) \ e^{i[(\ell+\frac{1}{2})\theta-\pi/4]}\}$$

$$\equiv (ik)^{-1} \ (2\pi \sin \theta)^{-\frac{1}{2}} \ [e^{i\pi/4} \ g(\theta) + e^{-i\pi/4} \ g(-\theta)]. \qquad (1)$$

Writing $f(\ell) = e^{2i\delta(\ell)}$, the stationary-phase evaluation of the sum, in the case of real $\delta(\ell)$, gives the usual semi-classical result,

$$g(\theta) = i \ \underset{n}{\Sigma} \ [\frac{\pi \mathcal{L}_n}{\delta''(\mathcal{L}_n)}]^{\frac{1}{2}} \ e^{i[2\delta(\mathcal{L}_n)-(\mathcal{L}_n+\frac{1}{2})\theta]} \ , \qquad (2)$$

valid for $\theta > 1/\mathcal{L}_n$. The sum is over the stationary-phase ℓ-values $\mathcal{L}_n$, which are given as a function of θ by the <u>real-ℓ</u> solutions of the trajectory-equation,

$$2\delta'(\mathcal{L}) = \theta. \qquad (3)$$

Keeping only the real-ℓ solutions is the ray-optics or refraction approximation. It omits all effects which allow flux to "leak" into shadow regions, by setting $g(\theta) = 0$ at angles which cannot be reached by classical trajectories, even though the actual $g(\theta)$ of Eq. (1) does not vanish there. However as θ moves beyond a rainbow angle, for instance, and into a rainbow shadow, the corresponding two $\mathcal{L}_n$'s do not "vanish" when they meet at $\theta = \theta_r$, but simply become complex, and several groups[3] have recently realized that the traditional semi-classical approach can be generalized to include flux-penetration into shadow regions simply by including the complex as well as the real solutions to Eq. (2). The essential point for a discussion of shadows is that if $\mathcal{L} = \mathcal{L}_R + i\mathcal{L}_I$, then $|e^{i\mathcal{L}\theta}|^2 = e^{-2\mathcal{L}_I\theta}$ describes the exponential shadow-edge (the tail of the Airy function, e.g.), with a "skin thickness" of $(2\mathcal{L}_I)^{-1}$. Thus it is the imaginary part $\mathcal{L}_I$ of a complex $\mathcal{L}$ which describes flux leakage into a shadow region (for both diffractive and refractive leakage). No real $\mathcal{L}$ (and hence no classical trajectory) exists for these angles, and $\mathcal{L}_I$ measures the amount by which the trajectory is "blurred".

With this inclusion of complex-$\mathcal{L}$ trajectories (and with $\delta(\ell)$ given more accurately than by the WKB approximation) the semi-classical formula (1) provides a useful approximation, even in shadow regions. The resulting $g(\theta)$ can then be interpreted as the sum of the positive-angle or repulsive trajectories (mainly Coulomb), and $g(-\theta)$ as the sum of the negative-angle or attractive ones (mainly nuclear). We have found that viewing $f(\theta)$ as their sum provides a simple understanding of many oscillatory interference patterns. As one example, from the fact that $e^{i\pi} = e^{-i\pi}$, we conclude that

$|g(-\pi)| = |g(\pi)|$. Hence the two amplitudes are guaranteed to interfere strongly at far back angles, and the result is the familiar glory oscillation seen there. Its analog at $\theta = 0$, where $g(\theta)$ and $g(-\theta)$ again cross, is the forward oscillation of heavy-ion transfer reactions.

3. RAINBOWS AND ℓ-SPACE LOCALIZATION

A rainbow is described by an extremum of the classical deflection function $\Theta(\ell)$. As such it determines a boundary at $\theta \approx \theta_r$ between a "bright" and "dark" (classically inaccessible) region of angle space. But since the extremum occurs at a specific ℓ, $\mathscr{L}_r$, a rainbow is of necessity <u>localized in</u> ℓ. The degree of localization is best characterized by $\Theta''(\mathscr{L}_r)$, the curvature of the deflection function at the extremum, and is essential in determining the "darkness" of the rainbow shadow, for a simple estimate shows that $[\Theta''(\mathscr{L}_r)]^{1/3}$ is approximately the skin depth of the shadow. <u>The broader the</u> <u>ℓ-width</u> of a <u>rainbow</u>, the <u>shorter</u> the <u>skin-depth of its shadow</u>, an example of the $\Delta\ell\ \Delta\theta$ uncertainty relation. In particular, a Coulomb rainbow maximum usually is much broader or "flatter" in ℓ than a nuclear rainbow minimum, so its shadow is correspondingly steeper.

If the rainbow extremum in $\Theta(\ell)$ is sufficiently pronounced, its two allowed trajectories can interfere to produce 1, 2 or more maxima, all on the bright side of θ_r. In Airy's familiar parabolic approximation to the extremum, the first maximum occurs at an angle θ_1, so located that the area beneath it in the Airy parabola, $A_1 = \int_{-\ell_1}^{\ell_1} [\theta_1 - \Theta(\ell)]d\ell$, is approximately $\pi/2$. The first minimum follows an area π further, and succeeding maxima and minima follow at intervals of π, as indicated in Fig. 2.

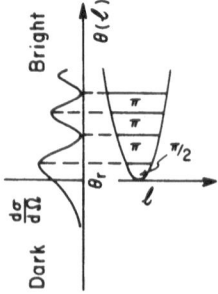

FIG. 2. Localization of maxima and minima of a symmetric rainbow.

4. REGGE RAINBOWS

Near an isolated elastic resonance, the energy-dependence of a phase shift in partial wave ℓ is given by the Breit-Wigner approximation

$$S(E,\ell) = e^{2i\delta(E,\ell)} = 1 - i\frac{\Gamma}{E-E_0(\ell) + i\ \Gamma/2}. \quad (4)$$

If the resonance is one member of a rotational band, so that a function $E_0(\ell)$ exists which correlates the energies of successive states along the band, and if their widths $\Gamma(\ell)$ depend only weakly on ℓ, the same expression can give the <u>ℓ-dependence</u> of $\delta(E,\ell)$ at a fixed energy E, within the approximation $E_0(\ell) \approx E_0(\ell_0) + (\ell-\ell_0) dE_0(\ell)/d\ell$. If, in particular, we choose ℓ_0 (in general not at an integer) so that $E_0(\ell_0) = E$, we have

$$e^{2i\delta(E,\ell)} = 1 + i\frac{\hat{\Gamma}}{\ell-\ell_0-i\ \hat{\Gamma}/2}, \quad (5)$$

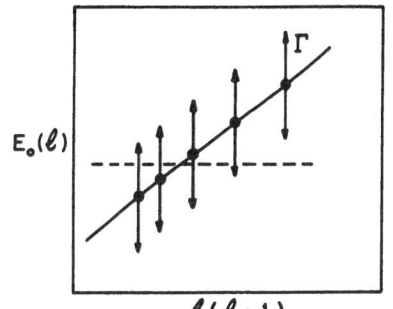

$\ell(\ell+1)$

FIG. 3. Overlapping resonances on a quasi-molecular rotational band.

a "Breit-Wigner-in-ℓ" approximation, with

$$\hat{\Gamma} = \Gamma \left(\frac{d\ell_o}{dE}\right) \geq 0. \tag{6}$$

Here $d\ell_o/dE$ is the rate at which the resonance moves along the band from one ℓ to the next as E is increased. Formally Eq. (5) becomes singular at the complex ℓ-value

$$\ell = \ell_o + i \, \hat{\Gamma}/2 \equiv \alpha_1 + i \, \alpha_2 \equiv \alpha, \tag{7}$$

and the singularity is known as a Regge pole. Since ℓ_o is a function of E, the pole "moves"

in the ℓ-plane as the energy increases, along a "Regge trajectory" which in our approximation is a straight line parallel to and above the real axis.

If the resonances on the band are sufficiently broad to overlap, a range of them will be excited at any energy E, centered at $\ell_o(E)$ with width $\Delta\ell = 2\alpha_2$, and in this way we again encounter a scattering phenomenon which is localized in ℓ-space.

What are the consequences of this ℓ-localization for the elastic scattering amplitude? All that can be said in general is that if the <u>entire</u> ℓ-dependence of $\delta(E,\ell)$ is given by the 1-pole approximation of Eq. (5), then[2a]

$$f(\theta) \sim P_\alpha(-\cos\theta) \sim \{e^{i[(\alpha+\frac{1}{2})\theta+\pi/4]} + e^{-i[(\alpha+\frac{1}{2})\theta+\pi/4]}\}/\sqrt{\sin\theta}. \tag{8}$$

Since $\alpha_2 > 0$, these are decaying and growing waves, which have equal amplitude at $\theta = \pi$, and diverge toward forward angles; the growing one is actually logarithmically singular at $\theta = 0$, where the above approximation fails. If $\alpha_2 \ll \frac{1}{\pi}$ the damping is small and the two waves remain nearly equal in magnitude all the way forward to $\theta = 0$; in particular, if $\alpha_2 = 0$ and α_1 is integral, they interfere over the entire angular range to produce the usual $P_\ell(-\cos\theta)$. But if $\alpha_2 \gg \frac{1}{\pi}$, they interfere (the glory effect) only at the back angles $\pi-\theta \lesssim 1/\alpha_2$; forward of that, $e^{+i\alpha\theta}$ dominates, to produce the exponentially-dropping cross section

$$|f(\theta)|^2_{Regge} \sim e^{-2\alpha_2\theta}/\sin\theta, \qquad \pi-\theta > 1/\alpha_2. \tag{9}$$

A physical understanding of this remarkable angular distribution can be given in two forms. One[2a] is based on the idea that a beam of projectiles incident at energy E will excite a band $\Delta\ell = 2\alpha_2$ of overlapping resonances in the rotational sequence. Assuming them all to couple equally to this incoming channel, they will be excited (coherently) with the amplitudes $f(\ell) \sim (\ell-\alpha)^{-1}$, to form a wave packet in ℓ, of width $\Delta\ell = 2\alpha_2$ [i.e., an angular pulse, $F(\theta)$] which propagates around the nucleus, decaying as it goes. The flux observed asymptotically is that which leaks outward from this surface propagation, and so decays in angle at the same rate the pulse itself does; for the decay-angle $(\alpha_2)^{-1}$ Fuller has proposed the apt name "life-angle". In agreement with the $\Delta\ell \, \Delta\theta$ uncertainty relation, a pulse with a small bandwidth α_2 will have a large life-angle $(\alpha_2)^{-1}$.

In this context it is very illuminating, as Fuller[4] has pointed out, to note that if the energy-spacing between resonances along the band is D, then their spacing in angular momentum is $\Delta \ell = 1 \overset{\sim}{\sim} (d\ell_o/dE)D$. Equation (6) then provides a very physical interpretation for α_2, given by

$$\hat{\Gamma} = 2\alpha_2 = \Gamma/D. \tag{10}$$

That is, α_2 is not only the inverse of the life-angle, but is also half the width-to-spacing ratio for successive resonances along the rotational band. Thus it is when $\Gamma/D << \frac{2}{\pi}$ that the isolated resonances are excited only one at a time, producing a "pulse" which extends entirely around the nucleus, and is simply $|P_\ell (\cos \theta)|^2$. It also means that Eq. (9) can be written

$$|f(\theta)|^2_{Regge} \backsim e^{-(\Gamma/D)\theta}/\sin \theta, \qquad \pi-\theta > D/\Gamma. \tag{11}$$

The alternative interpretation of this angular distribution is that it describes a rainbow shadow. The ℓ-dependence of the one-Regge-pole elastic phase shift is given by Eq. (5),

$$e^{2i\delta(E,\ell)} = \frac{\ell-\alpha^*}{\ell-\alpha}, \tag{5}$$

from which the deflection function is readily found to be the Lorentzian

$$\Theta(\ell) = 2\partial\delta/\partial\ell = \frac{-2\alpha_2}{(\ell-\alpha_1)^2 + \alpha_2^2}. \tag{12}$$

This has an attractive or negative-angle rainbow minimum at $\ell = \alpha_1$, and

$$\text{(Elastic Regge rainbow)} \qquad \theta_r = -2/\alpha_2. \tag{13}$$

Thus it is reasonable that the cross section should drop exponentially for $\theta < \theta_r$, but why does it not have the usual rainbow maxima on the bright side of θ_r? The reason is given by Fig. 4. It is readily found that the area inside a 1-pole Regge deflection function is exactly π. Hence in this sense it is too weak to have even a single minimum – which in this case is "lost" under the divergence of the cross section at $\theta = 0$ resulting from the pileup of classical trajectories there. A Regge rainbow is "all shadow".

These Regge-rainbow formulae all apply only to the phase shift given exactly by Eq. (5). However, Regge poles can (and do) occur in any scattering amplitude, where their influence is localized in ℓ, over the range $\Delta\ell \overset{\sim}{\sim} 2\alpha_2$, centered at ℓ_o. Hence if, semiclassically, there is an angular

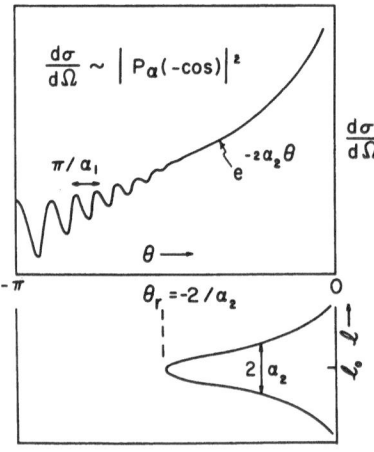

FIG. 4. One-Regge pole angular distribution and deflection function. Note back-angle glory oscillations.

region of a cross section which is dominated by this ℓ-range, we can expect to find it well-approximated by Eq. (9).

As an example, we return to the α+Zr scattering of Figs. 1. Accepting the interpretation of Goldberg et al.,[1c] in the resonant energy region below about 75 MeV the resonances along any rotational bands which may be present have $\Gamma/D < 2/\pi$, so that their life-angles, and hence the rainbow angle, exceed 180°. At these energies, no part of the angular range is classically inaccessible, and the cross section oscillates regularly with a period $\sim\pi/kR$ all the way from the (small) Coulomb grazing angle to 180°. By 75 MeV, the rainbow angle moves forward to perhaps 110°, and continues to decrease with energy, "exposing" more and more of the exponential rainbow shadow as it goes. From the slopes of these angular distributions, we can use Eq. (10) to extract Γ/D values of 6.9 at 99.5 MeV, 8.9 at 118 MeV and 10.5 at 141.7 MeV. These correspond to life-angles $(\alpha_2)^{-1}$ of 17°, 13°, and 11°, respectively, which describe very short-lived pulses, indeed. To be accurate, it should be noted that although we have applied the 1-pole formulae to this case, several poles are actually present. This can be seen most immediately at 141.7 MeV from the rainbow-shaped deflection function of the Goldberg potential,[1c] which has an area of 6π, suggesting 6 poles. And indeed, a search of the scattering amplitude of this potential by J. T. Londergan[5] using the Regge-pole search code developed by Wolter[6] revealed a number of poles in the neighborhood of $\ell_o \sim kR \sim 32$, with imaginary parts α_2 averaging around 5.

On the basis of these considerations, it is tempting to conjecture that <u>all</u> rainbows might be describable as clusters of Regge poles.

5. RAINBOWS AND REGGE POLES IN HEAVY-ION TRANSFER REACTIONS

The angular distributions for few-nucleon transfer between heavy ions undergo a now-familiar evolution from a grazing-angle peak at low energy to forward oscillations at high energy, which is nicely illustrated by the Brookhaven data[7] shown in Fig. 5 on the following page. As Dr. Garrett has discussed in his contribution to this conference, the forward oscillations arise from an interference between attractive and repulsive trajectories, i.e., between the amplitudes $g(\theta)$ and $g(-\theta)$ of Eq. (1). This is indicated in Fig. 6 (also on the following page), which shows $|g(\pm\theta)|^2$ for the dominant $M = 4$ component, as given by the DWBA calculation used in Fig. 5. The oscillations are deepest near $\theta = 0$, where $|g(\theta)| = |g(-\theta)|$, fade away at the grazing-angle peak near 20°, and, as the data seem to indicate, begin to return at larger angles where the two amplitudes again become comparable.[3c] Measurements at larger angles still should show a deepest minimum at about 70°; its exact position is very sensitive to the surface of the optical potential.

The relation of the oscillations to surface waves (Regge poles) thus becomes the question of how strongly the shape of $g(\theta)$ is affected by any Regge poles which may be present in the amplitude. Recent studies at Heidelberg and Wisconsin, in collaboration with R. C. Fuller, W. A. Friedman and G. W. T. Shuy, indicate that in many cases

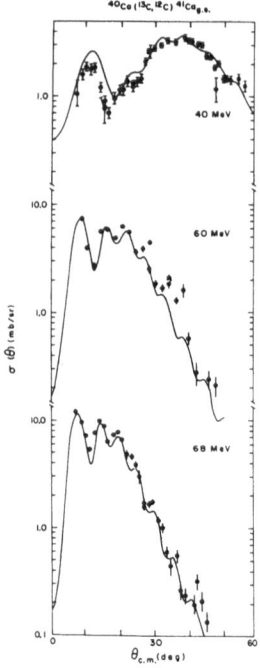

FIG. 5. Angular distributions and DWBA fits of the ^{40}Ca(^{13}C, ^{12}C)^{41}Ca (g.s.) transfer at several bombarding energies (Ref. 7).

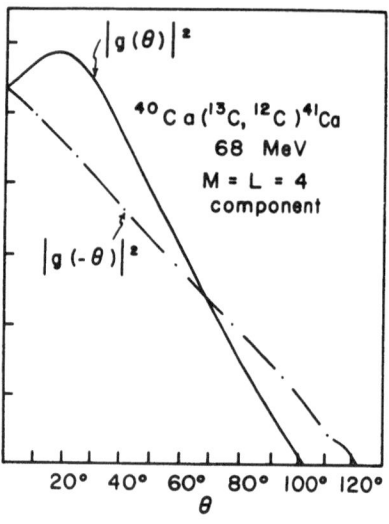

FIG. 6. $|g(\pm\theta)|^2$ (log scale) for the L = M = 4 component of the 68 MeV DWBA calculation shown in Fig. 5. Large oscillations in the cross section occur where the two curves cross.

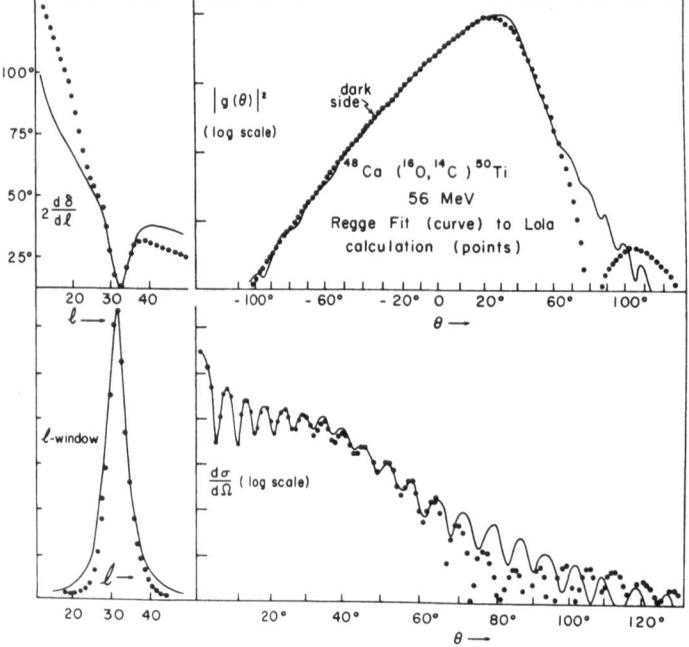

FIG. 7. Two-Regge-pole fit to the DWBA phase shifts and angular distribution of ^{48}Ca(^{16}O, ^{14}C)^{50}Ti at 56 MeV. $\alpha = 32 + 2.8i$.

(above all in well-Q-matched L = 0 transfers) $g(\theta)$ is in fact __dominated__ by two Regge poles, located, as they should be, at the positions of the leading poles of the entrance and exit channels. This is illustrated by Fig. 7, which shows a case of this type in which a simple two-pole approximation accurately fits the upper two orders of magnitude of the DWBA cross section, and hence the data, of Henning et al.[7] Both poles are at α = 32 + 2.8i, just where the Wolter code[6] located them in the entrance and exit channels. Physically this means that a pulse of overlapping quasi-molecular states (which in this case propagates through only 20° and has $\Gamma/D \stackrel{\sim}{\sim} 6$) so dominates the reaction dynamics that the cross section is insensitive to all details of the bound state but its spectroscopic factor and spin. In partial-wave terms, it means that the DWBA ℓ-window is nearly a Lorentzian function of ℓ, as Fig. 7 indicates, so that a Regge pole may provide a simple and physically-meaningful parametrization of the ℓ-localization of the amplitude. The possibility that this might be so was first suggested by Fuller and Dragun,[8] whose fit to the BNL data of Fig. 5 is shown in Fig. 8.

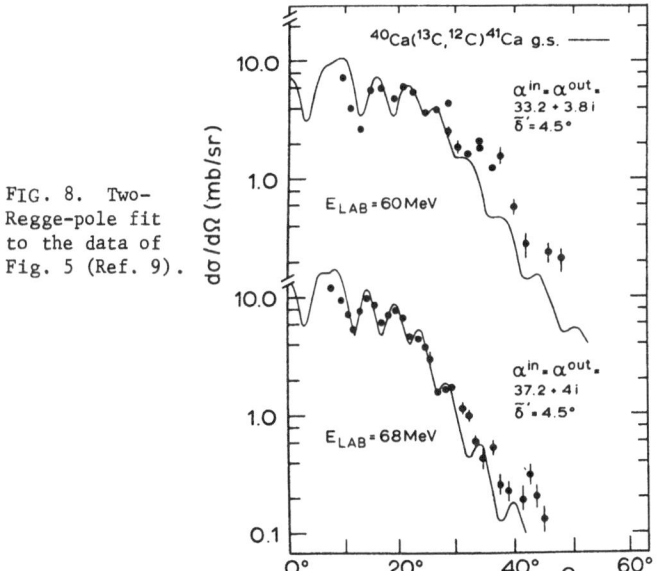

FIG. 8. Two-Regge-pole fit to the data of Fig. 5 (Ref. 9).

The further studies of Regge amplitudes currently in progress have concentrated so far on an attempt to understand the $g(\theta)$ functions of DWBA calculations, which are both simpler and more fundamental than the oscillatory cross sections they produce. One hint as to their nature is the nearly-exponential falloff of the left (small-angle) side of the grazing-angle peak of Fig. 7. We interpret it as a rainbow shadow, whose slope is directly correlated with the curvature of the rainbow-dip seen in the "deflection function" or ℓ-derivative $2\delta'(\ell)$ of the phase of the transfer amplitude $f(\ell) = \rho(\ell) \exp[-2i\delta(\ell)]$. The minimum or (nuclear) rainbow angle is at +12°, so all angles below this, including the entire negative-angle range, are classically forbidden (if we incautiously transfer the elastic-scattering language to reactions), and hence illuminated only by the surface creep wave which leaks beyond the bottom of the rainbow. In other words, all the forward oscillations result from the interference of a rainbow tail with itself.

We believe this to be the tail of a (double) Regge rainbow. A pure double Regge amplitude would have the very simple ℓ-dependence $f(\ell) = (\ell-\alpha)^{-2}$, which has a Lorentzian magnitude $\rho(\ell) = [(\ell-\alpha_1)^2 + \alpha_2^2]^{-1}$ and a negative Lorentzian phase derivative,

$2\delta'(\ell) = -2\alpha_2/[(\ell-\alpha_1)^2 + \alpha_2{}^2]$. In the stationary-phase approximation, this model yields the angular dependence:

$$g(\theta) = 0, \quad \theta > 0; \qquad g(\theta) = \pi\theta\,\exp[-i(\alpha+\tfrac{1}{2})\theta], \quad \theta < 0. \qquad (14)$$

This is a function with a single rainbow maximum at $\theta = -1/\alpha_2$. It has the correct rainbow tail on its left or dark side, and vanishes completely at positive angles on its "bright side", in agreement with the vanishing of $\delta'(\ell)$ there. Figure 9 shows the result of a discrete ℓ-sum (rather than integral) for this double-Regge $g(\theta)$,

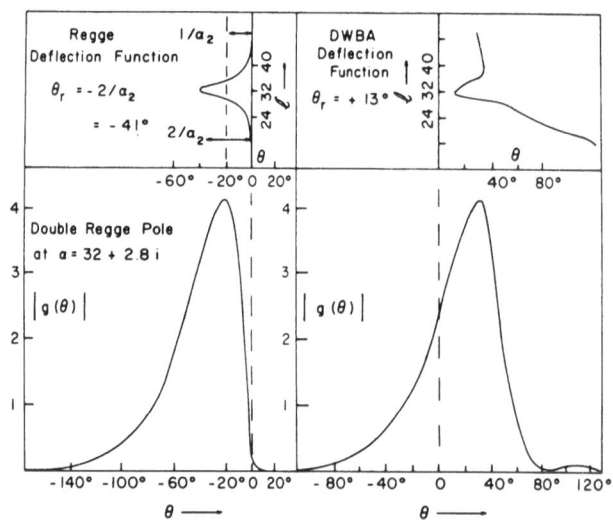

together with the same LOLA result that was plotted logarithmically in Fig. 7. The comparison provides a clear understanding of the qualitative features of the DWBA grazing-angle peak in this case, but we note that (a) the Regge peak occurs at $-20°$ while the LOLA one is at $+34°$, and (b) this simple Regge model does not fit the secondary (rainbow?) maximum of the LOLA amplitude at $+105°$. The Regge peak is shifted $+40°$ simply by including an obvious Coulomb back-

FIG. 9. Two-Regge-pole deflection function and $g(\theta)$ (linear scale), compared with the DWBA calculations of Fig. 7.

ground under the pole. The fit shown in Fig. 7 was obtained by arbitrarily shifting both $2\delta'(\ell)$ and $g(\theta)$ by $14°$ (as was also done in Ref. 9); this makes $2\delta'(\ell)$ too large in the large-ℓ Coulomb-dominated region, but a more reasonable "nuclear background" to $\delta(\ell)$ can be devised which provides a fit to $|g(\theta)|^2$ accurate for over 3 orders of magnitude. We have not, however, attained a fit over 4 orders of magnitude, which would be necessary to fit the secondary LOLA maximum and hence the accurate return of deep oscillations at large angles, such as that due to the curve-crossing seen in Fig. 6.

This Regge approach to a parametrized-phase-shift model of heavy-ion transfer reactions is still in its early infancy, and little is yet known about its applicability to other reactions, especially to the non-Q-matched ones. As some indication of what the challenges may be, Fig. 10 displays LOLA results for 1-, 2- and 4- nucleon transfer reactions out of ^{16}O+Ca entrance channels at 56 MeV, recently studied at Argonne, all using the potential of Ref. 8. All show an exponential falloff on the left side of the grazing-angle peak, and the corresponding rainbow-dip in the "deflection function". Similar dips (some of which show more structure) occur in the phase-derivatives for

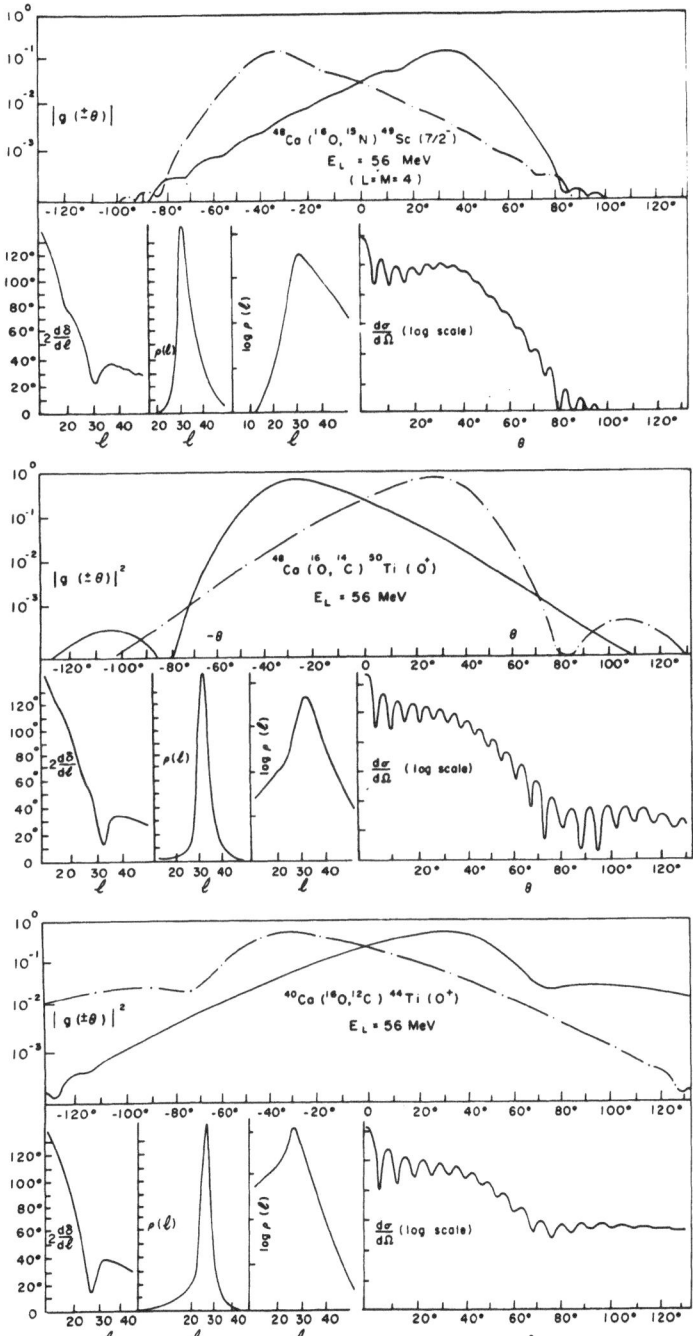

FIG. 10. Deflection function, ℓ-window (linear and log scales), $|g(\pm\theta)|^2$ and angular distributions from DWBA calculations for 1-, 2- and 4-nucleon transfers at 56 MeV, using the transparent-surface potential of Ref. 8. Note the greater symmetry of the ℓ-window in the 2-nucleon transfer case.

a large number of DWBA amplitudes studied, based on four different optical potentials, which suggests that rainbows, and perhaps Regge rainbows, will play a significant role in our understanding of many heavy-ion transfer reactions. This is consistent with the fact that all the potentials which have so far met with success in DWBA transfer studies have transparent surfaces and black interiors -- just the conditions necessary for enhancing a single peripheral surface wave or Regge pole.

But, as a final question, why <u>two</u> Regge poles? Simply because, as Fuller and Dragun[6] suggest, DWBA is adequate. It produces a transfer matrix element from a product of channel wave functions, and each channel wave function brings in its own pole. This argument is not entirely general, however, and the following refinement of it, based essentially on Austern's[8] "phase-averaging" argument, suggests that if the form-factor is sufficiently long-ranged, the two channel poles may be additive (simple poles) rather than multiplicative (which gives a double pole if the channels are nearly identical, as in the case of good Q-matching). To put it most simply, assume the form factor sufficiently long-ranged that the channel wave functions have their asymptotic form over the interesting range of integration, $\chi_c^{(+)} \sim e^{-ik_c r} + S_{cc}(\ell) e^{ik_c r}$. Then the DWBA radial integral has the form

$$\int \chi_{c'}^{(+)} F(r) \chi_c^+ dr = \int e^{-i(k_c+k_{c'})r} F(r) dr + S_{cc} S_{c'c'} \int e^{i(k_c+k_{c'})r} F(r) dr$$

$$+ S_{cc} \int e^{i(k_c-k_{c'})r} F(r) dr + S_{c'c'} \int e^{i(k_{c'}-k_c)r} F(r) dr.$$

$$(15)$$

The two elastic S-matrix elements each have simple Regge poles, which will be nearly coincident in the case of good Q-matching, giving $S_{cc} S_{c'c'}$ a double pole. If $F(r)$ is short-ranged, as in a 2-nucleon transfer amplitude, all four integrals will be comparable, and near a pole the double-pole term will clearly dominate. However, the first two integrals have "high frequency" exponentials, while the last two have "low-frequency" ones. Consequently, if the range of $F(r)$ is several wavelengths long (as might be the case in 1-nucleon transfer), the coefficient of $S_{cc} S_{c'c'}$ may become small enough by cancellation that only the single-pole terms survive. In this case a 1-pole parametrization would be more appropriate, and, since the first term in (15) is then also small, the transfer amplitude is simply a sum of the two elastic amplitudes. If they are nearly identical, the Regge zeros, as well as the poles, of the transfer amplitude should be near those of the elastic amplitude; evidence of this behavior has indeed been seen. Thus in Regge language it appears that the influence of the form factor on the reaction will enter through the positions and strengths of "Regge zeros" in the scattering amplitude $f(\ell)$, i.e., complex ℓ's at which $f(\ell) = 0$. Such zeros can produce dips in $|f(\ell)|$ for real ℓ, and may be related to double-humped ℓ-windows.

ACKNOWLEDGMENT

It is a pleasure to thank W. A. Friedman, G. W. T. Shuy and R. C. Fuller for useful conversations, and J. D. Garrett and D. G. Kovar for detailed information on their DWBA analyses.

REFERENCES

1. a) P. P. Singh, R. E. Malmin, M. High and D. W. Devins, Phys. Rev. Letters 23 (1969) 1124.
   b) L. W. Put and A. M. J. Paans, Phys. Letters 49B (1974) 266.
   c) D. A. Goldberg, S. M. Smith and G. F. Burdzik, Phys. Rev. C (to be published).

2. R. C. Fuller, Nucl. Phys. A216 (1973) 199; R. C. Fuller and I. Avishai, Nucl. Phys. A222 (1974) 365; N. M. Nussenzveig, Ann. Phys. (N.Y.) 34 (1965) 23.

3. a) J. Knoll and R. Schaeffer, Saclay preprints.
   b) W. E. Frahn, Extended Seminar on Nuclear Physics, International Atomic Energy Agency, Trieste, 1973.
   c) W. A. Friedman, K. W. McVoy and G. W. T. Shuy, Phys. Rev. Letters 33 (1974) 308.

4. R. C. Fuller, private communication.

5. J. T. Londergan, private communication.

6. T. Tamura and H. H. Wolter, Phys. Rev. C6 (1972) 1976.

7. P. D. Bond, J. D. Garrett, O. Hansen, S. Kahana, M. J. Levine and A. Z. Schwarzschild, Phys. Letters 47B (1973) 231.

8. W. Henning, D. G. Kovar, B. Zeidman and J. R. Erskine, Phys. Rev. Letters 32 (1974) 1015.

9. R. C. Fuller and O. Dragun, Phys. Rev. Letters 22 (1974) 617.

10. N. Austern, Ann. Phys. (N.Y.) 15 (1961) 299.

TWO-STEP PROCESSES

Hermann H. Wolter
Institut für Kernphysik, KFA Jülich
and
1. Phys. Institut der Universität Köln, Germany

Higher-order processes are examined as they are manifested in elastic and inelastic transfer reactions, particularly in the ^{16}O and ^{17}O scattering. It is demonstrated that the two-step transfer processes are very important. The dependence of such processes on the optical potentials is discussed.

1. INTRODUCTION

During the last years it has been learned, that the mechanism of reactions involving light projectiles is often much less direct than had commonly been thought[1]. It must then be expected, that two-step processes are even more important in heavy ion reactions, since the reaction proceeds slower, so that several interactions might take place during the encounter of the reaction partners. Furthermore because of the strong absorption heavy ion reactions are very surface localized. For a grazing path, however, the direct process may often be hindered by kinematical conditions[2] or because of the smallness of the form factor in the sensitive region[3]. Then indirect routes, i.e. two-step processes, will become important.

Two-step processes via inelastic routes were investigated in detail within the framework of the Coupled-Channels Born Approximation (CCBA). It was predicted, that such processes would strongly affect the forward angle cross section for inelastic scattering[4] and this was recently verified in an exact-finite-range calculation[5]. Secondly successive transfer processes have been suspected to contribute to two-nucleon transfer, because the two-nucleon transfer form factor is often smaller at the nuclear surface than the one-nucleon transfer form factor[3]. Indeed it has been calculated within the framework of the semiclassical theory that e.g. in the ^{94}Mo(^{18}O,^{16}O)^{96}Mo reaction the successive transfer contribution is about a factor four larger than simultaneous two-nucleon transfer[6].

There is a third process which in a way combines these two reaction mechanisms, namely elastic or inelastic transfer[7]. Here in the scattering of two heavy ions, which differ only by a light particle or cluster, this cluster may be scattered elastically or inelastically or may be exchanged between the heavy cores, possibly several times. Because of the identity of the cores these different contributions interfere and the resulting interference pattern is very sensitive to the contributing reaction mechanisms. Therefore, I believe it is useful in the following to study these processes in detail to get a feeling about higher order processes in heavy ion reactions.

2. COUPLED CHANNELS THEORY OF ELASTIC AND INELASTIC TRANSFER

First I shall briefly sketch a coupled channel theory of elastic and inelastic transfer, which we shall use to investigate the various one- and multi-step processes. These reactions have previously been treated by two other methods, which are included as special cases. One is the LCNO method of von Oertzen[8], which treats the case of $j \leq 1/2$ particle exchange and elastic scattering. On the other hand, when only the first order is important, the DWBA has been used esp. by Baur and Gelbke[9]. The present treatment is modelled after the Coupled-Reaction-Channels approach for multi-step transfer processes in light ions[10].

We consider the reaction A(B,B)A*, where A = B + x, i.e. B differs from A by some light cluster or nucleon x. We then use the system of coordinates given in Figure 1. We also use the no-recoil approximation $\vec{R}_1 \approx \vec{R}_2 \approx \vec{R}$, which is justified if x is small relative to B and the energy is not too high. For simplicity we assume 0^+ cores.

Let us define a channel wave functions as

$$\phi_{\ell j J}(\hat{R}, \vec{r}_1) = [Y_\ell(\hat{R}) \times \varphi_j(\vec{r}_1)]_{JM} \tag{1}$$

where φ_j is the wave function of x with respect to B. We then make the following ansatz for the total scattering wave function

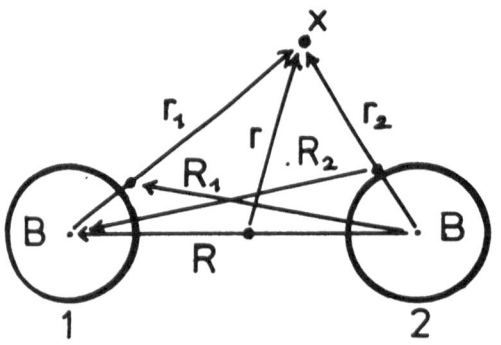

$$\psi^{(+)} = \frac{1}{R} \sum_{J\ell j} \chi_{\ell jJ}(R) \frac{1}{N} [\phi_{\ell jJ}(\hat{R},\vec{r}_1) + \phi_{\ell jJ}(-\hat{R},\vec{r}_1)]. \qquad (2)$$

This wave function is symmetric with respect to interchange of the iden-
tical cores B ($\hat{R} \leftrightarrow -\hat{R}$). N is a normalization factor. Inserting (1) into (2) we obtain explicitely

$$\psi^{(+)} = \frac{1}{R} \sum_{\ell jJ} \chi_{\ell jJ}(R) [2(1+(-)^\ell S(\hat{R}))]^{-1/2} \qquad (3)$$

$$\times [Y_\ell \times (\varphi_j(\vec{r}_1)+(-)^\ell \varphi_j(\vec{r}_2))]_{JM} ,$$

where the function

$$S(R) = \int dr^3 \; \varphi_j^*(\vec{r}_1) \; \varphi_j(\vec{r}_2) \qquad (4)$$

in the normalization factor arises because the wave functions of x at-
tached to one or the other core are not orthogonal. We shall at first
neglect all terms of order S and investigate its influence later.

Note that we take into account only the identity of the cores as
a whole but neglect all antisymmetrization effects of x with respect to
the nucleons in the core and between the cores. Clement et al.[11] have
shown that this is justified, if the wave function at the surface can
be well represented as a two-cluster wave function.

The Hamiltonian of the system

$$H = T_R + T_r + U(R) + V(r_1) + V(r_2) \qquad (5)$$

contains the interaction U between the cores (taken as the optical potential) and the interaction V between x and both of the cores (taken as the potential that binds x to B). We insert the trial wave function $\psi^{(+)}$ into the Schroedinger equation and project on a particular channel. We also use for short the index $\alpha = \{J\ell j\}$.

$$(\phi_\alpha | H-E | \psi^{(+)}) = 0 \tag{6}$$

The brackets denote integration over $\hat{R}$ and $\vec{r}$. Then we obtain the coupled equations

$$\{\frac{\hbar^2}{2\mu}(\frac{d^2}{dR^2} - \frac{\ell(\ell+1)}{R^2}) - U(R) + (E-\varepsilon_j)\} \chi_\alpha(R)$$
$$= \sum_{\alpha'} (V_{\alpha\alpha'}^{(+)}(R)+(-)^{\ell'}V_{\alpha\alpha'}^{(-)}(R)) \chi_{\alpha'}(R) \quad . \tag{7}$$

Here ε_j is the binding energy of x to B in the state φ_{jm}. The coupling potentials are given by

$$V_{\alpha\alpha'}^{(\pm)} = \theta_\alpha \theta_{\alpha'} (\phi_\alpha^*(\hat{R},\vec{r}_1) V(r_2) \phi_{\alpha'}(\hat{R},\vec{r}+\frac{\vec{R}}{2})) \quad . \tag{8}$$

$V^{(+)}$ represents the form factor for direct elastic or inelastic scattering and $V^{(-)}$ is the form factor for transfer of x between the cores given here in the no-recoil approximation. We have also introduced spectroscopic amplitudes θ_α to take into account the spreading of the single-particle strength. The phase $(-)^\ell$ multiplying the transfer term gives rise to the typical odd-even staggering of the reflection coefficients[8] and thus to the oscillations of the cross sections, which are encountered in elastic and inelastic transfer.

Let us briefly investigate the effect of the non-orthogonality of the basis wave functions. This occurs through the normalization factors in the wave function (3) as well as in non-vanishing overlaps in the projection (6). In analogy to the light ion case[12] or to semiclassical theory[6] one may show that in first order in the overlap an additional term is added on the rhs of eq. (8).

$$-\sum_{\alpha'\alpha''} (-)^{\ell''}(V_{\alpha\alpha'}^{(+)}+(-)^{\ell'}V_{\alpha\alpha'}^{(-)}) V_{\alpha'\alpha''}^o \chi_{\alpha'} \quad , \tag{9}$$

where

$$V_{\alpha'\alpha''}^o(R) = (\phi_{\alpha'}^*(\hat{R},\vec{r}_1)\phi_{\alpha''}(\hat{R},\vec{r}_2)) \quad . \tag{10}$$

In calculations this term is usually neglected and we shall see, that in our particular case V^o is small. Also in a two-step reaction of the

type pickup-stripping the non-orthogonality terms of the two steps can-
cel if the prior-post forms are used[12].

3. ELASTIC AND INELASTIC SCATTERING OF ^{16}O on ^{17}O

We shall use the above theory to investigate the importance of the
various reaction mechanisms. As our main example we shall take the elastic and inelastic scattering of ^{16}O on ^{17}O [13-16]. For this system, the
assumption of the model, namely inert O^{16} cores, good single-particle
wave functions, and neglect of recoil effects are probably well satisfied. An additional uncertainty in the treatment of heavy ion reactions
is the ambiguity in the optical potential. We shall therefore investigate the effects depending on two potentials which have both been fitted
to ^{16}O-^{18}O elastic scattering[17]. One is deep and strongly absorbing
($V = 100$; $W = 40$, $r_o = 1.2$, $a_R = 0.49$, $a_I = 0.32$), the other is shallow
and weakly absorbing ($V = 17$, $W = 2.6$, $r_o = 1.35$; $a_R = a_I = 0.49$).

We first consider the elastic scattering in Fig. 2. This had been

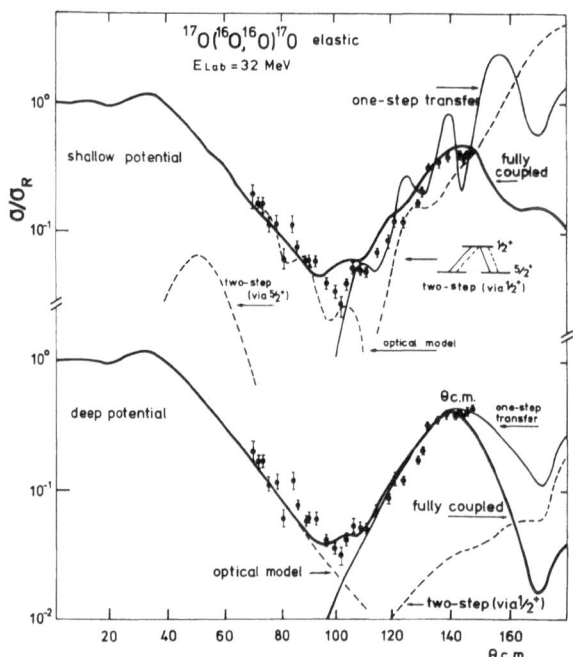

Figure 2:
Elastic scattering of
^{16}O on ^{17}O at E_{lab} =
32 MeV calculated with
a deep (below) and shal-
low (above) optical po-
tential (for parameters
see text). Given are
the contributions from
the elastic optical mo-
del scattering, from
the one-step transfer,
from the two-step path
via the excited state
and the two-step path
via transfer between
the ground states. The
result of the fully
coupled calculation is
given as the heavy full
curve.

done in DWBA in ref.[13] and in an approximate LCNO treatment in ref.[15].
The transfer is seen to strongly dominate at backward angles. For the
deep potential essentially only the one-step transfer contributes ex-

cept at extreme backward angles. For the shallow potential, however, the one-step transfer contribution alone is oscillating too strongly. Also higher order contributions, of which the one via the excited $1/2^+$ state is shown, are strong. Two-step transfer between the ground states is also strong, but is forward peaked and therefore dominated by the Coulomb scattering. However, when all the various contributions are summed up coherently, i.e. when a fully coupled calculation is done, one obtains again a smooth curve, which is very similar to the one for the deep potential and fits the data equally well. One should note that this effect is not proportional to the strength of the transfer, i.e. to the spectroscopic factors, which are taken as unity here. If one decreases the spectroscopic factor the backward cross section just decreases proportionally for the deep potential, but changes shape for the shallow one.

We may understand the reason for the different behaviour of the two potentials from inspecting in Fig. 3 the form factors for inelastic

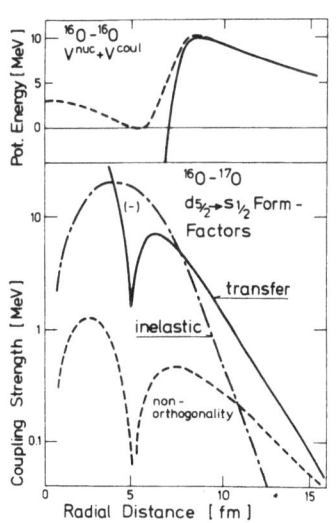

Figure 3:
Form factors for the ^{16}O on ^{17}O scattering. Above are the real parts of the deep and the shallow potentials used (for parameters see text). Below are the coupling potentials (eq. (8)) for inelastic excitation and transfer and the non-orthogonality term (eq. (10)) for the $d_{5/2} \rightarrow s_{1/2}$ transition. The transfer potentials for the $d_{5/2} \rightarrow d_{5/2}$ transition, which are not given here, are quite similar in shape and magnitude.

and the transfer coupling and the non-orthogonality term, all obtained from numerical integration. The shallow potential has substantial contributions from the interior, which can be checked by introducing various cutoff radii. Since the transfer form factor is rapidly increasing in the surface region, it is clear, why the shallow potential has so

much stronger one- and multi-step contributions. It is also seen that
the non-orthogonality term is small and will be unimportant in this
particular case.

We now turn to the inelastic scattering. From the magnitude of
the inelastic form factor in Fig. 3 we expect, that the transfer contribution is dominant. At higher energy the two contributions are well
separated in angle, the inelastic cross section being forward peaked,
the transfer one backward. The result of a DWBA calculation[14] is shown
in Fig. 4. The interference patterns in the theoretical curves are out
of phase with those observed in the data. It was found that only the

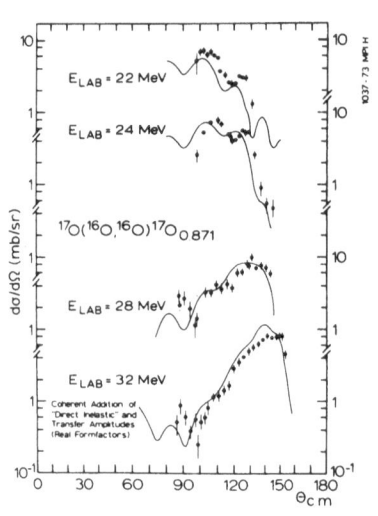

Figure 4:
Inelastic scattering cross section
calculated as a coherent sum of the
first order inelastic and transfer
contributions (from ref.[14]).

introduction of an arbitrary relative phase of 90° between the inelastic
and transfer contribution shifted the oscillations into agreement with
the data.

Fig. 5 shows the influence of the two-step contributions at 22 MeV
bombarding energy. It is seen that the two-step transfer is even slightly larger than the inelastic process. While the interference between the
one-step and inelastic contribution is out-of-phase with respect to the
data, the interference of the one- and two-step transfer produces the
correct phase of the oscillations.

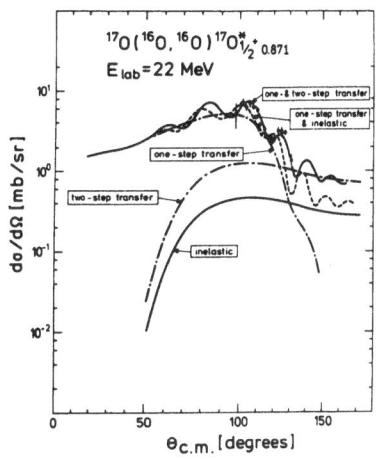

Inelastic scattering cross section at E_{lab} = 22 MeV. The contributions of the one-step and two-step transfer and of the inelastic excitation as well as the coherent sums are given. The deep optical potential is used (from ref.[16]).

Qualitatively this may be understood by writing the Green's function for the intermediate state in the two-step term as a principal value integral and an on-energy-shell term

$$G^{(+)}(\vec{R},\vec{R}') = P \int dE \, \frac{\chi_E^*(\vec{R})\chi_E(\vec{R}')}{E_o - E} + i\pi \, \chi_{E_o}^*(\vec{R}) \, \chi_{E_o}(\vec{R}')$$

Since the contribution of the intermediate state is expected to vary slowly with energy the principal value integral is strongly reduced by cancellation. Also since the contributions come from a narrow region in R the second term is approximately purely imaginary. We thus obtain the desired phase of 90° versus the inelastic coupling. Indeed inspection of the scattering amplitudes confirms this phase difference quite accurately.

In Fig. 6 we give the result of the fully coupled calculation including all transfers and the inelastic excitation to all orders. One obtains a reasonably good fit and reproduces the phase of the oxcillations consistently. The inclusion of the inelastic form factor, however, has shifted back the interference pattern somewhat as compared to the second order calculation in Fig. 5. However, in this model the microscopic inelastic form factor of eq. (9) is real, while it is well known that phenomenological complex form factors do much better for the description of inelastic scattering[18]. To account for this in the calculations of Fig. 6 the inelastic form factor has been multiplied by a

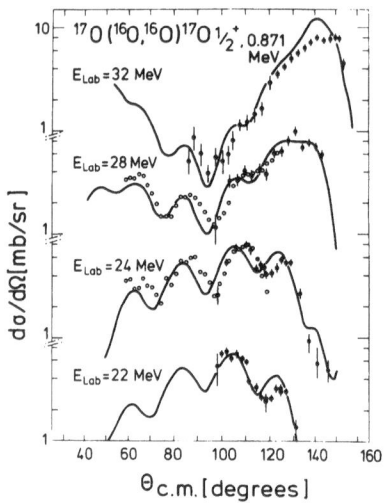

Figure 6:
Inelastic scattering cross section compared with theoretical curves obtained from a fully coupled calculation including inelastic excitation and transfer, (from ref.[16]) using the deep optical potential.

constant phase of 2C°. However it would be desirable to obtain independent information about the phase of the inelastic form factor. This could be done by investigating simultaneously the interference of the inelastic nuclear excitation with the Coulomb excitation at more forward angles.

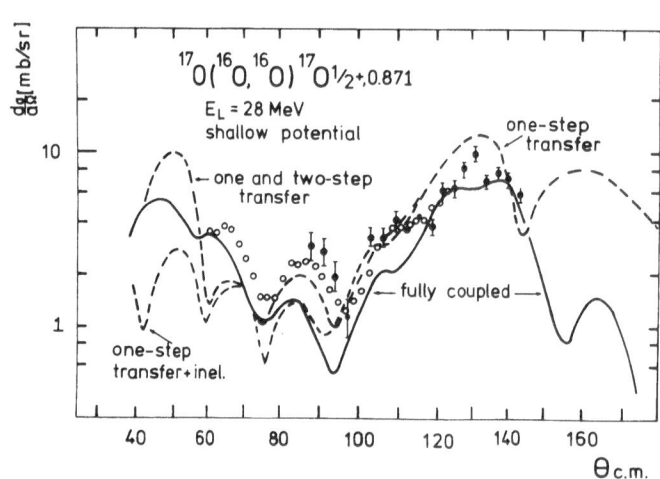

Figure 7:
Inelastic scattering at 28 MeV using the shallow optical potential. Given are the contribution of one-step transfer and the coherent sums of one-step transfer with two-step transfer and inelastic excitation, respectively. The full curve is the result of the fully coupled calculation.

In Fig. 7 we study the effect on the inelastic scattering of using the shallow potential. It is seen, that the one-step as well as the two-step transfer are very strong, especially at forward and backward angles. There is now no significant shift in the interference patterns of the one-step transfer with the inelastic excitation and the two-step transfer respectively. In the fully coupled calculations, however, the strong first and second order contributions are partly cancelled and one again obtains a reasonable fit. In Fig. 8 we compare the final fit ob-

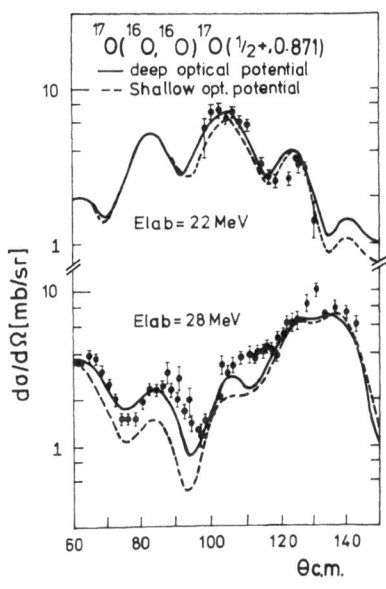

Figure 8:
Comparison of inelastic cross sections at 22 and 28 MeV, calculated with the deep and the shallow optical potentials (for parameters see text).

tained with the deep and the shallow potential at two energies. Even though the deeper potential gives better results above the Coulomb barrier, the fits are not widely different.

4. CONCLUSIONS

I have tried to demonstrate that in the scattering of ^{16}O on ^{17}O two- and multi-step processes are very important. However, they manifest themselves in a different way, depending on whether one uses a deep, strongly absorbing or a shallow transparent potential. For the deep potential the two-step processes are identified in the phase of the interference patterns for the inelastic scattering. For the shallow potential investigated here processes of every order are contributing significantly, which indicates, that these potentials should be used with

care if at all in heavy ion reactions. However, the fully coupled
calculation still reproduces the data.

Let me briefly remark on extensions of these investigations to
elastic and inelastic transfer involving exchange of two-nucleons, such
as the scattering of ^{16}O on ^{18}O. Higher-order contributions to this process would most likely not involve multiple exchange of the two neutrons, because the two-nucleon form factor is small at the surface[3].
Rather one should expect that successive transfer of neutrons involving
the intermediate $^{17}O + ^{17}O$ system is important, as in other work on two-
nucleon transfer reactions[6]. For the elastic scattering the LCNO description[19] and for the inelastic scattering a DWBA calculation[20] have
provided an adequate description of the data. However in the latter a
phenomenological form factor was used for inelastic excitation and a
two-neutron cluster form factor for the transfer. The phase relationship of these form factors derived from two different models is thus
undetermined, even though the calculations of ref.[20] obtained a fit
without any additional phase. Considering the importance and the information contained in the phases of interference structure, it would be
desirable to attempt a full coupled channels calculation of the $^{16}O +$
^{18}O and $^{17}O + ^{17}O$ systems using consistent microscopic form factors.
From such calculations one may then obtain structure information, as
e.g. demonstrated for the inelastic scattering of ^{30}Si on ^{32}S [21].

References

1) G.R. Satchler, Proc. Intern. Conf. on Nuclear Physics, Eds. J. de
   Boer and H.J. Mang, Munich (1973) Vol. 2, p. 569
2) R.J. Ascuitto, N.K. Glendenning, Phys. Lett. 45B (1973) 85
3) B. Nilsson, R.A. Broglia, S. Landowne, R. Liotta and A. Winther,
   Phys. Lett. 47B (1973) 189
4) N.K. Glendenning and R.J. Ascuitto, Symposium on Heavy Ion Transfer
   Reactions, Argonne Informal Report PHY-1973 B, p. 513;
   K.S. Low and T. Tamura, ibid. p. 655;
   P.J. Ascuitto and N.K. Glendenning, Phys. Lett. 48B (1974) 6
5) T. Tamura, K.S. Low and T. Udagawa, Phys. Lett. 51B (1974) 116
6) R.A. Broglia, U. Götz, M. Ichimura, T. Kammuri and A. Winther, Phys.
   Lett. 45B (1973) 23;
   U. Götz, M. Ichimura, R.A. Broglia and A. Winther, Proc. Intern.
   Conf. on Reactions Between Complex Nuclei, Eds. R.L. Robinson et al.,
   Nashville (1974).p. 74
7) A. Gobbi, Nashville Conference (1974) (see ref. 6), Vol. 2
8) W. von Oertzen, Nucl. Phys. A148 (1970) 529
9) G. Baur and C.K. Gelbke, Nucl. Phys. A204 (1973) 138
10) W.R. Coker, T. Udagawa and H.H. Wolter, Phys. Rev. C7 (1973) 1154
11) D. Clement, E.J. Kanellopoulos and K. Wildermuth, to be published
12) T. Udagawa, H.H. Wolter and W.R. Coker, Phys. Rev. Lett. 31 (1973)
    1507
13) C.K. Gelbke, R. Bock, P. Braun-Munzinger, D. Fick, K.D. Hildenbrand,

W. Weiss and S. Wenneis, Phys. Lett. 43B (1973) 284

14) C.K. Gelbke, G. Baur, R. Bock, P. Braun-Munzinger, W. Grochulski, H.L. Harney and R. Stock, Nucl. Phys. A219 (1974) 253

15) H.G. Bohlen and W. Nörenberg, Phys. Lett. 49B (1974) 227

16) G. Baur and H.H. Wolter, Phys. Lett. 51B (1974) 205

17) R.H. Siemssen, H.T. Fortune, A. Richter, J.W. Tippie, Phys. Rev. C5 (1972) 1839

18) G.R. Satchler, Phys. Lett. 35B (1971) 279

19) C.K. Gelbke, R.Bock and A. Richter, Phys. Rev. C9 (1974) 852

20) D. Kalinsky, D. Melnik, U. Smilansky, N. Trautner, Y. Horowitz, S. Mordechai, B.A. Watson, G. Baur and D. Pelte, Nashville Conference (1973) (see ref. 6), p. 12

21) B. Kohlmeyer, Thesis Marburg 1973.

APPROACHES TO THE INTERPRETATION OF HEAVY
ION TRANSFER REACTIONS AT HIGH ENERGIES

B. Buck
Department of Theoretical Physics
University of Oxford
Oxford OX1 3PQ, England

Abstract: Various theoretical approaches to the interpretation of recent transfer experiments on light nuclei are outlined.   Starting from some simple ideas on the reaction mechanism the observed highly selective nature of the reactions is discussed successively in terms of j-j coupled nucleon configurations, more sophisticated shell models of the SU 3 type and, finally, as possible evidence for cluster correlations in nuclei.

In this talk I will confine myself to discussing the possible implications of phenomena observed in multinucleon transfer reactions on targets of light nuclei in the mass region near ^{16}O. As a point of departure I take the recent experiments at Oxford [1] in which beams of light nuclei such as ^{12}C, with energies of about 10MeV/nucleon, were used to bombard various light elements. Transfers of one, two, three and four nucleons were observed. The angular distributions of the final detected ions tend to be rather featureless and are peaked sharply at forward angles. They are still rising at the smallest angles measured and so far they have provided little information.

The interesting thing about these reactions is their extreme selectivity. In each spectrum only a few states of the residual nucleus appear, over a wide range of excitation energies, and the states that are seen usually show up very strongly. The prominent peaks in the spectra indicate residual nuclear states with excitation energies ranging up to about 20 MeV and the corresponding reactions have moderate to large negative Q-values. A full explanation of these data requires consideration of both reaction theory and nuclear structure concepts. However, the basic physics of the transfer process can be treated very simply and the resulting picture gives interesting clues about the nature of the selectively populated states.

I will consider, for the moment, the mechanics of single nucleon transfer. The forward peaking of the detected residual ion suggests that the transfer occurs most probably in a grazing collision of the nuclear cores. Because of the high incident energy the de Broglie wavelength of the projectile is much smaller than the nuclear interaction radii and we may also suppose that the implied classical motion of the incident ion is not much perturbed by the grazing collision. Hence it is a reasonable approximation to assume that the orbit of the core of the projectile is a straight line traversed with a constant velocity v relative to the stationary target and that the effective impact parameter of the projectile is about equal to the sum of the radii of the colliding nuclei. This is sketched in Figure 1.

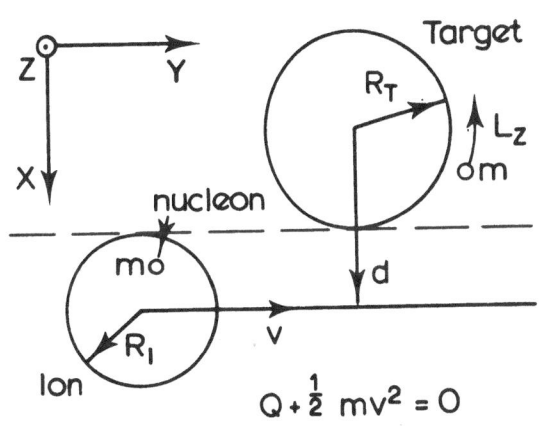

Incident ion with 10 MeV / nucleon
v = velocity of 10 MeV nucleon

$$Q + \tfrac{1}{2} mv^2 = 0$$

$$L_z = m v R_T$$

$$\text{Impact Parameter} = d = R_T + R_I$$

L($\hbar$)	TARGET	SHELL
1 - 2	^{12}C	p + (sd)
2	^{16}O	(sd)
3	^{40}Ca	(pf)

For Transfer of A Nucleons

$$\sum_{i=1}^{A} \ell_z(i) = L_z = L = \sum_{i=1}^{A} \ell_i$$

Fig. 1: Simplified picture of single particle transfer and table of favoured final state L-values for one nucleon transfer to various targets.

Let us now suppose that a nucleon m is initially bound somewhere near the surface of the incident ion and is eventually captured by the target into a single particle orbit of angular momentum L, with z-component L_z perpendicular to the reaction plane. Some simple arguments based on this idealised picture lead to three important conclusions if we assume that the energy and angular momentum of the transferred nucleon are approximately conserved.

To begin with the nucleon of mass m is bound in the incident ion with some energy ε_1 and it also has an average energy of translation $\frac{1}{2}mv^2$ due to the motion of the projectile. Finally, the nucleon gets captured by the stationary target into an orbit of energy ε_2 and it will then have lost its overall translational motion. The residual ion continues with the same velocity v, but with energy reduced by about $\frac{1}{2}mv^2$. Approximate energy conservation for the transferred nucleon gives the condition $\varepsilon_2 = \varepsilon_1 + \frac{1}{2}mv^2$. But the difference ($\varepsilon_1 - \varepsilon_2$) between the initial and final bound state energies of the nucleon is just the reaction Q-value. Hence we find the selection rule

$$Q + \tfrac{1}{2}mv^2 \approx 0 \qquad\qquad 1)$$

The nucleon will be transferred to the target most easily when it is in the region where the two nuclei just overlap. That is, at the moment of transfer, the nucleon will be at distance R_T from the centre of the target and will be moving with average velocity v perpendicular to the line joining the nuclei. Thus the angular momentum of the nucleon about the centre of the target will be mvR_T just before the collision and will be $\hbar L$ just after the transfer. Angular momentum conservation then yields the relation

$$\hbar L \approx mvR_T. \qquad\qquad 2)$$

Finally, the diagram suggests that the reaction should proceed most efficiently when the nucleon happens to be in the reaction plane. Under these conditions it is clear that the z-component of angular momentum should be as large as possible and we conclude that

$$L_z \approx L. \qquad\qquad 3)$$

Detailed semi-classical calculations support these conclusions except that the third one is not quite as good as we should like for a clear interpretation of the several nucleon transfer data. Equation (1)

shows that for high energy projectiles the largest transfer probabilities
should occur for reactions with reasonably large negative Q-values, so
that the nucleon is in general less bound in the final state than it was
in the incident ion. Evaluation of equation (2) for projectiles with
10MeV/nucleon and for various targets gives results for the optimum L-
values of the final orbits as shown in Figure 1. The two results together indicate that, whatever the target, the reaction mechanism at
these energies favours nucleon transfer into orbits of highest L-value
in the first unfilled shells above the Fermi surface, e.g. p and d levels
in ^{12}C, d orbits for ^{16}O and f levels in the ^{40}Ca region.

The conclusion (3) suggests that the states selectively populated
by transfer of several nucleons have high spins. If each nucleon i is
transferred with its z-component of angular momentum $l_z(i)$ equal to its
total angular momentum l_i, then for several nucleons the total transferred $L_z = \sum_i l_z(i) = \sum_i l_i$. Since the orbital L of the final state
must at least be equal to L_z we expect that

$$L \approx \sum_i l_i \qquad\qquad 4)$$

Finally, since the individual l_i have the largest values compatible with
the nuclear shell structure, we have grounds for believing that the many
particle transfer states have high spins and are constructed out of
stretched configurations.

This interpretation of the observed strong preference for high spin
final states is not altogether convincing on closer analysis. To uphold
rule (4), for example, we must assume that all nucleons are transferred
on the same side of the nucleus and this suggests that we really need
some additional hypothesis about the structure of the final states and
about possible correlations between the nucleons in the transfer process
itself. One possibility is that the nucleons are all transferred together in a real or virtual bound state of internal relative motion so that
the process can be regarded effectively as the transfer of a single
'particle' with two, three or four times the nucleon mass. The above
simple picture of the transfer mechanism may then be applied to the
centre of mass of this heavier particle and the L-value of the final
state will be correspondingly greater than for one nucleon. I will
come back to this idea later.

For the moment I will just make the working hypothesis of stretched
configurations for the selectively excited states and see what is implied.
Generalising slightly to include the intrinsic spins of the nucleons

I assume that the nucleons tend to go into the levels with highest j in the first unfilled shells and couple to the highest total spin consistent with those j-values. Clearly there are only a few such states in each residual nucleus and they correspond quite well with the observed strong peaks. For transfers on to nuclei near ^{16}O the preferred orbits are Od levels ($j = \frac{5}{2}$ and $\frac{3}{2}$) and in the case of ^{12}C a nucleon may also go into the $p\frac{1}{2}$ level. For various numbers of nucleons transferred into d levels we get the scheme shown in Figure 2

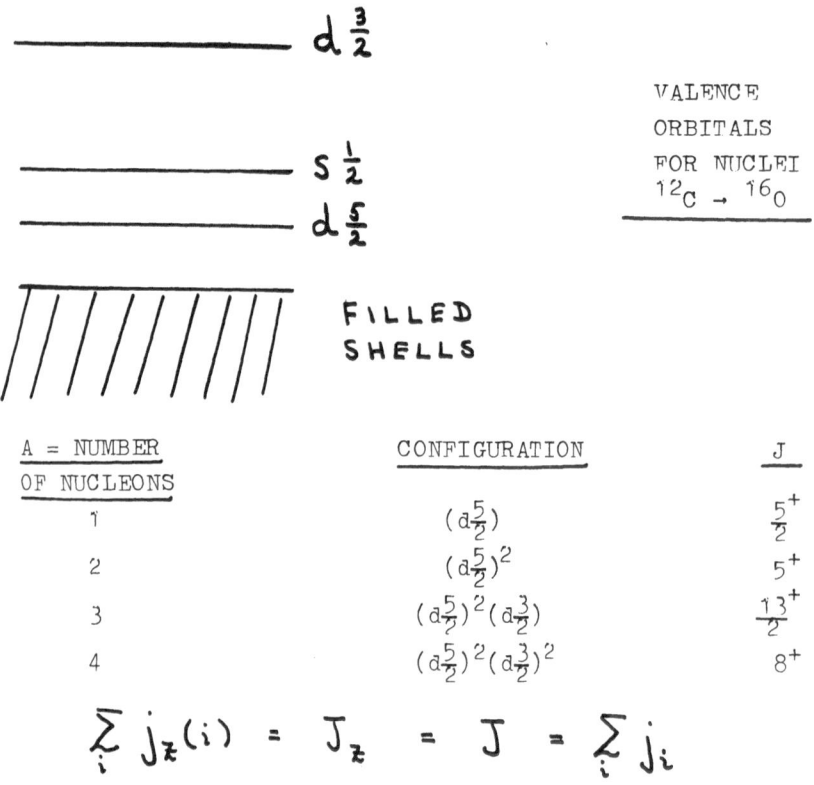

A = NUMBER OF NUCLEONS	CONFIGURATION	J
1	$(d\frac{5}{2})$	$\frac{5}{2}^+$
2	$(d\frac{5}{2})^2$	5^+
3	$(d\frac{5}{2})^2(d\frac{3}{2})$	$\frac{13}{2}^+$
4	$(d\frac{5}{2})^2(d\frac{3}{2})^2$	8^+

$$\sum_i j_z(i) = J_z = J = \sum_i j_i$$

Fig. 2: Examples of stretched configurations reached by transfers into the (sd) shell.

Once there is a neutron and a proton in the $(d\frac{5}{2})$ level, the Pauli principle does not allow us to add another nucleon with $j_z = j = \frac{5}{2}$, so the next transferred nucleons tend to go into the $(d\frac{3}{2})$ orbit. If the third and fourth particles went into $(d\frac{5}{2})$ their orbits would have to be tilted away from the reaction plane. For some reason this seems to be much more highly forbidden than is suggested by the semi-classical calculations mentioned earlier. This kind of stretched coupling scheme is extremely useful for interpreting the data and stimulating further experi ments. In addition the picture is supported by the results of shell-model calculations, using realistic residual interactions, which show that the excitation energies of the stretched configurations are close to those of the selectively populated levels.

I should also mention here some other evidence for the principle of maximal spin coupling in these high energy transfer reactions. First, consider one neutron transfer on to the nucleus ^{13}C whose structure can be described as a $p\frac{1}{2}$ neutron outside a core of ^{12}C. The transferred neutron goes preferentially into the $d\frac{5}{2}$ orbit. Since the $p\frac{1}{2}$ neutron is essentially a spectator it can couple both ways with the captured neutron to yield the states with configurations $[(p\frac{1}{2}d\frac{5}{2}): J = 2^-, 3^-]$. These levels are both excited strongly. However, in two neutron transfer on to ^{12}C, in which one neutron goes into the $p\frac{1}{2}$ level and the other into $d\frac{5}{2}$, only the maximal spin state with $J = 3^-$ is seen strongly and there is no trace of the $J = 2^-$ state.

Another example concerns levels in mass 15. In three nucleon transfer on to ^{12}C we should expect to see two strongly excited levels with maximal spin configurations $[\{(d\frac{5}{2})^2 5^+\}(p\frac{1}{2}): J = \frac{11}{2}^-]$ and $[\{(d\frac{5}{2})^2 5^+\}(d\frac{3}{2}): J = \frac{13}{2}^+]$. They are observed as prominent peaks at about 13 and 15 MeV excitation energy. On the other hand, in (np) transfer on to ^{13}C the spectator $p\frac{1}{2}$ nucleon can couple two ways with the maximal spin state $[(d\frac{5}{2})^2: J = 5^+]$ of the transferred pair and we should thus excite the configurations $[\{(d\frac{5}{2})^2 5^+\}(p\frac{1}{2}): J = \frac{11}{2}^-, \frac{9}{2}^-]$. Hence both reactions show two strong peaks, but only one in common. This turns out to be the one at 13 MeV. We can then immediately assign the other states as $J = \frac{13}{2}^+$ in the first reaction and $J = \frac{9}{2}^-$ in the second.

We see that the principle of maximal spin coupling for the transferred nucleons is well supported and is useful as a spectroscopic tool. But it is very unsatisfactory that the semi-classical reaction model predicts only a rather moderate preference for high spin coupling, at least when we consider the nucleons to be transferred essentially independently. As indicated earlier, the most natural additional hypothesis is that the particles are transferred as a correlated cluster, in the lowest state of internal motion, with the centre of mass of the cluster carrying all the transferred energy and angular momentum. If, for example in the reaction ^{12}C(^{12}C, ^{10}C)^{14}C the two neutrons are transferred as a cluster with internal quantum numbers $l = 0$, $S = 0$, $T = 1$ (a dineutron) and the centre of mass of the dineutron goes into various high L-orbits about the ^{12}C target, we should expect to see the state with spin $J = 3^-$, but the state with $J = 2^-$ would be ruled out as being of unnatural parity. This seems to be the only reasonable way to account for the complete absence of the 2^- state in the spectrum.

Of course, the dineutron is not a particularly good spatially correlated structure since it is not bound, but this fact is in itself of some interest because of another very striking feature of the multinucleon transfer data. It has been found that single nucleon, (np), (npp) and (2n2p) transfer reactions all have forward angle cross sections of about 1 mb/steradian while (nn), (pp), (3n) and five nucleon transfer cross sections are about fifty times smaller. The relevant numbers are shown in Figure 3.

REACTION	$d\sigma/d\Omega$
n or p	1 mb/st
(np)	1 mb/st
(nn) or (pp)	20 μb/st
(2p1n)	.5 mb/st
(3n)	3 μb/st
(2p2n)	1 mb/st
5 nucleon	20 μb/st

Fig. 3: Table of cross sections for transfer reactions on light nuclei induced by heavy ions at incident energies of 5-10 MeV/nucleon.

The only apparent difference is that the latter groups can not form bound states by themselves, so there is a suggestion that strong multinucleon transfers actually proceed by transfer of deuteron, helion or α-particle like clusters, again with the centres of mass of the bound groups carrying all the transferred quantum numbers. Nevertheless, the (2p) and (2n) transfers are also highly selective, even though they have much smaller cross sections, which implies that there are still some residual correlations present in these reactions; that is, it appears to be true that the centres of mass of these virtually bound states carry all the transferred information.

These observations in turn suggest that the selectively excited states themselves may have a structure describable, in some sense, as a correlated cluster orbiting the target nucleus. However it is very difficult to analyse the results of the j-j coupled shell-model calculations of these final states so as to make apparent the existence of such spatially localised clusters or to show the relations between states of different spins which would be implied by such a cluster picture. In particular, the shell model calculations are always done in rather severely truncated model spaces which are not big enough to allow the nucleons to correlate spatially in a marked way. Furthermore, the mixing of a few low lying shell model configurations will give only a rather poor description of the surface peaked wave functions of a true cluster state.

Fortunately there is another way of doing the structure calculations which lends itself to a cluster analysis. This is to use many particle basis states classified by means of the SU3 model. In essence, this model uses single particle harmonic oscillator orbitals to build up configurations which have a deformed density distribution in a space fixed coordinate frame. Such states do not have a good angular momentum, but one can project from a single deformed intrinsic state to construct a whole band of related states which do have good J-values. A shell model calculation with such a basis shows that the high spin states for two, three and four particles in the (sd) shell are predominantly as shown in Figure 4.

NUCLEONS	$(\lambda\mu)$	J	(j-j) OVERLAP
2	4,0	5	$1 \times (d\frac{5}{2})^2$
3	6,0	13/2	$\sqrt{\frac{4}{5}} \times [(d\frac{5}{2})^2 d\frac{3}{2}]$
4	8,0	8	$\frac{4}{5} \times [(d\frac{5}{2})^2 (d\frac{3}{2})^2]$

$$\Psi(\lambda 0) \longrightarrow \phi_{n\ell}(\underline{R})\chi_{00}(\underline{r})$$

where $2n + \ell = \lambda$
$\underline{R} \rightarrow$ C.M. coordinate of cluster
$\underline{r} \rightarrow$ internal coordinates of cluster
E.G. For 4 nucleons obtain 5 states with same internal α-cluster structure

(8,0) STATE					High spin J = 8
ℓ 0 2 4 6 8					only special case
n 4 3 2 1 0					of cluster state.

Fig. 4: Overlaps of stretched configurations with SU3 basis states and cluster decomposition of SU3 representations.

We observe that the SU3 states have large overlap with our simple maximal spin configurations. Now this is very interesting because it may be shown, by transforming the oscillator coordinates, that the SU3 states in turn have large overlap with wave functions appropriate for clusters. We can think of the state as a cluster of nucleons, in its lowest internal level, moving so that its centre of mass carries all the orbital angular momentum relative to the centre of the core nucleus. This is indicated in the figure for the case of four nucleons. The (8,0) SU3 intrinsic state is analysed into a band of α-cluster levels with various spins. Of course, the low spin members of the band should not appear very strongly in the (2p2n) transfer reactions at high energies precisely because these reactions are selective for high spin.

With this kind of structure in mind we may now attempt to calculate transfer amplitudes and decay widths for quasi-molecular final nuclear states which appear, for example, as α-clusters orbiting a core and we now see that there may be whole sequences of states with very similar intrinsic structure which include high spin states only as special cases. But we now get into difficulties because our original harmonic oscillator single particle basis states have the wrong asymptotic form for describing α-clusters out in the nuclear surface. If we insist on using SU3 states we are forced to try admixing quanta from higher shells in order to make the internal cluster structure more spatially correlated and at the same time increase the amplitude of the centre of mass wave function in the nuclear surface region. Finally, there is the complicated and ambiguous problem of matching bound state wave functions from the structure calculation with the asymptotic free state wave functions necessary to describe α-decay.

However, since we know that these four particle cluster states seem to have large spectroscopic factors for transfer of α-like structures and, in the case of the well-known rotation bands of ^{16}O and ^{20}Ne, large widths for emitting real α-particles, we can try a different approach. This involves describing such states as being just a real α-particle bound to, or resonating with, a core nucleus through some effective potential well. The cluster and core are assumed to be in their ground states as in free space and exchange effects between cluster and core will be incorporated in the effective well. The orbital blocking effects of the Pauli principle can be satisfied by a careful choice of the quantum numbers of the motion of the cluster relative to the core as already indicated for the SU3 model in which the α-clusters have their lowest nucleon states in the (sd) shell. I assume that the n and l of the centre of mass wave function of the cluster are related to the n_i and l_i quantum numbers for the individual nucleons by the formula

$$2n + \ell = \sum_{i=1}^{c} (2n_i + \ell_i)$$

where c is the number of cluster nucleons.

In shell model terms, each nucleon carries at least $(2n_i + l_i)$ quanta of excitation appropriate to the lowest available oscillator levels. Then with d particles present I assume that they correlate so that the internal motion carries zero quanta and all the excitation quanta go into the centre of mass motion. Thus for four nucleons, with lowest states in the (sd) shell, each carrying $2n_i + l_i = 2$ quanta, I calculate states of the centre of mass of the cluster which have $2n + l = 8$. This gives five states with $l = 0, 2, 4, 6$ and 8, which is very convenient for describing the lowest rotation bands of ^{16}O and ^{20}Ne.

This scheme works very well if the effective α-nucleus potential is taken to be proportional to the folding of the densities of the α-particle and the core nucleus as determined by electron scattering experiments. For states of ^{20}Ne we have

$$V(\underline{r}) = V_0 \int \rho_\alpha (\underline{r} - \underline{s}) \rho_0 (\underline{s}) d^3\underline{s}$$

where ρ_α and ρ_0 are the densities of the α-particle and the ^{16}O ground state, which may, for example, be chosen in the form

$$\rho(r) \sim (1 + \alpha r^2) \exp(-\beta r^2)$$

with α and β determined by experiment. The folding integral may then be evaluated analytically, while the depth V_0 can be found by locating one of the ^{20}Ne states at the correct binding energy.

It is now straightforward to calculate the energies of the other bound states and to find the position and widths of the resonances. The results for the first positive and negative parity bands of ^{20}Ne, which have $2n + l = 8$ and 9 respectively, are shown in Figure 5.

The widths of the narrow resonances are obtained as the energy intervals for the elastic phase shift to go from $\pi/4$ to $3\pi/4$. Since the nuclear levels are calculated just as single particle states of an α in a potential well it is also easy to compute the B(E2) values for γ- decay as the α-particle jumps from orbit to orbit. The theoretical results for α and γ decay widths are compared with experiment in Figure 6. We see that there is good overall agreement.

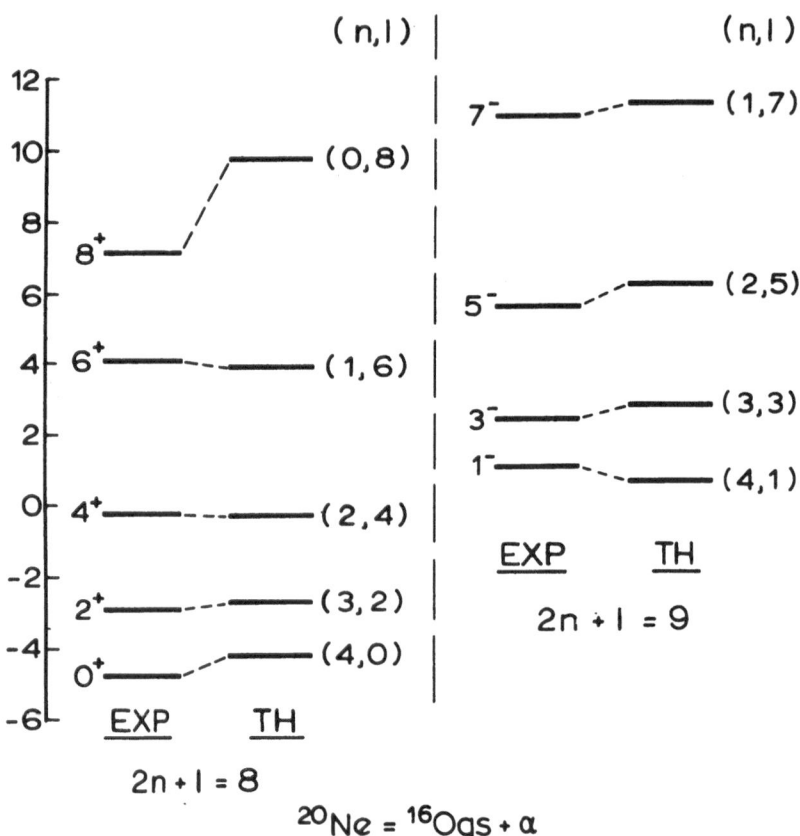

Fig. 5: Comparison of theory and experiment for rotation bands in ^{20}Ne with KΠ = 0+ and 0-. Theoretical spectra were calculated using the folded potential and the scale shows the binding energy of the α-particle relative to Oxygen 16.

J^{Π}	ALPHA WIDTHS	
	Γ_{α}^{exp} (KeV)	Γ_{α}^{calc} (KeV)
6^+	0.110	0.210
8^+	0.035	0.108
1^-	> 0.013	0.021
3^-	8	6.7
5^-	141	81
7^-	280	183

GAMMA DECAYS

TRANSITION	B(E2)exp	B(E2)calc $(e^2 fm^4)$
$2^+ \rightarrow 0^+$	57.3	57.3
$4^+ \rightarrow 2^+$	71.0	71.9
$6^+ \rightarrow 4^+$	66.0	60.0
$8^+ \rightarrow 6^+$	24.0	34.6

Fig. 6: Alpha widths and B(E2) values for various states in ^{20}Ne.

Similar calculations have been done for the rotation bands in ^{16}O described as ^{12}C + α and the same idea has been applied to explain the properties of some high spin states in mass 15 in terms of a triton orbiting the ^{12}C ground state [2].

Finally I should mention a very intriguing theoretical result. Various kinds of parameterisation of the densities have been tried, but the folded potential usually turns out to have a roughly Gaussian shape and the single particle energies of the states in a band with given 2n + l follow an l(l+1) law very accurately. To investigate this behaviour I have calculated, with Dr. E. Tomusiak, all the bound state levels in a very deep potential of pure Gaussian shape, which has bound bands up to 2n + l = 12. The results are shown in Figure 7. The really striking thing is that all the levels fall into nearly perfect rotational bands and that every band has the same moment of inertia. This remarkable result may have some bearing on the widespread occurrence of rotational characteristics in nuclei.

To sum up, I believe that the many particle – many hole states selectively populated in multinucleon transfer at high energies may be well described in terms of cluster correlations and that heavy ion reactions are a powerful tool for investigating such collective effects.

References:

1) N. Anyas-Weiss, J.C. Cornell, P.S. Fisher, P.N. Hudson, A. Menchaca-Rocha, D.J. Millener, A.D. Panagiotou, D.K. Scott, D. Strottman, D.M. Brink, B. Buck, P.J. Ellis and T. Engeland, Physics Reports 12C, No. 3 (1974).

2) B. Buck, C.B. Dover and J.P. Vary, Annals of Physics (to be published).

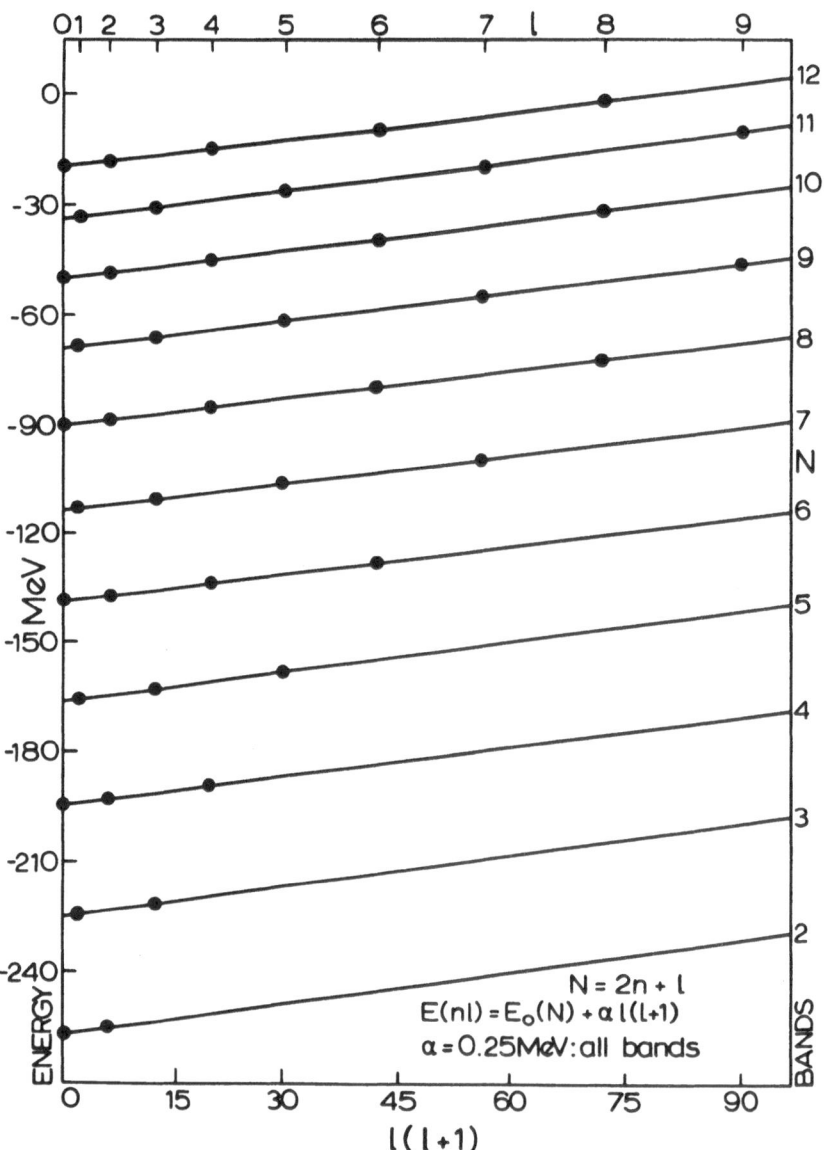

Fig. 7: Spectrum of bound energy levels in a deep Gaussian potential.

THE ANALYSIS OF HIGH ENERGY HEAVY-ION TRANSFER REACTIONS*

D. K. Scott
Lawrence Berkeley Laboratory
University of California
Berkeley, California 94720

ABSTRACT

   The regions of validity of quantal and semiclassical theories are discussed
for high energy transfer reactions with heavy-ions. After demonstrating the equi-
valence of the two formalisms, they are applied quantitatively to predict spectroscopic factors, the energy dependence and Q dependence of single nucleon transfer
on lead. The power of semiclassical theory for making wide surveys is shown. The
discovery of simple cluster states in multinucleon transfer reactions on lighter
nuclei is discussed using semiclassical theory. Finally the quantal analysis of
new effects due to multistep processes in high energy heavy-ion reactions is presented.

1.   Introduction

   The rapid advances in experimental heavy-ion physics have sparked off a

remarkable inventiveness among theoreticians in developing new reaction theories and

interpretations of the data. These range from simple qualitative semiclassical in-

sights to formal semiquantal theories, and to a bewildering variety of approximations

to make the exact quantum mechanical theories amenable to calculations. In two years

the field has evolved from the viewpoint that heavy-ion reactions were beautifully

simple, to one implying that the proper interpretation is of staggering complexity.

This conference may show us that the simplicities are still there after all. We

should not be too elated at the successes of the simple theories, nor despair at the

failures of the complicated theories. The lesson to be learned from the last few years

and from this conference was learned long ago on the English Public School playing

fields: "It matters not who won or lost, but how they played the game." This talk

is about some of the games that have been played in interpreting high energy heavy-

ion transfer data from quantum mechanical and semiclassical approaches. Some general

aspects of these approaches and the effect on differential cross sections are discussed

in the next section. In section 3, both methods are applied to single nucleon trans-

fer data on ^{208}Pb, which serves as a standard of calibration and of comparison for

different theories. This is followed in section 4 by a discussion of transfers on

lighter nuclei, where the semiclassical approach is on less sure ground, but where

we show that it can give physical insight, and suggest interesting experiments. In

section 5 we discuss multistep processes at high energy, for which so far only the quantum mechanical calculations have been done; the results are sufficently exciting to make us hope that the development of a semiclassical theory will be forthcoming in order to make surveys of the type discussed in sections 3 and 4. Our conclusions are presented in section 6.

2. General Semiclassical and Quantum Mechanical Aspects of High Energy Data

A characteristic feature of high energy heavy-ion reactions is that the wave length of relative motion (λ) is very short compared to the nuclear radii. Typically for ^{16}O ions of 10 MeV/nucleon on ^{208}Pb, $\lambda \approx 0.2$ fm compared to $R_1 + R_2 \approx 13$ fm. This localisation leads to the concept of a well-defined classical trajectory. It is important to remember, however, that the classical picture arises from the inter- ference of a large number of waves[1], and therefore the concept of a trajectory is valid only under the condition that the beam contains a sufficiently large number of orbital angular momenta ($\Delta\ell$), which however must still be small compared to the grazing orbital angular momentum (ℓo). Only then exists the possibility of defining phase shifts and Legendre Polynomials as smooth functions of ℓ, and the replacement of quantized sums by integrals. An example is shown[2] in fig. 1, for the reaction of 100 MeV ^{18}O ions on Sn, where S_ℓ the amplitude in the outgoing channel is plotted as a function of ℓ. In this case the semiclassical condition might be satisfied, with ℓo $\approx 57\,\hbar$ and $\Delta\ell \approx 15\,\hbar$ (evaluated at 1/e of the maximum).

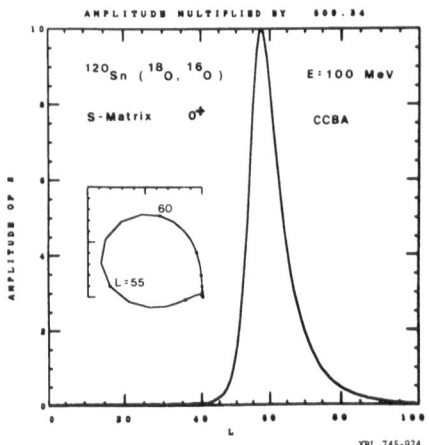

AMPLITUDE MULTIPLIED BY 100.34

^{120}Sn (^{18}O, ^{16}O)   E = 100 MeV

S-Matrix   0⁺   CCBA

60

L = 55

XBL 745-924

Fig. 1. The amplitude of the S-matrix for the ground state transition in the re- action ^{120}Sn(^{18}O,^{16}O)^{122}Sn at 100 MeV. The inset shows S plotted in the com- plex plane, its real part on the x- axis. The positions of ℓ = 55 and 60 are marked.

The orbit concept also requires that the uncertainty in the angle of scattering $\Delta\Theta$, should be small compared to Θ. Now we can write $\Theta \approx \frac{\Delta p}{p}$, where p is the incident momentum and the change is Δp. So $\Theta \approx \int \frac{F dt}{p}$, where F, the force acting on the particle over the region "a", is V/a, and dt $\approx$ a/v, so $\Theta \approx$ V/E. When the particle passes through the region "a" the uncertainty in the transverse momentum is $\delta p \approx \hbar/a$, and so $\Delta\Theta \approx \hbar/ap$; and finally $\Theta/\Delta\Theta \approx \frac{Va}{\hbar v}$, which we require to be much greater than unity. This condition is better satisfied the lower the energy. On the other hand the condition $\lambda \ll R_1 + R_2$ is fulfilled better at highincident energy. It is important therefore to have a more general criterion for the degree of "class- icality" of a reaction.

We write the scattering amplitude,

$$f(\Theta) = \frac{1}{2ik} \sum (2\ell+1)\, \eta_\ell\, e^{2i\delta_\ell}\, P_\ell(\cos\Theta) \tag{1}$$

and the reaction amplitude, assuming the peripheral nature of the reaction, as a Gaussian distribution[3] (justified by the output of "exact" DWBA calculations e.g. see fig. 1)

$$\eta_\ell = \eta_{\ell o}\ \ \mathrm{EXP}\ \left[-\frac{(\ell-\ell o)^2}{(\Delta\ell)^2}\right] \tag{2}$$

$P_\ell(\cos\Theta)$ is replaced by the asymptotic expression valid for large ℓ, and $\sin\Theta > 1/\ell$,

$$P_\ell(\cos\Theta) \approx [1/2(\ell+1/2)\,\pi\,\sin\Theta]^{-1/2}\ \ \sin[(\ell+1/2)\Theta + \pi/4] \tag{3}$$

For δ_ℓ we make the Taylor expansion:

$$\delta_\ell = \delta_{\ell o} \pm \left(\frac{d\delta}{d\ell}\right)_{\ell=\ell o}(\ell-\ell o) + 1/2\left(\frac{d^2\delta}{d\ell^2}\right)_{\ell=\ell o}(\ell-\ell o)^2 + \ldots\ldots\ldots \tag{4}$$

On account of the WKB classical relationship[4] for the scattering angle Θ_ℓ corresponding to partial wave ℓ,

$$\Theta_\ell = 2\frac{d\delta_\ell}{d\ell} \tag{5}$$

we can write

$$\delta_\ell = \delta_{\ell o} \pm \frac{\Theta_o}{2}(\ell-\ell o) + 1/4\left(\frac{d\Theta\ell}{d\ell}\right)_{\ell=\ell o}(\ell-\ell o)^2 + \ldots\ldots\ldots \tag{6}$$

where Θ_o is the classical angle of deviation for the tangetial trajectory (not necessarily purely Coulomb). Substituting in (1) and converting the summation to an integral gives

$$\frac{d\sigma}{d\Omega} = |f(\Theta)|^2\ \alpha\ \mathrm{EXP}\left\{\frac{-(\Theta-\Theta_o)^2}{(\Delta\Theta)^2}\right\} + \mathrm{EXP}\left\{\frac{-(\Theta+\Theta_o)^2}{(\Delta\Theta)^2}\right\} + \left(\begin{array}{c}\mathrm{INTERFERENCE}\\\mathrm{TERM}\end{array}\right) \tag{7}$$

For the sake of historical accuracy it is worth noting that the interference term was present in the early treatments of high energy heavy-ion transfer[3,5,6] theories but it was always averaged over, because the data were too crude at that time. Kahana et al.[7] interpret the differential cross section as the interference of two classical distributions centered at the physical angle (Θ_o) and the unphysical $(-\Theta_o)$. Here we discuss only the term

$$\frac{d\sigma}{d\Omega}\ \alpha\ \mathrm{EXP}\left[\frac{-(\Theta-\Theta_o)^2}{(\Delta\Theta)^2}\right], \tag{8}$$

since so far no high energy data has revealed the interference oscillations. They remain a challenge to experimental ingenuity. This equation describes a symmetric distribution of width[8]:

$$(\Delta\Theta)^2 = \frac{2}{(\Delta\ell)^2} + \frac{1}{2}\left(\frac{d\Theta_\ell}{d\ell}\right)^2 (\Delta\ell)^2 \tag{9}$$

Thus for small $\Delta\ell$ the distribution is broad due to quantal dispersion, and if $\Delta\ell$ is very large the distribution is also broad due to "dynamic" dispersion. The minimum value of $(\Delta\Theta)$ is obtained for

$$\Delta\ell = \sqrt{2\left(\frac{d\ell}{d\Theta}\right)_{\ell o}} = \sqrt{\eta} \;\; \mathrm{cosec}\left(\frac{\Theta_o}{2}\right), \tag{10}$$

where η is the Sommerfeld parameter, using the classical result $\ell = \eta \cot\left(\frac{\Theta}{2}\right)$. (11)

Then
$$(\Delta\Theta)_{MIN} = \frac{2}{\sqrt{\eta}} \;\; \sin\left(\frac{\Theta_o}{2}\right). \tag{12}$$

If $(\Delta\ell)^2 \gg \mathrm{cosec}^2\left(\frac{\Theta_o}{2}\right)$ we have a classical situation, and the width of $\Delta\Theta$ increases with $\Delta\ell$, but if $(\Delta\ell)^2 \ll \eta\,\mathrm{cosec}^2\left(\frac{\Theta_o}{2}\right)$, we have a quantum situation and as $\Delta\ell$ increases, $\Delta\Theta$ decreases[9].

If we take as an example the data of fig. 1, $\Delta\ell$ is derived independently from a quantum mechanical calculation using a correct form factor and optical parameters pertinent to elastic scattering[2]. The value of 225 for $(\Delta\ell)^2$ is approximately the same as $\eta\,\mathrm{cosec}^2(\Theta/2) \approx 200$. These data correspond therefore to the region of the minimum $\Delta\Theta$, and are in the transitional region between classical and quantum descriptions. A consideration of the data from both viewpoints is likely to be instructive. Further if we consider the single nucleon transfer data[10,11] on Pb shown in fig. 2, at 98 MeV, the classical conditions are better satisfied. On the figure we show the value of $(\Delta\Theta)_{MIN}$ and the value of Θ_o predicted from the equation

$$\mathrm{cosec}\left(\frac{\Theta_o}{2}\right) = \frac{2E\,(R_1+R_2)}{z_1 z_2 e^2} - 1 \tag{13}$$

It is interesting that this formula predicts that the grazing angle, taken as an average over the initial and final orbits, should move to large angles with increasing excitation energy (as E in the final channel decreases). This effect is observed in fig. 2(a) for neutron transfer, and is predicted by DWBA for the proton transfer in fig. 2(b), although in this case the position of the experimental peak is in fact constant. This disagreement, which is greatest for the case of lowest angular momentum transfer has been discussed by von Oertzen[12]. If we write the initial and final angular momenta

$$\ell_i = \eta_i \cot\left(\frac{\Theta_o^i}{2}\right), \quad \ell_f = \eta_f \cot\left(\frac{\Theta_o^f}{2}\right) \tag{14}$$

then the requirement that $\ell_i \approx \ell_f$, together with the fact that $\eta_f < \eta_i$ in proton stripping, implies $\Theta_f < \Theta_i$, and that the absorption and hence the position of the classical maximum are determined primarily by the _initial_ orbit. This effect has been reproduced by (somewhat artificial) adjustment of optical parameters[13,14]. Probably there are subtleties of the heavy-ion potential as yet unaccounted for.

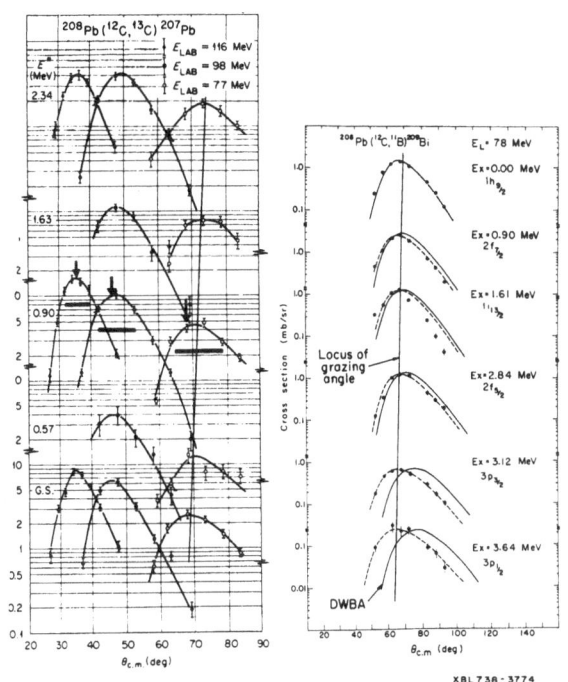

XBL 738-3774

The characteristic feature of the distribution function derived by expanding δ_ℓ to second order in $(\ell - \ell_o)$ is that it is symmetrical about Θ_o. The data in fig. 3 for one and two nucleon transfer on Nd, from Berkeley illustrate another effect. The one nucleon transfer data has a symmetric peak of width somewhat larger than the estimate of eq. 12, and in fact this reaction meets the semi-classical criteria. On the other hand the peak for two nucleon transfer is considerably broadened and asymmetric. Shown in the figure are two DWBA analyses using the optical model parameters[15,16] of Table 1. For one nucleon transfer the predictions are almost identical, whereas there is a factor of ten between the predictions at forward angles for two nucleon transfer. Glendenning and Ascuitto[17] have discussed how the sharper fall-off of the two nucleon form factor, makes the forward cross section particularly sensitive to close trajectories, and consequently they provide a probe of the nuclear edge, and of the relationship between the imaginary and the real potentials

Fig. 2. (a) differential cross sections for the ^{208}Pb(^{12}C, ^{13}C)^{207}Pb reaction leading to single hole states at incident energies of 77, 98, and 116 MeV. The bold arrows denote the grazing angle predicted by eq. 13. The locus of this angle as a function of energy is indicated. The bold horizontal lines are the minimum FWHM of the distributions, predicted from eq. 12. (b) differential cross sections for the ^{208}Pb(^{12}C, ^{11}B)^{209}Be reaction at 78 MeV. The dotted lines are drawn through the data points, and the solid curves are DWBA predictions.

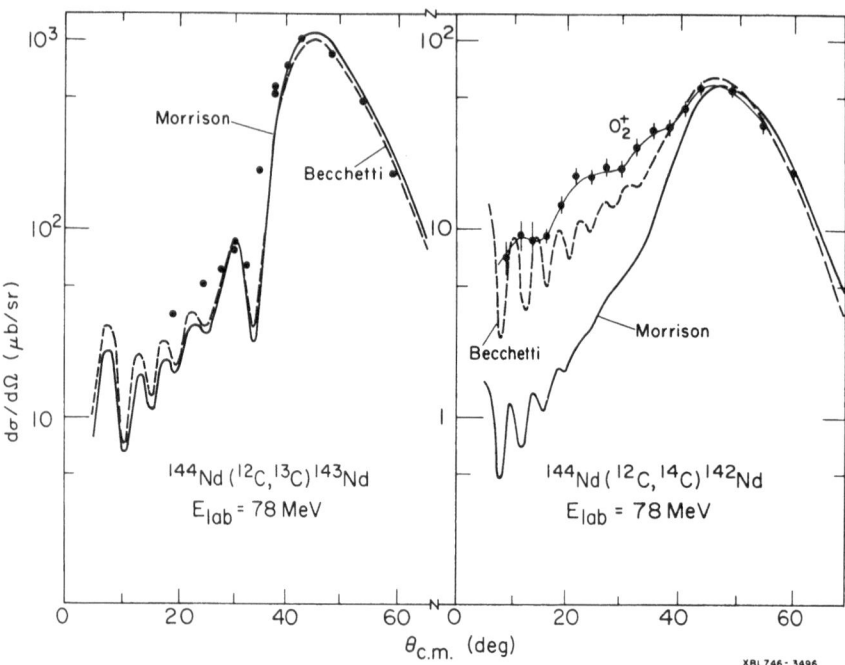

Fig. 3. Differential cross sections for the reactions $^{144}Nd(^{12}C,^{13}C)^{143}Nd$ and $^{144}Nd(^{12}C,^{14}C)^{142}Nd$ at 78 MeV. Two sets of DWBA calculations are shown using the optical model parameters of Table 1. These give almost identical results for one nucleon transfer, but are makedly different for two nucleon transfer. (The calculation for $(^{12}C,^{14}C)$ with the Morrison potential used different real and imaginary diffuseness of 0.49 and 0.6 fm. In the $(^{12}C,^{13}C)$ these parameters would lower the forward cross section by approximately a factor of 2, but does not affect our conclusions in the text).

Table 1. Optical model parameters used in the analysis of one and two nucleon transfer reactions on ^{144}Nd.

	V	W	r_o	a	r_c
Morrison (ref. 16)	-100	-40	1.22	0.5	1.2
Becchetti (ref 15)	- 40	-15	1.31	0.45	1.2

How do we interpret such an asymmetry semiclassically? Recently it has been shown[18] that if we expand δ_ℓ to third order,

$$\delta_\ell = \delta_{\ell o} + \left(\frac{\Theta_o}{2}\right)(\ell-\ell_o) + \ldots + \frac{1}{3}\beta(\ell-\ell_o)^3 \qquad (15)$$

the resultant $|f(\Theta)|^2$, takes on the form shown by $|g(\Theta)|^2$ in fig. 4, where the cross section for the physical scattering becomes tipped to more forward angles, and takes on the form of the Airy function. The effect on the deflection function of a third order term in δ_ℓ is

$$\Theta_\ell = 2\frac{d\delta_\ell}{d\ell} = \Theta_o + \ldots + \beta(\ell-\ell_o)^2 \qquad (16)$$

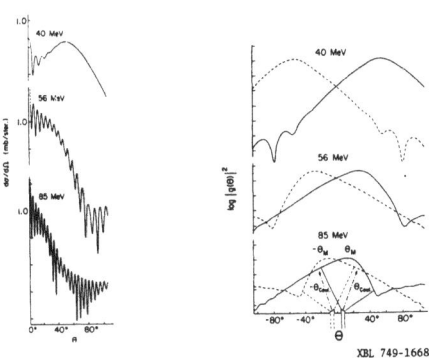

XBL 749-1668

i.e. it adds a parabolic dip, deviating the deflection angle for trajectories passing through the ℓ-window. The physics of this "refraction" is obviously closely related to the optical potential. The interesting insights will come from such associations of the parameters of the semiclassical expansion with the optical model of the heavy-ion interaction. Fig. 4 shows how at high energy the overlap of the two asymmetric distributions leads to a modulation of the <u>diffractive</u> oscillations. It is also interesting to note the existence of gross oscillations in fig. 3 for two nucleon transfers, which, in the light of past experiences, we should perhaps not ignore.

Fig. 4. On the right are shown single-slit diffraction-refraction patterns from DWBA calculations for the reaction ^{48}Ca(^{16}O,^{14}C)^{50}Ti (shown on the left). The two distributions correspond to the grazing angle distributions centered at the physical angle Θ_M and the unphysical $-\Theta_M$. (compare eq. 7). At 85 MeV the distributions take on an asymmetric form. (see the discussion of of eq. 15).

Finally it is amusing to note that kinks, and modulations in deflection functions and phase shifts are not entirely new in the subject of heavy-ion physics. Fig. 5 illustrates an early calculation[5] for the multinucleon transfer reaction (Ne^{20}, Na^{24}) on Au, where the discontinuty between the initial and final orbits leads to an unusual average deflection function in the region where the reaction amplitude for multinucleon transfer is concentrated. The phase shift as a function of ℓ develops inflexions and the differential cross section becomes double peaked.

3. <u>DWBA and Semiclassical Theories Applied to High Energy Transfer Data on Pb</u>

Before discussing the formal DWBA and semiclassical theories applied to high energy transfer data, it is worthwhile recalling the main features of the data which stimulated their development. The work of the Berkeley group for $(^{16}O, ^{15}N)$ on ^{208}Pb from 104 to 216 MeV is shown[19] in fig. 6. The variety of pure single particle states excited makes this reaction a standard of comparison and calibration of different reaction theories. The striking feature is the dominance of the $j_f = 1_f + 1/2$ state at low energy and the equality of $1_f \pm 1/2$ at the highest energy, the understanding of which brought about a revolution in reaction theories for heavy-ions. From an intuitive viewpoint the effect is easily understood as the overcoming of the orbital velocity by the velocity of the transferred particle due to the projectile motion (see fig. 6), whereas for low projectile velocity a smooth transition selects a final $j_f = 1_f + 1/2$ from an initial $j_i = 1_i - 1/2$.

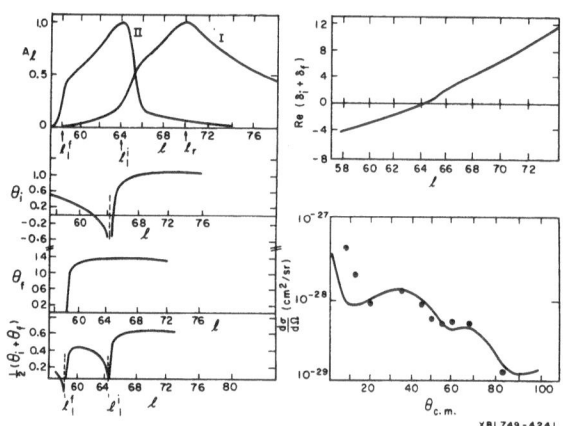

Fig. 5. The top figure shows the amplitude for single nucleon (I) and multinucleon (II) transfer induced by ^{20}Ne on ^{197}Au at $E_{CM} = 127$ MeV. The multinucleon transfer $(^{20}Ne, ^{24}Na)$ is peaked at smaller ℓ-values. Underneath are the deflection functions for the initial and final channels, together with the average $1/2(\Theta_i + \Theta_f)$. The corresponding phase shift is plotted at the top right. The differential cross section below has a double maximum, due to the discontinuities between the initial and final channels of the multinucleon transfer.

The formalisms of DWBA and SC theory will now be described to show their relationship. Their quantitative agreement is compared by applying these theories to various features of the data, such as spectroscopic factors, energy dependence and Q-dependence.

3.1 DWBA Formalism

The relevant vector diagram for the reaction $A(a,b)B$ with $a = b + x$ and $B = A + x$ is shown in fig. 7. For single nucleon transfer the transition probability involves the six-dimensional integration over $\underline{r}_i$, $\underline{r}_f$

$$T^{DWBA} = \int d\underline{r}_f \, d\underline{r}_i \, \chi_f^* \, (\underline{k}_f, \, \underline{r}_f) \, \phi_B^* \, (\underline{r}_2) \, V\phi_a(\underline{r}_i) \, \chi_i \, (\underline{k}_i, \underline{r}_i) \tag{17}$$

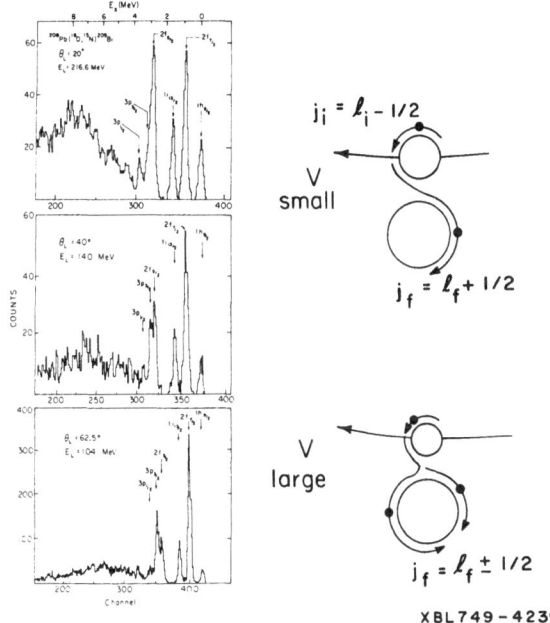

XBL749-4239

Fig. 6. Energy spectra for the reaction ^{208}Pb $(^{16}$O,^{15}N$)^{209}$Bi at 104, 140 and 216 MeV. At the lowest energy the reaction favors $j_f = \ell_f + 1/2$ over $j_f = \ell_f - 1/2$, whereas at the highest energy they are equal. The figure at the right explains the phenomenon in terms of orbit matching at low and high velocity.

The χ's are distorted waves, and the ϕ's represent wave functions of the relative motions of the nucleon x bound to the cores A or b. In general one must also include spectroscopic amplitudes Θ_i, Θ_f for the decomposition A → b + x, B → A + x. In the post representation $V = V_{bB} - U_{bB}$, the difference between the total interaction in the final channel and the average interaction of the optical potential, and is approximately $V_{bx}(\underline{r}_1)$; in the prior representation, it is $V_{Ax}(\underline{r}_2)$. A plethora of methods exist for approximating this integral (see the discussion in refs. 20, 21), the most drastic of which is the "no-recoil" approximation obtained by ignoring the difference between $\underline{r}_f$ and $\underline{r}_i$. In order to show the relationship to the semi-classical theory we derive this approximation by setting $\underline{r}_f \approx \frac{A}{B} \underline{r}$ and $\underline{r}_i \approx \underline{r}$. Then

$$T^{DWBA} = \int \chi_f^*(\underline{k}_f, \frac{A}{B}\underline{r}) \; F(\underline{r}) \; \chi_i(\underline{k}_i, \underline{r}) \; \underline{dr} \qquad (18)$$

$$F(r) = \int \phi_B^*(\underline{r} + \underline{r}') \; V_{bx}(\underline{r}') \; \phi_a(\underline{r}') \; \underline{dr}' \qquad (19)$$

and we obtain two three-dimensional integrals. The effect of this approximation can be seen[22] by expanding the distorted waves,

$$\chi(\underline{k}, \underline{r} + \underline{\delta r}) = e^{\underline{\delta r} \cdot \nabla} \; \chi(\underline{k},\underline{r}) \qquad (20)$$

$$\approx e^{\underline{\delta r} \cdot K(\underline{r})} \; \chi(\underline{k},\underline{r}) \; ,$$

where $K(\underline{r})$ is the local momentum vector at point $\underline{r}$. Then eq. (17) reduces to the form of eq. 18 with F replaced by

$$F(\underline{r}) = \int e^{iP(\underline{r}) \cdot \underline{r}'} \phi_B^*(\underline{r} + \underline{r}') \, V_{bx}(\underline{r}') \, \phi_a(\underline{r}') \, \underline{dr}' \tag{21}$$

and

$$P(\underline{r}) = \frac{x}{B} K_f(\underline{r}) + \frac{x}{a} K_i(\underline{r}) \tag{22}$$

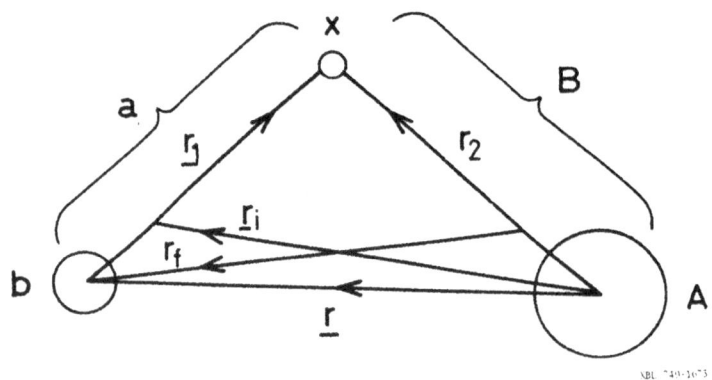

Fig. 7. Vector diagram for the reaction A(a,b)B with a = b + x
and B = A + x. The relative coordinates of the colliding
nuclei in the initial and final channels are $\underline{r}_i$, $\underline{r}_f$, and $\underline{r}$
is the relative seperation of the cores. The coordinate of
the transferred particle x in the incident and residual nuclei
is represented by $\underline{r}_1$, $\underline{r}_2$.

or in terms of velocities, since

$$\underline{k}_f = \frac{(A + x)b}{A + x + b} \frac{\underline{v}_f}{\hbar} \quad \text{and} \quad \underline{k}_i = \frac{(b + x)A}{A + x + b} \frac{\underline{v}_i}{\hbar} \tag{23}$$

$$P \approx \frac{x(A\underline{v}_i + b\underline{v}_f)}{\hbar(A + x + b)} \approx \frac{x\bar{v}}{\hbar} , \tag{24}$$

were $\bar{v}$ is an average of the initial and final velocities.

The classical picture shows the main contribution to the reaction came from
distances of order sum of the nuclear radii, so therefore $\underline{r}' \approx R_a$, the projectile
radius. The recoil term introduces additional angular momentum transfers of order
$\underline{P} \cdot \underline{R}_a$ which is the angular momentum carried by the transferred particle due to the
projectile motion at the surface. It also allows unnatural party terms through, for
example, the first term in the expansion

$$e^{i\underline{P} \cdot \underline{r}'} = 1 + i\underline{P} \cdot \underline{r}' + \ldots\ldots\ldots \tag{25}$$

The corresponding selection rules[22] for a transition in the no-recoil approximation   from a state $(l_1 j_1)$ to $(l_2 j_2)$ are:

$$|l_1 - l_2| \leqslant \Delta l \leqslant l_1 + l_2 \; ,$$

$$|j_1 - j_2| \leqslant \Delta l \leqslant j_1 + j_2 \; , \tag{26}$$

$$(-1)^{\Delta l} = (-1)^{l_1 + l_2} \; .$$

The last rule is relaxed when the additional transfers are permitted in the full recoil treatment. For example in a transition from p 1/2 $\rightarrow$ f 7/2, $\Delta l = 3,4$ and from p 1/2 $\rightarrow$ f 5/2, $\Delta l = 2,3$. In the no recoil approximation only 4 and 2 are permitted which has the effect of enhancing the f 7/2 state at the lower energy (see fig. 6) since high angular momentum transfers are favored.

Many techniques have been developed for the evaluation of the DWBA integral without making the no-recoil approximation. They have been reviewed recently by Glendenning and Nagarajan[21] and by Blair et al.[20] We shall defer a comparison of the results until we also develop the basic semiclassical transition amplitude.

3.2   Semiclassical Theory for High Energy Reactions

Under the conditions outlined in section 2, it was shown that for many reactions, there is a well-localized trajectory. Then the transition amplitude can be evaluated by integration of the quantum mechanical matrix element along the orbit.[23,24] Using the same notation as for the DWBA, the nucleon x starts out at t = - ∞ in a bound state of the potential V_1 provided by the moving core b. We must calculate the probability that the nucleon transfers to a bound state of the potential V_2 of the core A at t = ∞.

$$T^{SC} = \frac{1}{\hbar} \int < \psi_B \, V \, \psi_a > \, dt. \tag{27}$$

where the wave functions refer to bound states of the particle in __moving__ potentials. In the transformation to a stationary frame:

$$< \psi_B \, V \psi_a > \; = \; F(t) \; \text{EXP} \left\{ \frac{-i}{\hbar} \, [Q + \frac{1}{2} \, x \dot{\underline{r}}^2] \right\} \; , \tag{28}$$

$$F(t) \; = \; e^{\frac{i}{\hbar} \, x \dot{\underline{r}} \cdot \underline{r}'} \; \phi_B [\underline{r}(t) + \underline{r}'] \; V(\underline{r}') \; \phi_a (\underline{r}') \; d\underline{r}' \; . \tag{29}$$

Here Q is the reaction Q value modified by the change in Coulomb energy $(z_1^f z_2^f - z_1^i z_2^i) e^2 / R$ in charge transfer; $\underline{r}(t)$ is the relative separation of the cores, and $\underline{r}'$ is the coordinate of x relative to core b.

Obviously F(t) in eq. (29) is closely related to F(r) in eq. (21) in view of eq. (24), while the phase factor $(Q + 1/2) \, x \, \dot{\underline{r}}^2$ replaces the distorted wave integral in eq. (18). This becomes clear when the reaction Q-value is expanded to first order in $(\underline{k}_f - \underline{k}_i)$ and the mass of the transferred particle x:

$$Q = \delta \left(\frac{\hbar^2 k^2}{2\mu} \right) \approx \hbar v_i (k_f - k_i) - \frac{1}{2} \hbar (v_i k_i) \frac{\delta\mu}{\mu} , \tag{30}$$

where μ is the reduced mass. Setting $\frac{\delta\mu}{\mu} \approx \frac{x(b - A)}{b + A}$ leads to

$$Q \approx \hbar v_i (k_f - k_i + \frac{x}{b} k_i) - \frac{1}{2} x v_i^2 \tag{31}$$

$$\therefore \hbar v_i (k_f - k_i) \approx Q + \frac{1}{2} x v_i^2 \tag{32}$$

and $\quad r.(k_f - k_i) \approx \frac{t}{\hbar} (Q + \frac{1}{2} x v_i^2),$

which relates the phase factor of the distorted waves to that of the semiclassical expression. The two evaluations of the transition probability essentially contain the same physics as required by the correspondence principle. We shall now see how well they compare in describing the experimental data.

3.3 Comparison of SC and DWBA for High Energy Single Nucleon Transfer on Pb

In principle the DWBA and semiclassical integrals can be evaluated exactly, but to economize on computing time a number of approximations have been developed.[20,21] For the DWBA Nagarajan has used a first order expansion[25] of the recoil phase factor in eq. (25), a method extended by Kahana and Baltz to higher orders[26]. More accurate methods are discussed by Elbaz et al.[27] The expansion to first order can seldom be justified at high energies as the expansion parameter usually exceeds unity. The semiclassical formulation has the advantage that the recoil term can be included exactly if the bound state wave function is approximated by a Hankel function[24]. Other novel techniques have been the direct evaluation of the multidimensional integral using Montecarlo techniques[28], or expansion of the distorted waves in a plane wave series to achieve separation of coordinates[29]. Low and Tamura point out the saving in computer time of using interpolation to evaluate the slowly varying form factor from points calculated on a coarser mesh than the rapidly varying distorted waves[13,30].

Some comparisons of these methods for the reaction $^{208}Pb(^{16}O, ^{15}N)^{209}Bi$ are shown in Tables 2 and 3. The semiclassical calculations were made by integrating along a hyperbolic orbit, corresponding to a grazing collision, since differential cross sections have not been treated so far. (This would require taking into account the effect on the trajectory of the nuclear and absorptive potentials[31]). The tables show that the methods of including recoil give an agreement within a factor of approximately two for the spectroscopic factors. The Oak Ridge group have also made a study of ^{11}B induced reactions on Pb using the "exact" approach and conclude[14] that effects due to optical potentials, normalisations, finite-range and non-locality may

Table II. Spectroscopic factors for the reaction $^{208}Pb(^{16}O,^{15}N)^{209}Bi$ at 104 MeV.

Method \ State	$h_{9/2}$	$f_{7/2}^*$	$i_{13/2}$	$f_{5/2}$	$p_{3/2}$	$p_{1/2}$
No-recoil[a]	4.80	1.00	0.83	4.00	1.15	3.50
Nagarajan[a]	1.32	1.00	0.80	1.12	1.28	0.82
Tamura[b]	1.29	1.00	-	0.92	0.79	0.61
Semiclassical[c]	0.71	1.00	1.12	1.10	1.26	0.78
Exact[d]	2.60	1.00	0.96	1.48	1.48	0.60-1.00

*Spectroscopic factors are normalized to unity for the f 7/2 state.

[a] D. G. Kovar et al. Phys. Rev. Lett. 30 (1973) 1075

[b] Ref. 30;

[c] Refs. 24 and 33

[d] Ref. 19.

Table III. Spectroscopic factors for the reaction $^{208}Pb(^{16}O,^{15}N)^{209}Bi$ at 140 MeV.

Method \ State	$h_{9/2}$	$f_{7/2}^*$	$i_{13/2}$	$f_{5/2}$	$p_{3/2}$	$p_{1/2}$
No-recoil[a]	8.00	1.00	0.83	6.67	1.00	6.20-10.0
Nagarajan[a]	2.77	1.00	0.84	1.35	1.04	1.47
Tamura[b]	1.42	1.00	-	0.81	0.79	0.83
Semiclassical[c]	0.96	1.00	0.74	1.00	1.49	2.17
Exact[d]	2.08	1.00	0.87	1.04	0.96	0.60-1.30
$(^3He,d)$[e]	0.89	1.00	0.84	1.02	0.96	0.63-0.80

*Spectroscopic factors are normalized to unity for the $f_{7/2}$ state

[a] D. G. Kovar et al, Phys. Rev. Lett. 30 (1973) 1075

[b] Ref. 30

[c] Refs. 24,33

[d] Ref. 19

[e] B. H. Wildenthal et al., Phys. Rev. Lett. 19 (1967) 960.

contribute of order 20%. So within the present state of the art the simple semi-classical theory does fairly well. We now discuss its use for performing surveys, impossible to contemplate with the DWBA formulations.

We return to the energy variation, observed by the Berkeley group, shown in fig. 8. Detailed finite range DWBA calculations are not yet available. However the calculations apparently do not produce the correct energy dependence of the cross sections.[19,32] There is however a difficulty of making wide surveys, and varying optical parameters, owing to the prohibitive cost. The semiclassical calculations in post and prior forms are shown[33] in fig. 9 which gives a good overall representation of the data, including the equality of the f 7/2, f 5/2 cross sections at 200 MeV.

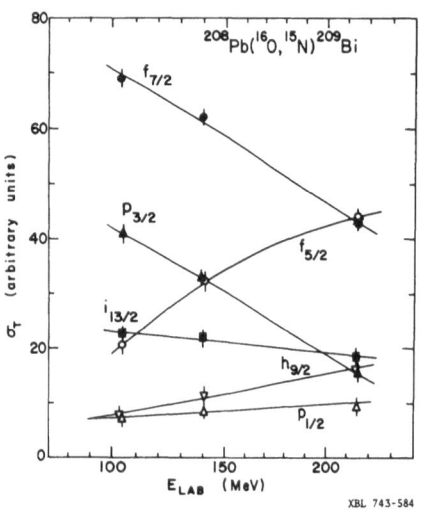

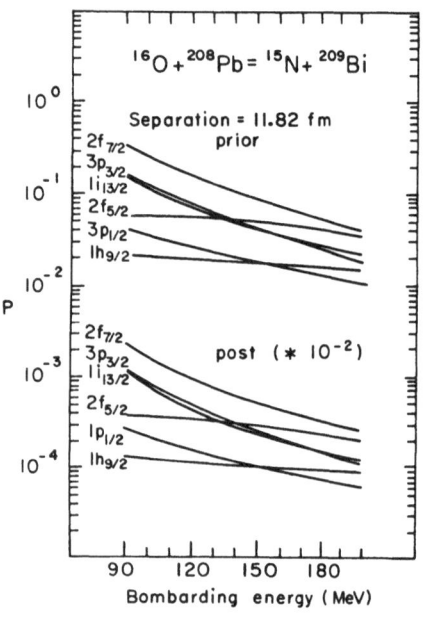

Fig. 8. Observed energy dependence of the cross sections of single particle states populated in the ^{208}Pb(^{16}O,^{15}N) ^{209}Bi reaction (the summed cross section of all measured angles is plotted).

Fig. 9. Theoretical energy dependence of the semiclassical transition probability for the ^{208}Pb(^{16}O,^{15}N)^{209}Bi reaction, evaluated in the prior and post representations.

Both formulations have been used to study the Q-dependence of heavy-ion transfers. Buttle and Goldfarb have shown that close to the Coulomb Barrier, the optimum Q corresponds to equal distances of approach before and after transfer, as expressed by the relation[34]

$$Q_{opt} = \frac{Z_3 Z_4 - Z_1 Z_2}{Z_1 Z_2} E_{cm} \tag{33}$$

This relation predicts Q_{opt} = -11 MeV for the $Pb^{208}(^{16}O,^{15}N)$ reaction at 104 MeV. At this energy the proper matching conditions should balance Q to the angular momentum, mass and energy transfer. The calculations using no-recoil DWBA[11] and semi-classical[33] theory are shown in fig. 10, where we see both theories predict Q opt $\approx$ -6 MeV. The semiclassical calculation shows that higher ℓ-transfers peak at more negative Q-values, an effect which is not apparent in the DWBA calculations.

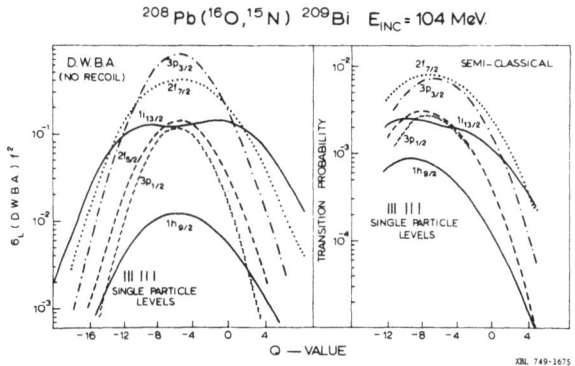

Fig. 10. Theoretical calculation of the Q-dependence for the $^{208}Pb(^{16}O,^{15}N)^{209}Bi$ reaction (a) using DWBA theory and (b) semiclassical theory. In (a) the form factor ($F(\underline{r})$ in eq. 19) was calculated with the binding energy of the state fixed at the value for the actual single particle level, whereas in (b) the binding energy was allowed to change with Q-value.

The Q matching conditions are more transparent if we make an approximate evaluation of the space and time integrals in eq. 28,29. Then we find the probability for a transition[35] from a state $(\ell_1\lambda_1)$ to $(\ell_2\lambda_2)$ is

$$P(\lambda_2\lambda_1) \propto P_o(R) |Y_{\ell_1}^{\lambda_1}(\tfrac{\pi}{2},0) \ Y_{\ell_2}^{\lambda_2}(\tfrac{\pi}{2},0)|^2 \ \text{EXP} [- (\tfrac{R \ \Delta k}{\sigma_1})^2 - (\tfrac{\Delta L}{\sigma_2})^2], \qquad (34)$$

where

$$\Delta k = k_o - \lambda_1/R_1 - \lambda_2/R_2 \qquad (35)$$

$$\Delta L = \lambda_2 - \lambda_1 + \tfrac{1}{2} k_o(R_1 - R_2) + Q_{eff} \ R/\hbar v \qquad (36)$$

$$Q_{eff} = Q - (z_1^f z_2^f - z_1^i z_2^i)e^2/R. \qquad (37)$$

In our previous notation $k_o = \frac{xv}{\hbar}$, the recoil term; $P_o(R)$ depends on the radial wave functions and the distance of closest approach. The widths σ_1 and σ_2 are not precisely known but uncertainty principle estimates suggest $\sigma_1 \approx \pi$ and $\sigma_2 \approx \sqrt{\gamma R}$ where $\gamma^2 = 2x \ \epsilon/\hbar^2$ and ϵ is some average of the binding energy of the transferred particle in the initial and final states. For a large transition probability

Δk, $\Delta L \approx 0$, which are the generalized kinematic conditions replacing $Q_{eff} \approx 0$ for sub-coulomb transfer. The conditions correspond to conservation of linear and angular momentum in the reaction. The total transition probability can be calculated by summing over the final magnetic substates λ_2, and averaging over the initial λ_1, weighted by angular momentum coefficients coupling the nuclear spin and angular momenta.

An application of this simple theory for [11]B induced reactions on [208]Pb at 114 MeV can be found in ref. 36. Here the increased recoil allows equal population of $j_f = \ell_f \pm 1/2$ states at lower energy. This is confirmed by the full semiclassical calculation[33] and agrees fairly well with an exact quantum mechanical calculation[14]. In the subsequent discussion of lighter nuclei we shall make use of the simple version of the semiclassical theory in eq. 34-37.

4. Theory of High Energy Reactions on Lighter Nuclei

When we bombard lighter nuclei with heavy-ions, the effects of recoil increase, and are particularly dramatic for multinucleon transfer. An estimate of the importance is obtained from the condition that the angular momentum carried by the recoil momentum at the target surface be greater than one unit, or for single nucleon transfer,

$$E_L > 20 \frac{A_P}{A_T^{2/3}} \tag{38}$$

where A_P, A_T are the projectile and target mass numbers[37]. For single nucleon transfer induced by ^{12}C on ^{12}C this limit is set at 45 MeV, and for multinucleon transfer at much lower energy. In three nucleon transfer at 10 MeV/nucleon the associated transfer is ≈ 6 $\hbar$. The effect is twofold: it leads to the population high spin states; and it leads to damping of diffraction patterns in the angular distributions. This is an important point as it forces us to look for other signatures of states rather than the customary differential cross sections.

To see this, it is useful to discuss a classical wave optics analogy[37]. The projectile wave scatters from a circular slit of radius R (due to localization). As it scatters it transfers angular momentum, by virtue of the momentum transfer $q = \underline{k}_i - \underline{k}_f$. (see fig. 11). In a classical model this can only change the angular momentum vector in the Z-direction, setting up L complete de Broglie wavelengths around the ring locus. The interference of waves from two characteristic spectral points A,B, will depend not only on the path difference 2d, but also on the intrinsic phase difference. For constructive interference we require:

$$2d = n\lambda \tag{39}$$

if A, B have the same relative phase, i.e. $L = 0,2.....$, but for odd L transfer of $1,3,5 ...$

$$2d = (n + \frac{1}{2})\lambda \tag{40}$$

So odd L transfers will be out of phase with even transfers.

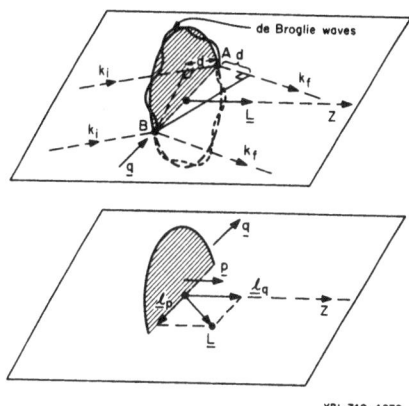

XBL 749-4272

Fig. 11. A semiclassical picture of de Broglie
waves on the ring locus is illustrated at
the top. Waves that emanate from A have a
path difference of 2d over those from B.
The bottom figure illustrates the angular
momentum ℓq arising from momentum transfer
q along the ring locus, and ℓp arising from
the recoil effect. The total angular
momentum transfer is L = ℓq + ℓp.

We have seen that recoil imparts additional momentum p, in the Z-direction,
and an angular momentum ℓp perpendicular to Z, so that then the final L is made up
of ℓq and ℓp, both of which take on a variety of odd and even values as Θ changes.
But only ℓq determines the phase of the diffractive oscillations that emanate from
the ring locus. Thus the effect of ℓp is to allow even and odd ℓq and consequently
in phase and out-of-phase oscillations contribute, damping the oscillatory cross-
section. (compare the selection rules of eq. 26).

A striking case[38] is the $^{12}C(^{14}N, ^{13}N)^{13}C$ reaction shown in fig. 12 in which
the selection rules of eq. 26 allow $\Delta\ell$ = 0 and $\Delta\ell$ = 1 for the ground state transition,
whereas the transition to the s 1/2 state allows only $\Delta\ell$ = 1. As predicted above,
the ground state diffraction pattern is damped, but the excited state is oscillatory
albeit out of phase with the finite-range recoil calculation. It is an open question
at present whether some other processes are competing here since in general the
finite-range code "Lola" has been highly successful in its application to light
nuclei[39]. It is clear however that recoil effects in high energy heavy-ion transfer
reactions make the differential cross sections unpromising signatures in general of
the properties of nuclear states. To confirm this we show in fig. 13, the collected
differential cross-sections for one, two[40] and three nucleon transfers[41] induced by
heavy-ions of approximately 10 MeV/nucleon. The dependence on q, the linear momentum
transfer was explained by a simple recoil theory[42], using harmonic oscillator bound
state wave functions, a Gaussian interaction and distorted waves replaced by amplitude

modulated plane waves, that are then evaluated using the ring locus technique.

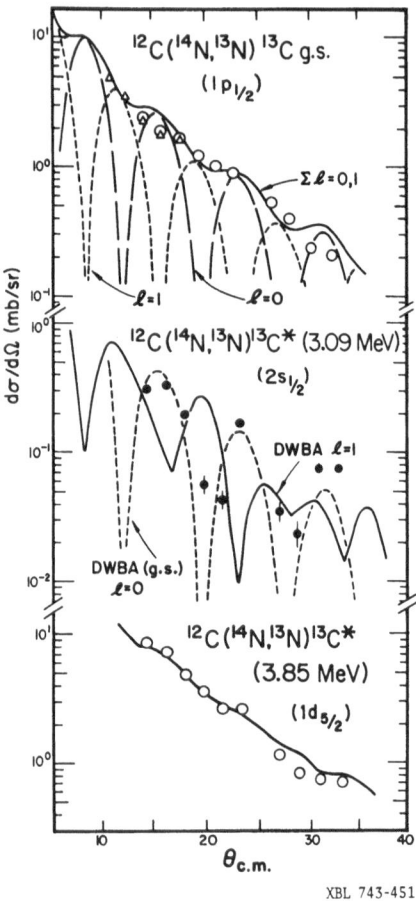

XBL 743-451

Fig. 12. Differential cross sections for the reaction $^{12}C(^{14}N,^{13}N)^{13}C$. The ground state distribution illustrates the damping of diffraction oscillations by the superposition of $\Delta \ell = 0$ and 1 transfers in the recoil DWBA calculation. The $s_{1/2}$ transition is highly oscillatory, since only one $\Delta \ell = 1$ is allowed. As discussed in the text, the phase appears better reproduced by $\Delta \ell = 0$. The DWBA agrees well with the $d_{5/2}$ transition at the bottom.

Then
$$\frac{d\sigma}{d\Omega} \, \alpha \, q^{-3} \, \text{EXP}\left(-p^2 a^2 / 6\right) , \tag{41}$$

if $L > 1$ and $q \gg 1/a$, where a is the range of the bound state wavefunction, and p is the recoil momentum.

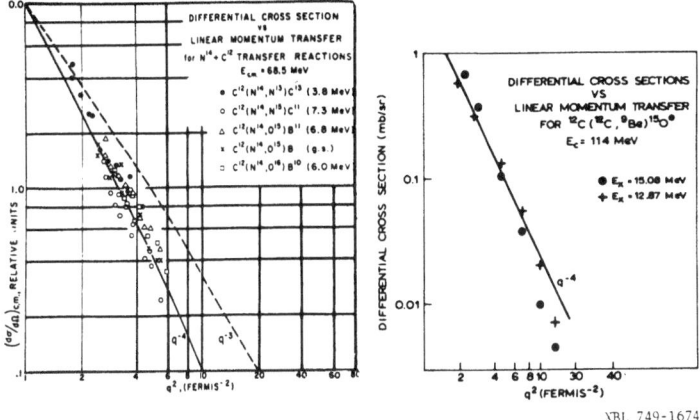

XBL 749-1674

Fig. 13. Collected differential cross sections for one, two and three nucleon transfer reactions on light nuclei induced by heavy-ion beams of approximately 10 MeV/nucleon. The data are plotted against the square of the linear momentum transfer q to remove kinematic differences. The theoretical lines q^{-3} and q^{-4} are based on an approximate recoil DWBA calculation.

The simplicity of the distributions has its compensations however. The forward rising of the cross-section suggests that the transfers occur during a grazing collision of the cores. Because of the high incident energy we may also suppose that the motion of the incident projectile is not much perturbed by the collision, so that the orbit is a straight line, with an impact parameter equal to the sum of the radii. We can then use the simple theory outlined in eq. 34-37, and compare differential cross sections at forward angles, as shown by the following considerations.

If P is the transfer transition probability, then as in the theory of Coulomb excitation we write, for cases where $\eta \gg 1$,

$$(d\sigma/d\Omega) = (d\sigma/d\Omega)_{el} \times P \tag{42}$$

Also we have,

$$\sigma = 2\pi \int \left(\frac{d\sigma}{d\Omega}\right) \sin \Theta d\Theta = \frac{2\pi}{k^2} \int P(L) dL$$

where L is the angular momentum of relative motion. The last equation is valid if the grazing angular momentum, $kR \gg 1$ and if ΔL, the contributing angular momentum range is $\gg 1$. This equation does not require $\eta \gg 1$. The main contribution to the first integral comes from Θ near the maximum of $\sin \Theta \frac{d\sigma}{d\Omega}$. If data on the monotonic distributions are taken near this maximum, then

$$\sigma \propto \frac{d\sigma}{d\Omega} \tag{43}$$

since the angular distributions are similar for different final states. Similarly the main contribution to the second integral comes from L near the critical value, L_c where $P(L_c)$ has a maximum. Thus we expect, as in eq. 42, that

$$\frac{d\sigma}{d\Omega} \propto P(L_c) \qquad (44)$$

will hold approximately at forward angles, even though η is not large compared to unity.

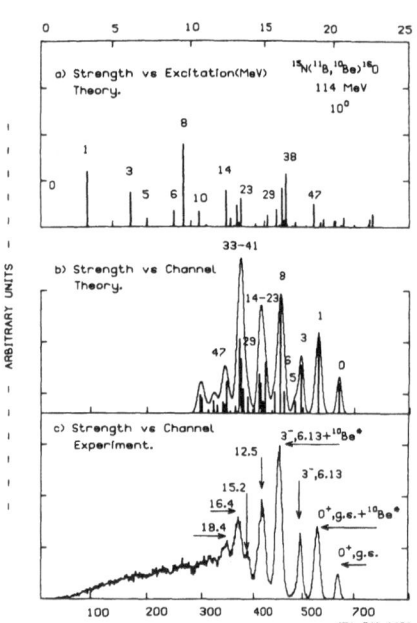

Fig. 14. (c) shows the experimental spectrum for the $^{15}N(^{11}B,^{10}Be)^{16}O$ reaction at 114 MeV; which selectively excites particle-hole-strength. (a) shows the calculated stength using shell-model spectroscopic amplitudes and the simpli-fied semiclassical theory of eq. 34. In (b) the theoretical predictions are folded with the experimental resolution to produce a theoretical spectrum.

An application of eq. 44 combined with shell-model spectroscopic amplitudes for a survey[43] of the reaction $^{15}N(^{11}B,^{10}Be)$ ^{16}O at 114 MeV is shown in fig. 14. This reaction is expected to excite preferentially the particle-hole strength in ^{16}O, formed by coupling a p 1/2, d 5/2, s 1/2 or d 3/2 particle to the ^{15}N core. Examples are the g.s., the 6.13, 3^- state, the 8.87, 2^-, the T = 1 quartet 1^-, 0^-, 2^-, 3^- analogues of ^{16}N, centered at 12.5 MeV, the 2_3^- at 15.2 MeV etc. In addition ^{10}Be can be formed in its low lying particle stable states, but particularly the 3.37 MeV 2^+ state. The theoretical spectrum was calculated as it would appear in an experimental spectrum, each peak having a Gaussian shape of area proportional to the theoretical strength and a width equal to the average experimen-tal value. Over 50 states were included in the calculation which automatically genera-ted the theoretical spectra. A satisfactory representation of the data is obtained.

If this method can be used for spectroscopy even within a precision of a factor of 2 or 3, it furnishes us with a powerful technique for making wide surveys in a way impossible to imagine with more elaborate theories.

More exciting of course, is the extension of this programme to two, three and four nucleon transfers. The Oxford group have shown[41,44] how these high energy heavy-ion reactions can selectively populate simple "single particle" states of ^3He or ^4He, due to the apparent preference for transfer of spatially localized clusters.[45] An example is shown in fig. 15 where it is suggested that the strongly excited states

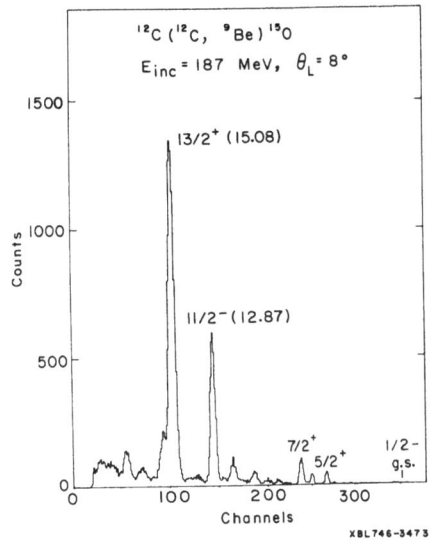

Fig. 15. Energy spectra for the ^{12}C $(^{12}C,^9Be)^{15}O$ reaction at 187 MeV showing the selective excitation of postulated high spin states $13/2^+$ and $11/2^-$.

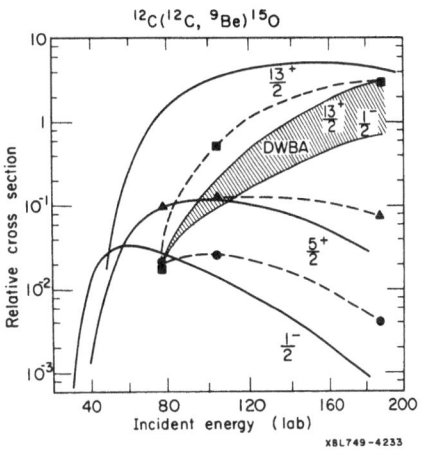

Fig. 16. The energy variation of the cross section for states excited in the $^{12}C(^{12}C,^9Be)^{15}O$ reaction. The solid curves are the predictions of semiclassical theory (no spectroscopic factors included) For comparison the no-recoil DWBA predictions are shown for the $1/2^-$ and $13/2^+$ states, in which there is no strong enhancement of the high spin states.

have $J^\pi = 13/2^+$ and $11/2^-$. A recent calculation using a folded potential model[46] for ^{3}He outside the ^{12}C core gives good agreement with the experimental locations for L = 6 and L = 5, ^{3}He orbitals. Table 4 gives a comparison between experiment and theory, using the methods described above, with SU_3 cluster spectroscopic amplitudes.[41] (see also the previous lecture by B. Buck).

Table IV. Comparison of experimental cross sections for the $^{12}C(^{12}C,^9Be)^{15}O$ reaction at 114 MeV with theoretical cross sections evaluated using a semiclassical theory for the reaction and three nucleon cluster spectroscopic amplitudes.

STATE	σ THEORY	σEXPT.
G.S. $1/2^-$	0.01	≈ 0
5.24, $5/2^+$	0.10	0.12
6.79, $3/2^+$	0.003	≈ 0
7.28, $7/2^+$	0.33	0.28
$(9.08)^a$, $9/2^+$	0.76	? STATES
$(10.80)^a$, $7/2^+$	0.29	? NOT IDENTIFIED
12.87^c, $11/2^-$	1.00^b	1.00
15.08, $13/2^+$	2.16	1.16

(continued)

TABLE IV (continued)

---

[a]Theoretical excitation energy

[b]Data normalized to unity for 12.87 MeV state

[c]These states were identified as $11/2^-$ and $13/2^+$ by comparison with the theory.

As new states of this type are located in other regions of the periodic table, it will be important to have methods of inferring their properties. The similarity of the distributions to different states can be turned to further advantage by studying the energy variation of the reaction at one, or a few, forward angles. The experimental variation for some states in the $^{12}C(^{12}C,^{9}Be)^{15}O$ reactions are shown in fig. 16, from recent work at Berkeley, together with the predictions of the semiclassical theory, which gives some agreement with the high energy trend, and confirms the high spin assignment of the $13/2^+$ state. For comparison we show the no-recoil DWBA predictions, using optical model parameters from a fit to the elastic scattering data at 104 MeV. The prediction was arbitrarily normalized to the $13/2^+$ state, in order to emphasize the small enhancement over the low spin $1/2^+$ state. Better agreement with the detailed shapes could be obtained using variations of the optical potentials, but the enhancement of 10^3 for $13/2^+$ compared to $1/2^-$ could only be reproduced in the DWBA calculations by including recoil. The enhancement of high spin states due to the recoil effect was demonstrated by Dodd and Greider long ago[47]. The advantages of the semiclassical theory is its ability to make wide surveys rapidly, without any unknown parameters to vary.

Many of the interesting states discovered, and awaiting discovery, in heavy-ion reactions are such high-spin states, often of small binding energy or even unbound. In the limit of small final binding energy, defined by

$$\chi_2 R_1 \ll 1 , \tag{45}$$

where $\quad \chi_2 = \sqrt{\dfrac{2m\varepsilon_2}{\hbar^2}}$, ε_2 is the binding energy of m in the final state

and R_1 is the radius of the projectile, Nagarajan has shown[48] that the reaction proceeds almost entirely through the recoil momentum transfer, and that the six-dimensional integration of the DWBA in eq. 17 separates into:

$$T \propto \int \underline{dr}_1 \, e^{-i\underline{k}_R \cdot \underline{r}_1} \, V(\underline{r}_1) \, U_{\ell_1}(\underline{r}_1) \int \underline{dr} \, e^{i\underline{q} \cdot \underline{r}} \, h_{\ell_2}^*(i\chi_2 r) \, Y_{\ell_2}^{\lambda_2}(\underline{r}) \times \Theta(\underline{r}) \tag{46}$$

Here $U_{\ell_1}(r_1)$ is the initial radial wave function, and the final weakly bound wave function is approximated by a Hankel function. Further

$$\underline{q} = \underline{k}_i - \underline{k}_f \tag{47}$$

$$\underline{k}_R = \frac{x}{B} \underline{k}_f + \frac{x}{a} \underline{k}_i \tag{48}$$

and $\theta(r)$ is an amplitude modulation on the plane waves to simulate distorted waves. Since eq. 48 essentially defines the recoil momentum of eq. 22, eq. 46 shows that the final result is a product of a zero-range DWBA integral with a radial integral correcting for recoil effects. The correspondence of eq. 46 with the form of the simple semiclassical theory of eq. 34 is also transparent. The spectroscopy of these new correlations in nuclear motion is likely to become a promising area of investigation with the new high energy, high resolution accelerators. The approximate forms of the exact theories outlined here will enable us to make surveys to see where the interesting regions of investigation lie.

5. Multistep Processes

In this final section, I wish to discuss some aspects of multistep processes in heavy-ion reactions at high energy. In contrast to the direct transfer reactions, so far the analyses have been done only with the coupled channels Born approximation (CCBA). However the results are sufficiently interesting and suggestive of what might be done in the future with heavier projectiles, to make us hope that soon a coupled channels semiclassical theory will be available to enable us to make the same wide surveys that have been possible for one-step direct reactions at high energies.

It has taken a long time to establish the presence of multistep processes in heavy-ion reactions, although they have been well studied in light-ion reactions. Such a case is two neutron pick-up on ^{144}Nd leading to two types of 2^+ states in ^{142}Nd. A spectrum - actually for the heavy-ion case we are about to discuss - is shown in fig. 17. The 2_2^+ state is a two neutron hole state in the N = 82 closed shell, i.e. a removal type quadrupole vibrational state, excited strongly in two neutron pick-up. The 2_1^+ state on the other hand, is dominantly a particle-hole quadrupole vibration which is forbidden in the direct pick-up. This state can be populated via the indirect routes, involving inelastic scattering shown in fig. 18. The CCBA and DWBA calculations[49] are compared for the (p,t) data in fig. 19, in which the CCBA calculation gives better agreement in magnitude for the 2_1^+ state (the forward angle phase is also better reproduced). Apart from a change in the magnitude, the effect of multistep processes is not very dramatic in this case. The same reaction induced by heavy-ions[50] is illustrated in fig. 20. Here the $0_1^+, 0_2^+$, and 2_2^+ all have the well-known, bell-shaped distribution, which is the characteristic of direct reactions with heavy-ions. The 2_1^+ state has no such "semi-classical" maximum, but is flat and rises at forward angles. Tamura and Low have suggested[51] how this forward rise, due to indirect processes, can be understood on a semiclassical trajectory picture. In fig. 21, we see that a nucleus deviated from its spherial shape (by for example inelastic scattering) for a given impact parameter makes a closer collision and thereby is deflected to more forward angles by the nuclear force. These results have been reproduced by CCBA calculations[50]. This example is a good "control" case for demonstrating indirect effects, because the direct 2_2^+ distribution must be reproduced simultaneously.

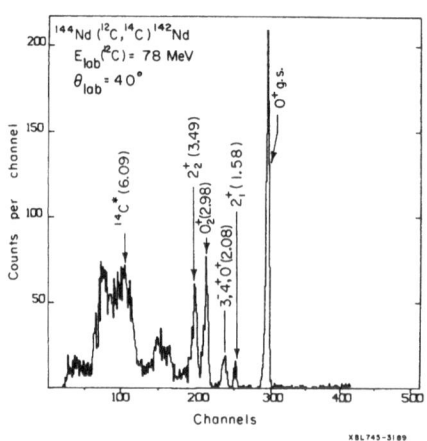

Fig. 17. Spectrum for the $^{144}Nd(^{12}C,^{14}C)^{142}Nd$ reaction at 78 MeV, showing the population of two different types of 2^+ states; the 2_1^+ state is a particle-hole vibration, forbidden in direct pick-up and the 2_2^+ which is more strongly excited is a two-neutron hole state.

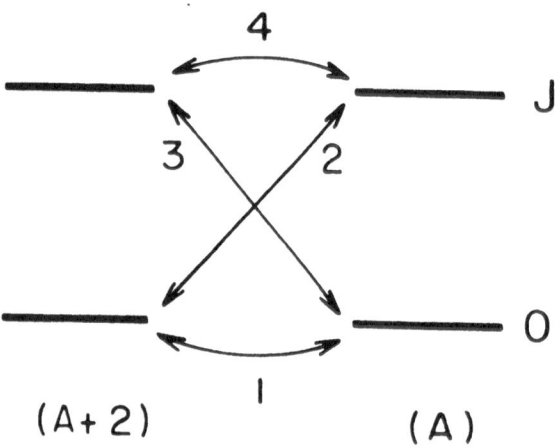

XBL 739-1244

Fig. 18. Illustration of direct and indirect routes in transfer reactions. In two neutron pick-up to the state J route 2 is direct and in stripping 3 is direct. The opposite sign of 2 and 3 can lead to opposite interference characteristics with indirect routes of which 4 and 1 are branches.

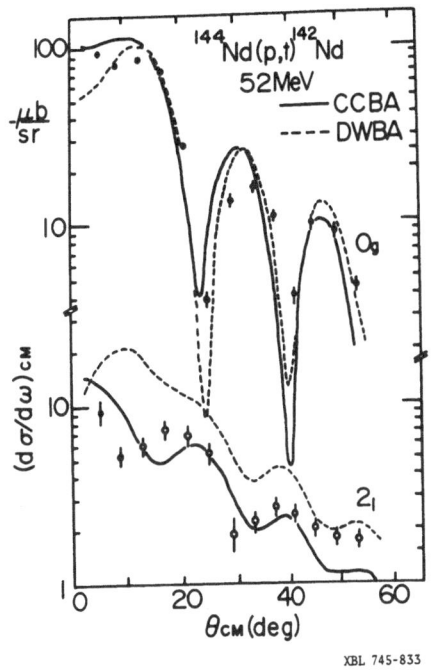

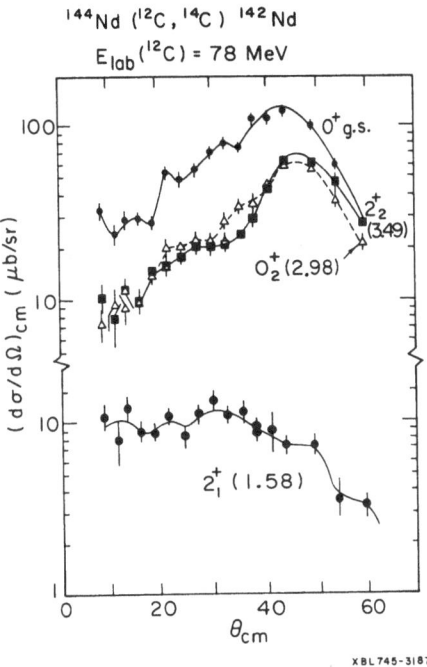

Fig. 19. Comparison of DWBA and CCBA calculations for the ^{144}Nd(p,t)^{142}Nd reaction. The CCBA gives better agreement in magnitude and phase for the weakly excited 2_1^+ state.

Fig. 20. Differential cross sections for the ^{144}Nd(^{12}C,^{14}C)^{142}Nd reaction at 78 MeV. The $0_{g.s.}^+$,0_2^+ and 2_2^+ distributions all exhibit the semiclassical maximum characteristic of a direct reaction, whereas the weakly excited 2_1^+ has no clear peak and rises at forward angles.

Using this signature of the indirect transition, we can study an interesting effect which has not, to my knowledge, been studied in light-ion reactions. The interference between direct and indirect modes to the lowest 2^+ vibrational states in the Sn isotopes is predicted to be constructive in the pick-up (^{16}O,^{18}O) reaction and

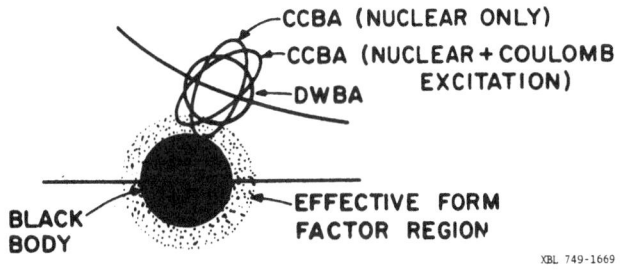

Fig. 21. Semiclassical interpretation of forward rising differential cross sections in indirect transitions, due to the closer interaction of nuclei deformed from their spherial shape.

destructive in the $(^{18}O, ^{16}O)$ stripping reaction[52]. Data for these reactions[53] at the same center of mass energy are shown in fig. 22. The cross sections for the ground states are identical as required by time-reversal invariance and exhibit the classical maximum, as does the cross-section for the 2^+ state in pick-up. The 2^+ in stripping however is flattened in the same way as the data on niodymium. The theoretical calculations are shown in fig. 23. Here the interference is sufficiently strong to produce a dip which does not appear in the data, but the magnitude of this effect is sensitive to deformations, nuclear structure and optical parameters. The figure also shows the result of neglecting higher order effects in the calculation, when <u>all</u> the distributions assume the same form. In a recent study of the $^{186}W(^{12}C, ^{14}C)^{184}\underline{W}$ reaction at 70 MeV, the interference effects <u>do</u> produce a dip in the differential cross-section[54]. Finally it is interesting to note that for the Sn case, the effect is one of higher energies, since at 72 MeV, it disappears[55]. Heavy-ion reactions are rich in possibilities for the study of interference effects, and clearly we are seeing only the beginning of an area of research which may yield spectacular results in the future, as heavier projectiles are involved.

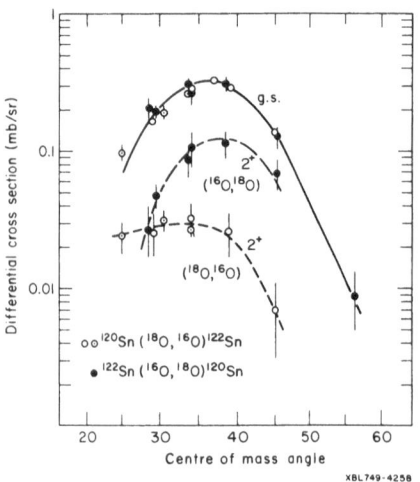

Fig. 22. Differential cross sections for the reaction $^{120}Sn(^{18}O, ^{16}O)^{122}Sn$ and $^{122}Sn(^{16}O, ^{18}O)^{120}Sn$ at the same center of mass energies of 89 MeV. Destructive interference between direct and indirect routes to the 2^+ state in stripping leads to an anomalous distribution.

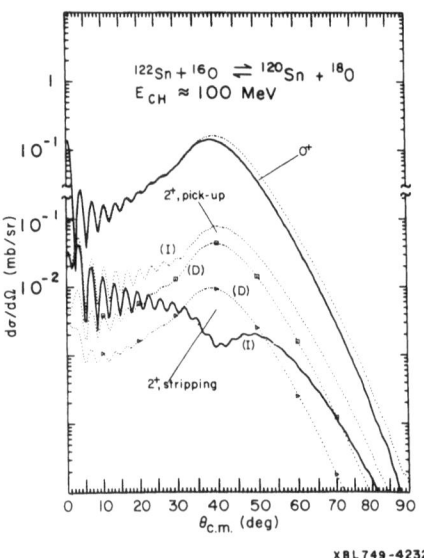

Fig. 23. Theoretical predictions for the $^{120}Sn(^{18}O, ^{16}O)^{122}Sn$ and $^{122}Sn(^{16}O, ^{18}O)^{120}Sn$ reactions. For the 2^+ states both CCBA calculation (I) and DWBA calculation (D) are compared.

6. Conclusion

In this talk, I have presented some of the approaches to the analysis of high energy transfer reactions with heavy-ions, concentrating on concrete achievements at the present time rather than on the possiblities in sight on the horizon. First we discussed the use of simple semiclassical concepts in the physical interpretation of various features of differential cross sections. Reliance on full quantal theories in this area often lead to what Glendenning has called[2] the "experimental approach" in which a parameter, of the optical potential for example, is varied and its effect catalogued and noted. A more formal semiclassical theory was compared with the exact and approximate DWBA theories for reactions where the semiclassical assumptions were well fullfilled. The power of the method in making rapid surveys of effects such as energy and Q-dependence was demonstrated. In this way one can determine what are the interesting areas to investigate experimentally. This was followed by a discussion of transfers on lighter nuclei where heavy-ion reactions have been used to give us information on new states and correlations in nuclear motion. The final discussion of the indirect effects in multistep processes at high energy gives us a glimpse of new areas of research unique to heavy ions in which semiclassical theories are likely to play an increasingly important role.

This discussion has left untouched the many technical developments in progress which are transforming the semiclassical approach from a crude approximation to a refined theory. These methods calculate a semiclassical scattering amplitude which gives a proper account of interference and diffractive phenomena in heavy-ion physics[31]. These approaches can go beyond the geometric optics limit towards wave optics allowing for complex trajectories which describe quantal phenomena in terms of classical quanities[56,57]. In this spirit the semiclassical approach becomes more and more useful at higher incident energies. In the high energy limit the Glauber or eikonal approaches[58] will become possible, and there have already been some successes in this direction[59,60,61].

Since we are now entering an era of higher energies and higher mass projectiles, we are likely to encounter situations with an enormous number of partial waves, hundreds or even thousands. Here the semiclassical approach may become our only hope of making theoretical progress. Or more aesthetically we shall probably have to combine semi-classical and quantal theories, using quantum mechanics to calculate the S matrix up to the region of critical angular momentum where the nuclear field causes rapid changes in the deflection function, with classical methods in the wider region where S varies smoothly[2].

I wish to thank the experimental and theoretical heavy-ion groups at Berkeley and Oxford for supplying me with many ideas and data.

References

*Work performed under the auspices of the U. S. Atomic Energy Commission.
1) K. W. Ford and J. A. Wheeler, Ann. Phys. (N.Y.) 7 (1959) 259
2) N. K. Glendenning, Proceedings of the Conference on the Interactions of Complex Nuclei (Nashville, 1974) Vol 2, to be published.
3) V. M. Strutinskii, Sov. Phys. JETP 19 (1964) 1401
4) L. D. Landau and E. M. Lifshitz, Quantum Mechanics. (Pergamon Press Ltd, London, 1958)
5) J. Grabowski, B. N. Kalinkin and N. F. Markova, Nuc. Phys. 65 (1965) 294
6) F. A. Gareev, J. Grabowski and B. N. Kalinkin Sov. J. of Nuc. Phys. 5 (1967) 85
7) S. Kahana, P. D. Bond and C. Chasman, Phys. Lett. 50B (1974) 199
8) V. M. Strutinskii, Phys. Lett. 44B (1973) 245
9) P. J. Siemens and F. D. Becchetti, Phys. Lett. 42B (1972) 389
10) J. S. Larsen, J. L. C. Ford, R. M. Gaedke, K. S. Toth, J. B. Ball and R. L. Hahn, Phys. Lett. 42B (1972) 205
11) D. G. Kovar, Symposium on Heavy-Ion Transfer Reactions (Argonne, 1973) p. 59
12) W. von Oertzen, Symposium on Heavy-Ion Transfer Reactions (Argonne, 1973) p. 675
13) K. S. Low and T. Tamura, Phys. Lett. 48B (1974) 285
14) J. L. C. Ford, K. S. Toth, G. R. Satchler, R. M. Devries, R. M. Gaedke, P. J. Riley and S. T. Thornton, Phys Rev. (C), to be published.
15) F. D. Becchetti, D. G. Kovar, B. G. Harvey, J. Mahoney, B. Mayer and F. G. Pühlhofer Phys. Rev. C6 (1972) 2215
16) G. C. Morrison, J. de. Physique 32, 11-12 (1971)
17) R. J. Ascuitto and N. K. Glendenning, Phys. Lett. 48B (1974) 6
18) W. A. Friedman, K. W. McVoy and G. W. T. Shuy, Phys. Rev. Lett. 33 (1974) 308
19) D. G. Kovar, F. D. Becchetti, B. G. Harvey, D. L. Hendrie, H. Homeyer, J. Mahoney and W. von Oertzen, Nuclear Chemistry Annual Report (1973) LBL-2366, p 94
20) J. S. Blair, R. M. Devries, K. G. Nair, A. J. Baltz and W. Reisdorf, to be published.
21) N. K. Glendenning and M. A. Nagarajan, LBL-2378 Preprint (1974) and to be published.
22) G. R. Satchler, Symposium on Heavy-Ion Transfer Reactions (Argonne, 1973) p 145 and references therein.
23) R. A. Broglia and A. Winther, Physics Reports 4C (1972) 153 and refs. therein.
24) D. M. Brink, P. N. Hudson and M. Pixton, to be published.
25) M. A. Nagarajan, Nuc. Phys. A196 (1972) 32; A209 (1973) 485
26) A. J. Baltz and S. Kahana, Phys. Rev. C9 (1974) 2243
27) E. Elbaz, J. Meyer and R. S. Nahabetian, Nuc. Phys. A205 (1973) 299
28) B. F. Bayman and D. H. Feng, Nuc. Phys. A205 (1973) 513
29) L. A. Charlton, Phys. Rev. Lett. 31 (1973) 116; Symposium on Heavy-Ion Transfer Reactions (Argonne, 1973) p 161
30) T. Tamura and K. S. Low, Phys. Rev. Lett. 31 (1973) 1356
31) R. A. Broglia, S. Landowne, R. A. Malfliet, V. Rostokin and A. Winther, Phys. Reports C11 (1974) 1
32) D. G. Kovar, private communication, 1974
33) P. N. Hudson, D. Phil Thesis (Oxford, 1973)
34) P. J. A. Buttle and L. J. B. Goldfarb, Nuc. Phys. A115 (1968) 461
35) D. M. Brink, Phys. Lett. 40B (1972) 37
36) N. Anyas-Weiss, J. Becker, T. A. Belote, J. C. Cornell, P. S. Fisher, P. N. Hudson, A. Menchaca-Rocha, A. D. Panagiotou and D. K. Scott, Phys. Lett. 45B (1973) 231
37) K. R. Greider in Nuclear Reactions Induced by Heavy-Ions, ed. by R. Bock and W. R. Hering (North Holland, Amsterdam, 1970) p 217
38) R. M. Devries, M. S. Zisman, J. G. Cramer, K-L Liu, F. D. Becchetti, B. G. Harvey, H. Homeyer, D. G. Kovar and W. von Oertzen, Phys Rev. Lett. 32 (1974) 680

39) R. M. Devries, Phys. Rev. C8 (1973) 951

40) J. Birnbaum, J. C. Overley and D. A. Bromley, Phys. Rev. 157 (1967) 787

41) N. Anyas-Weiss, J. C. Cornell, P. S. Fisher, P. N. Hudson, A. Menchaca-Rocha, D. J. Millener, A. D. Panagiotou, D. K. Scott, D. Strottman, D. M. Brink, B. Buck, P. J. Ellis and T. Engeland, Physics Reports C, to be published.

42) L. R. Dodd and K. R. Greider, Phys. Rev. 180 (1968) 1187

43) A. Menchaca-Rocha, D. Phil. Thesis (Oxford, 1974) and A. Menchaca-Rocha et al.,to be published.

44) D. K. Scott, P. N. Hudson, P. S. Fisher, N. Anyas-Weiss, C. U. Cardinal, A. D. Panagiotou, P. J. Ellis and B. Buck, Phys. Rev. Lett. 28 (1972) 1659

45) D. Kurath, Lectures at International School of Physics, Enrico Fermi, Varenna 1974, to be published

46) B. Buck, C. B. Dover and J. P. Vary, to be published.

47) L. R. Dodd and K. R. Greider, Phys. Rev. Lett. 14 (1965) 959

48) M. A. Nagarajan, LBL-2918 (Preprint 1974) and to be published.

49) K. Yagi, K. Sato and Y. Aoki, T. Udagawa and T. Tamura, Phys. Rev. Lett. 29 (1972)

50) K. Yagi, D. L. Hendrie, L. Kraus, C. F. Maguire, J. Mahoney, D. K. Scott, Y. Terrien, T. Udagawa, K. S. Low and T. Tamura, to be published.

51) K. S. Low and T. Tamura, Symposium on Heavy-Ion Transfer Reactions (Argonne, 1973) p 655

52) R. J. Ascuitto and N. K. Glendenning Phys. Lett. 45B (1973) 85

53) D. K. Scott, B. G. Harvey, D. L. Hendrie, U. Jahnke, L. Kraus, C. F. Maguire, J. Mahoney, Y. Terrien, K. Yagi and N. K. Glendenning, to be published.

54) K. A. Erb, private communication, 1974

55) H. G. Bohlen and H. J. Körner, private communication, 1974

56) J. Knoll and R. Schaeffer, Lectures at the International Center for Theoretical Physics (Trieste, 1973) to be published.

57) R. Malfliet, this Symposium.

58) R. J. Glauber, Lectures in Theoretical Physics, Boulder, Vol 1 (Interscience Publishers, N. Y. 1959) p 315.

59) A. Dar and Z. Kirzon, Phys. Lett. 37B (1971) 166

60) R. da Silveira J. Galin and C. Ngo, Nuc. Phys. A159 (1970) 481

61) B. Buck, private communication, 1974

FRICTION MODEL FOR HEAVY-ION COLLISIONS AND ITS APPLICATION TO HEAVY ION FUSION AND DEEP INELASTIC REACTIONS

D. H. E. Gross, H. Kalinowski and J. N. De[+]

Hahn-Meitner-Institut für Kernforschung Berlin GmbH

Bereich Kern- und Strahlenphysik

Berlin-West, Germany

Abstract

Heavy ion collisions leading to fusion or deep inelastic direct reactions are described by a simple classical model. The ions are assumed to move along classical trajectories under the influence of frictional forces. With the use of only two parameters fixed simultaneously for all systems the systematics of the existing data on fusion cross sections and deep inelastic collisions could be reproduced. Especially the model gives a simple explanation of the focussing effect observed in Kr-induced deep inelastic collisions. The amount of energy loss observed in these collisions is qualitatively reproduced but quantitatively still differs considerably.

1) Introduction

For collisions between heavier ions, a large part of the reaction cross-section goes into the deep inelastic collision channels. The individuality (its mass and charge) of the projectile and target are more or less conserved in these reactions, but a large amount of energy is brought into internal excitations. These collisions are fast and though the two colliding nuclei do not have time to equilibrate all their degrees of freedom, yet excitations to energies of more than 100 MeV are possible. At such high excitations the level densities are so large that only a statistical treatment seems to be promising. In describing the collision dynamics, we have to take into account the non-equilibrium processes like diffusion of nucleons from one ion into the other, dissipation of energy from the relative motion into the intrinsic degrees of freedom etc. Since the total degrees of freedom are rather innumerable, suppression of many of them necessitates the introduction of models with certain macroscopic parameters for the de-

---

[+] On leave of absence from Saha Institute of Nuclear Physics, Calcutta/India.

scription of the colliding system. To describe the dissipation of energy, we exploit the classical concept of friction.

In heavy ion collisions, the de Broglie wavelengths of relative motion are much smaller than the geometrical sizes of the system. An average over many unresolved degrees of freedom is made in the experiments also. Therefore, a classical treatment of the collision dynamics may not be unreasonable. Moreover, a test of our simple-minded classical picture of the process seems to be worthwhile.

The model we want to propose here was first introduced in 1973 by Gross, Kalinowski and Beck[1] and is up to now the only one that synthesizes fusion cross-section systematics and the general trends of deep inelastic collision data in a quantitative way. This differs essentially from the models of Bass[2] and Lefort[3] which treat the compound formation process only.

Let us sketch the basic idea first.
In fig. (1.1) we show the potential including the centrifugal term for $^{40}Ar + ^{232}Th$. Indicated schematically are the trajectories of radial motion for different initial angular momenta. (It should be said clearly that these trajectories have no relation with the realistic calculations, especially the realistic friction will not act at such large distances).
If we follow a classical trajectory with a large angular momentum $L \gg L_{CF}$, it will keep always outside of the effective potential barrier made by the sum of the nuclear, Coulomb and centrifugal potentials. For smaller angular momenta L but still $L > L_{CF}$ it will go through the region where friction is working. Thus a particle moving along such an orbit will lose energy (more for smaller L than for larger L) and will be counted as a very inelastic direct collision.

There is a lower bound L_{CF} of L-values below which the trajectories do not come out of the interaction region any more. Particles moving along these will lose some energy before they reach the effective potential barrier, but still they have enough energy to surmount it and then come into the attractive region of the effective potential (potential valley). Due to their increasing radial velocity there it will lose even more energy, too much actually to come out of the barrier again. These particles are captured and fuse.

Particles moving along trajectories with turning points close to the top of the effective barrier will remain some time in this region of small radial forces and thus spiral round. In some cases like the collision of $^{40}Ar + ^{232}Th$ at 400 MeV the friction stops many orbits with quite different angular momenta just in that critical region. In these cases there is a large cross-section for collisions leading to negative scattering angles. Wilczynsky[4] had guessed such a behaviour from the experimental data as shown in figs. (1.2a and b), the famous thumb print of Wilczynski.

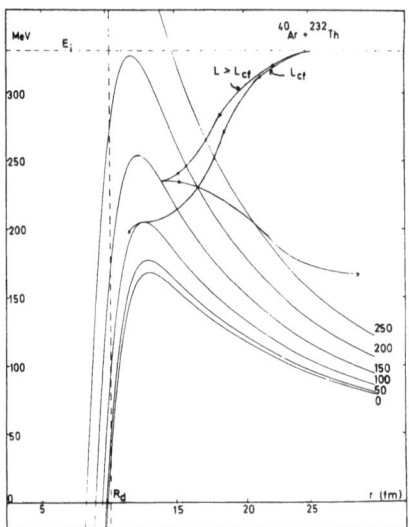

Fig. 1.1: Effective potential.

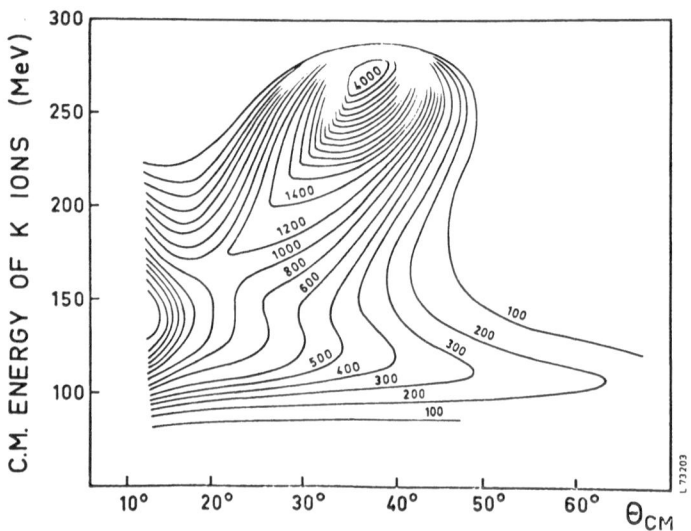

Fig. 1.2a: ^{40}Ar+^{232}Th → K, E = 388 MeV, counts v.s. E_{CM}^f, θ_{CM}.

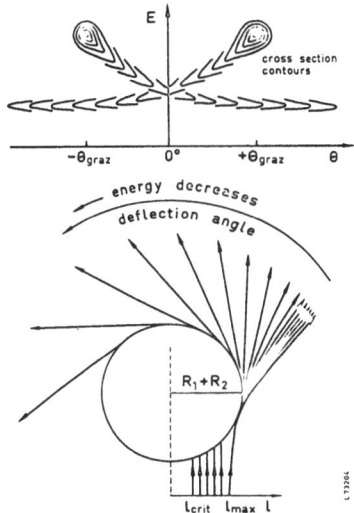

Fig. 1.2b: Schematic interpretation of the results in fig. 1.2a.

It is impressive to compare these drawings with the results of our calculation (fig. 1.3)!

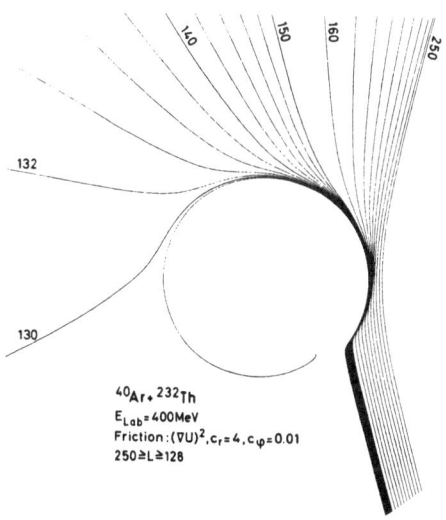

$^{40}Ar + ^{232}Th$
$E_{Lab} = 400 MeV$
Friction: $(\nabla U)^2, c_r = 4, c_\varphi = 0.01$
$250 \geq L \geq 128$

Fig. 1.3: Theoretical trajectories for the same system.

2) Microscopic Aspects of Nuclear Friction

The microscopic understanding of nuclear friction is at the moment
quite rudimentary. There are mainly two physically different theories
for the dissipation effects in ion-ion collisions: The Landau-Zener
single nucleon jumping model by Glas and Mosel[5] and the perturbation
formula given by Beck and Gross[6].

In the Landau-Zener model[5] it is assumed that the diabatic
single nucleon levels given for example by a two center shell-model
calculation are crossing at some distance between the two ions. Each
crossing point is assumed to be sufficiently far away from the others
and the perturbation is assumed to be small compared to the energy-
differences of neighbouring levels. Under these assumptions the transition probability (jump probability) is given by the Landau-Zener
theory. The physical picture is then that as the two ions approach each
other the nucleons will jump to higher and higher states and thus energy will be transferred from the motion of the two ions into the intrinsic excitations leading to a frictional damping of the relative motion.
Glas and Mosel claim that there is a characteristic critical distance
$R \approx 1 \cdot \sum A^{1/3}$ inside of which many levels cross consequently leading to
strong friction and outside of which there is more or less elastic
(frictionless) motion. Thus this model may serve as a foundation of
the concept of a critical distance, which was introduced quite successfully by the Orsay group around M. Lefort[3] to explain the systematics
of fusion cross-sections.

The Landau-Zener model was already criticised long time ago by
Wilets[7]. We will give a quite similar argument here:
The distance between neighbouring crossings is in the case of $^{16}O + ^{16}O$
about (or less than) 1 fm. In order to distinguish between these crossings the energy of the incoming beam must have an uncertainty of
$\Delta E \gtrsim v \hbar / \Delta R$, where v is the mean velocity of the relative motion. But
the crossings must not only be resolved spatially but also energeti-
cally. Assuming the distance between neighbouring levels to be
$\Delta E_\ell \lesssim$ 10 MeV we find from the condition $\Delta E_\ell > \Delta E$ that the energy
of relative motion of the two ion must be: $T \ll \frac{1}{2} m \left(\Delta R \cdot \Delta E_\ell / \hbar \right)^2 \lesssim$ 10 MeV,
which may even be much smaller because we certainly overestimated ΔR
and ΔE_ℓ. Therefore for energies larger than this, as we usually have
in the experiments of deep inelastic heavy ion collisions, the assumption of independent two level crossings and jumpings is questionable.

The perturbation theory of nuclear friction[6] gives a development of the classical equation of motion including frictional forces
from the full many body quantum scattering-theory of two colliding

heavy ions, which is rigorous up to second order in the nondiagonal interaction of the two nuclei. Of course, the validity of perturbation theory for such strongly interacting processes as deep inelastic heavy ion collisions is quite doubtful. However, the use of a second order expression for the friction tensor in a classical equation of motion is not just second order perturbation theory for the whole collision process, because the friction is infinitely iterated in calculating the frictional path.

Let us sketch here the main stream of this theory.

We start with the full Hamiltonian H of the two colliding ions, which we split into a relative motion, an intrinsic and an interaction part:

$$H = H_0^{rel} + H_0^{intr.} + V \tag{2.1}$$

$$H_0^{rel} = -\frac{\hbar^2}{2M} \Delta_R + U(R) \tag{2.2}$$

Then we construct the formal solution of Liouville's equation for the time dependent density matrix $\varsigma(t)$ of the total system:

$$i\hbar \frac{\partial}{\partial t} \varsigma(t) = [H, \varsigma(t)] \tag{2.3}$$

with the boundary condition:

$$\lim_{t \to -\infty} \| \varsigma(t) - \varsigma_0(t) \| = 0 , \quad \varsigma_0(t) = |P_i(t), 0\rangle \langle P_i(t), 0| \tag{2.4}$$

Here $|P_i(t)\rangle$ is the intrinsic noninteracting wave packet of relative motion and $|0\rangle$ the ground state of the separated nuclei.

By calculating the time dependent expectation values of the relative distance R we get a classical equation of motion corresponding to (2.3) by Ehrenfest's theorem:

$$M\langle \ddot{R}_v \rangle_t + \langle \frac{\partial}{\partial R_v} U(R) \rangle_t + \langle \frac{\partial}{\partial R_v} V \rangle_t = 0 \tag{2.5}$$

Here the third term $\langle \frac{\partial}{\partial R_v} V \rangle_t$ is the (negative) force coming from the coupling of the relative motion to the intrinsic degrees of freedom. This term can be calculated to second order of V and is found to be:

$$\langle \frac{\partial}{\partial R_v} V \rangle_t = -\langle | \nabla_v \{ V \frac{P}{H_0 - E_i} V \} | \rangle_t$$

$$+ 2\pi\hbar \sum_\mu \langle \frac{\partial}{\partial R_v} V \frac{P}{H_0^{intr} - E_0^{intr}} \delta(E_i - H_0) \frac{\partial}{\partial R_\mu} V \rangle_t^s v_\mu(t) \tag{2.6}$$

with $\langle AB \rangle^s = \frac{1}{2} \{ \langle AB \rangle + \langle BA \rangle \}$ (2.7)

The first term is the usual second order correction to the conservative potential $U(R)$, whereas the second term is of the form $\sum_\mu c_{\nu\mu}(t)\, v_\mu(t)$, a frictional force proportional to the velocity $v_\nu(t)$.

Simple estimates[8] show that the main origin of the friction is the excitation of 1p-1h states in one nucleus by the moving edge ($\frac{\partial}{\partial R_\nu} V_{sp}(R)$) of the single particle potential V_{SP} of the other nucleus. The friction tensor $c_{\nu\mu}(t)$ is anisotropic having the largest component c_{rr} in the radial-radial direction, whereas the tangential-tangential component $c_{\varphi\varphi}$ is much smaller.

In order to become more familiar with the concept of friction in nuclear scattering we shall formally relate it to the other well known dissipative quantity, the imaginary part W of the optical model potential. The latter one controls the dissipation of flux from the elastic channel due to the excitation of the intrinsic degrees of freedom. In the same way friction controls the dissipation of energy from the relative motion due to the excitation of the intrinsic degrees of freedom: The energy loss ΔE per unit time is given by

$$\frac{dE}{dt} = -\sum_{\nu\mu} c_{\nu\mu}\, v_\nu\, v_\mu \approx \langle \Delta E \rangle \frac{dN}{dt} = \langle \Delta E \rangle \frac{2}{\hbar} W$$ (2.8)

The same physical mechanism connects $c_{\nu\mu}$ with W!

3) Input into the Model

The equation of motion to be solved is

$$M\ddot{R}_\nu + \nabla_\nu U(R) + \sum_\mu c_{\nu\mu}(R)\dot{R}_\mu = 0 , \qquad M = \frac{M_1 M_2}{M_1 + M_2}$$ (3.1)

Clearly such an equation of motion discribing only the degrees of freedom of relative motion of the centers of mass of the two nuclei is an oversimplification. It can only be assumed to work well when the two nuclei do not overlap too much, otherwise additional degrees of freedom like compression modes, neck formation, deformations, mass and charge exchange have to be considered explicitly. The hope of the present investigation is that the decision between a collision leading to fusion or to deep inelastic reactions is made at such distances already and that the main features of deep inelastic collisions are determined there.

3.1) The potential.

Several possibilities for the ion-ion potentials suggested in the li-

terature are shown in fig. (3.1). They are quite different, representing of course our lack of detailed knowledge about the ion-ion potential.

The liquid drop potential is discussed in detail in the talk given by H. Krappe[9]. In a somewhat less sophisticated form it was used
by Bass[2] for explaining the fusion cross-section systematics.

A detailed investigation of the liquid drop potential of [9]
shows that it cannot explain the experimental fusion cross section in
a model like eq. (3.1) with any type of friction. One has to go so far
inside the touching distance $R_{d,K\tau}= 1.16 \sum A^{1/3}$ that one has to take
explicitly into account other degrees of freedom like neck formation,
mass and charge asymmetries, quadrupole and octupole deformations, compression modes and possibly a host of others. Besides the complicated
problem of treating the corresponding coupled equations of motion, there are many unknown but important quantities in such a model.

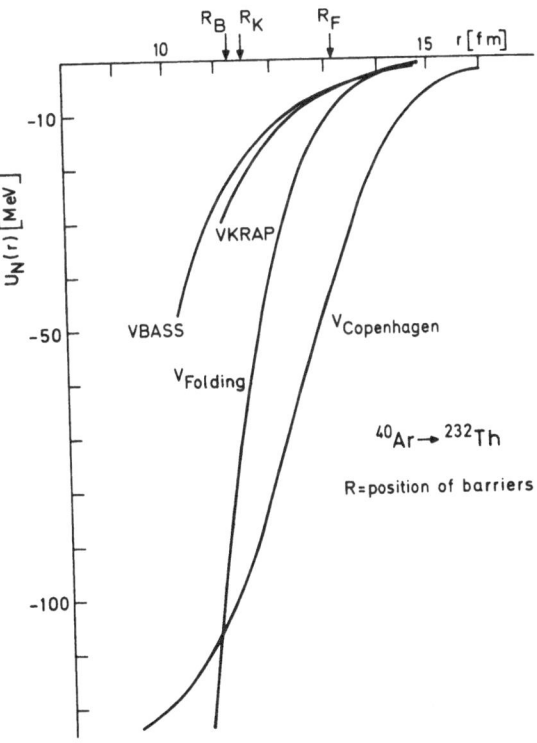

Fig. 3.1: Different ion-ion potentials (nuclear part only).

One needs not only the conservative forces but also the inertial and
frictional forces for each degree of freedom. Nobody knows anything
definite about these. Moreover all these may depend even on the temperature of the system, when it is excited to 100 MeV or more.

If we assume the liquid drop potential to be the right one (and
there are serious doubts about it), then we have to take into account
all relevant degrees of freedom; but the treatment becomes too complicated and the only hope is then to find some effective potential,
which can be used together with some friction in eq. 3.1 to produce
fits to the experimental data. Maybe the results presented here have
to be understood that way, may be they are more fundamental. We cannot
judge this at the moment.

The Bass potential, which is based on exactly the same physical
arguments as the Krappe-Nix potential, was quite successful for explaining the fusion data. The difference between the two curves shows that
the input parameters used in that potential are not so basic and unambiguous as one might think. Especially the critical distance $R_{d,Bass}=$
$1.07 \Sigma A^{1/3}$ which plays the decisive role to distinguish between fusion
and non fusion trajectories is in no way specified, neither by the potential itself nor by the underlying liquid drop theory. Following
Krappe one has to expect that inside $R_{d,Kr} = 1.16 \Sigma A^{1/3}$ or even more
far out other degrees of freedom like the ones mentioned above, become important.

The potential by the Copenhagen[10] group is based also on the
liquid-drop model but apparently quite different! With this potential
it is not possible to reproduce s-wave barriers which are near to the
experimental ones[9]. In that respect it is certainly only a type of an
effective potential, which in conjunction with some friction may produce the experimental data. That it is able to reproduce the main
systematics of fusion and deep inelastic data is still to be proved.

The folding potential used by us is given by:

$$U(R) = \tfrac{1}{2} \{ U_{12}(R) + U_{21}(R) \}$$

$$U_{12}(R) = \int d^3r \, V_1(R-r) \varrho_2(r) + \int d^3r \, d^3r' \, \varrho_{p1}(R-r') \frac{e^2}{|r'-r|} \varrho_{p2}(r) \qquad (3.2)$$

where $V_1(r)$ is the average potential acting between a nucleon and
nucleus 1, which is given by[11]

$$V_\Lambda(r) = V_0 \left\{ 1 + \exp\left(\frac{r - R_{V\Lambda}}{a_V}\right) \right\}^{-1}, \quad V_0 = -50 \text{ MeV}, \quad a_V = 0.65 \text{ fm}, \quad R_{V\Lambda} = 1.25 \cdot A_\Lambda^{1/3} \text{ [fm]} \quad (3.3)$$

The mass and charge densities $\mathcal{S}_2(r)$ and $\mathcal{S}_{p_2}(r)$ respectively are given by[11)

$$\mathcal{S}_2(r) = \frac{A_2}{Z_2} \mathcal{S}_{p_2}(r) \quad , \quad \mathcal{S}_{p_2}(r) = \mathcal{S}_{o_2} \left\{ 1 + \exp\left(\frac{r - R_{d2}}{a_d}\right) \right\}^{-1}$$

$$a_d = 0.54 \text{ fm} \quad , \quad R_{d2} = 1.12 A_2^{1/3} - 0.86 A_2^{-1/3} \text{ [fm]}, \quad \int d^3 r \, \mathcal{S}_{p_2}(r) = Z_2 \quad (3.4)$$

This potential is more attractive than the liquid drop potential of ref.[9)] discussed above. Certainly it makes sense only a few fm outside the distance of touching half densities. In our calculations there was no orbit which we had to follow to a distance less than $R_d + 2$ fm. The deepest potential seen in the calculations was less than 30 MeV deep.

In table (3.1) a comparison of the experimental fusion barriers with the predictions of the model is given. We see that the theoretical fusion barrier lies up to 80 MeV higher than the theoretical s-wave barrier.

3.2 The Friction Tensor
The friction is zero at large separations and increases as the two nuclei approach each other. The nature of the expression (2.6) suggests an ansatz of the following form:

$$c_{rr}(r) = C_r \, g(r) \quad , \quad c_{\varphi\varphi}(r) = C_\varphi \, g(r)$$

$$g(r) = \left(\nabla U_N(r) \right)^2 \quad , \quad U_N(r) = U(r) - U_{Coul}(r)$$

$$(3.5)$$

with c_r and c_φ to be free parameters. The best fit to all fusion data and to the deep inelastic scattering data was given by

$$c_r = 4 \left[\frac{10^{-23} \text{ sec}}{\text{MeV}} \right] \quad \text{and} \quad c_\varphi = 0.01 \left[\frac{10^{-23} \text{ sec}}{\text{MeV}} \right]$$

$$(3.6)$$

	$FB^{exp}_{c.m.}$ [MeV]	$CB^{th}_{c.m.}$ [MeV]	$FB^{th}_{c.m.}$ [MeV]	$\Delta=FB^{th}-CB^{th}$	Ref.
$^{12}C \rightarrow {}^{205}Cl$	56.0	54.6	55.7	1.1	16)
$^{12}C \rightarrow {}^{209}Bi$	57.0	55.8	57.0	1.2	16)
$^{16}O \rightarrow {}^{205}Tl$	77.0	72.0	74.3	2.3	16)
$^{32}S \rightarrow {}^{24}Mg$	28.3±2 %	26.0	26.6	0.6	17)
$^{32}S \rightarrow {}^{27}Al$	29.7±2 %	27.9	28.4	0.5	17)
$^{32}S \rightarrow {}^{40}Ca$	43.5±2 %	41.9	42.8	0.9	17)
$^{32}S \rightarrow {}^{58}Ni$	59.5±2 %	56.8	58.1	1.3	17)
$^{40}Ar \rightarrow {}^{164}Dy$	134,129	129.4	136.8	7.4	18)
$^{84}Kr \rightarrow {}^{72}Ge$	147.0	129.9	137.1	7.2	19)
$^{84}Kr \rightarrow {}^{116}Cd$	204.0	184.6	201.8	17.2	19)
$^{84}Kr \rightarrow {}^{232}Th$	332.0 (transfer)	317.0	395.7	78.7	20)

Tab. 3.1: FB = experimental and theoretical fusion barriers in the
CM-system, CB^{th} = theoretical Coulomb (s-wave) barrier. In
the case of $^{84}Kr + {}^{232}Th$ only the threshold for 6-9 nucleon
transfer has been determined. The experimental fusion barrier must be much higher.

$c_{\tau\tau\,(\varphi\varphi)}(r)$ has the dimension $\mathrm{MeV\cdot sec/fm^2}$.

With this choice of only two free parameters c_{τ}, c_{φ} and keeping them fixed for all masses and energies we obtained the results given in chapters 4 and 5.

In fig. (3.2) we show for the case $^{40}\mathrm{Ar}+^{238}\mathrm{U}$ the nuclear potential $U_N(r)$ and the radial friction $c_{rr}(r)$. The points R_1 and R_2 are the turning points of the orbit starting with the critical angular momentum for 400 MeV (R_1) and 300 MeV (R_2), respectively.

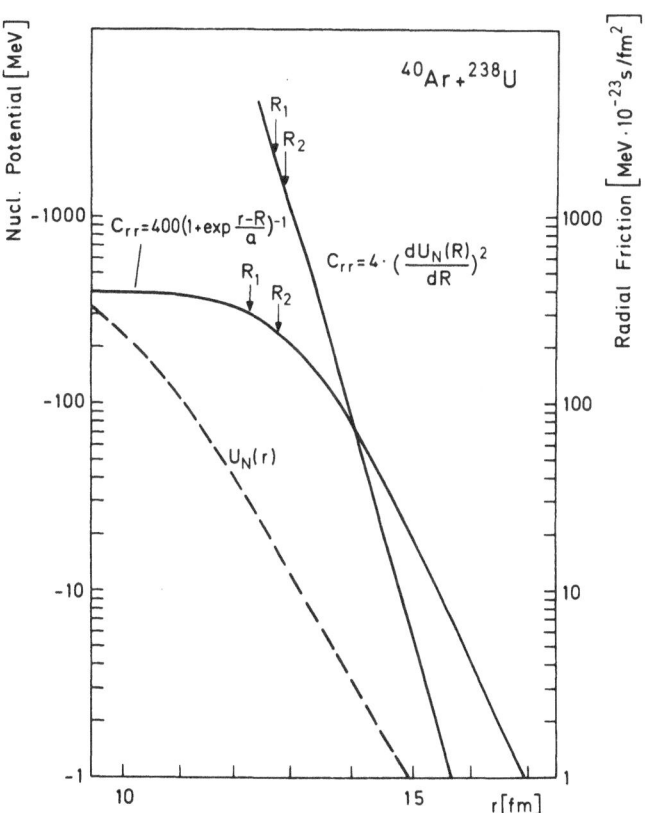

Fig. 3.2: Comparison of the radial friction coefficient and the nuclear part $U_N(r)$ of the ion-ion potential.

In the calculation of Gross and Kalinowski[12] we used a form-factor for the friction tensor which was different from the one in (3.5). It was a Fermi-type distribution with a radius $R = R_d + 2.7$ fm and a diffuseness $a = 0.65$ fm. It is shown also in fig. (3.2). This one was very successful in reproducing all light and medium heavy cases but failed for the Kr induced reactions and could not reproduce also the features of the deep inelastic collision cross-sections. The main reason for this failure was that it is only depending on the distance between the two nuclear surfaces irrespective of the fact that the overlap between two heavy nuclei at given distance is larger than that between lighter ones at the same distance. This fact is taken into account by the friction we use now (eq. 3.5), which is depending on the overlap itself. We are convinced that there might be another formfactor for the friction by which a better fit to the fusion data may be achieved. But we believe that the present calculation sufficiently proves the power of the model already.

4) Results for the Fusion Cross-Section

The table (4.1) (at the end) gives the results of our model with the choice of the two free parameters c_γ, c_φ as given in chapter (3.2). The potentials were computed individually using the formula (3.2) and the parameters in (3.3) and (3.4). We see that in most of the cases the experimental critical angular momenta are sufficiently well reproduced inspite of the fact that our friction parameters c_γ, c_φ (eq. 3.6) are constant and independent of whether we have a light system like $^{12}C + ^{27}Al$ or a very heavy one like $^{84}Kr + ^{238}U$.

5) Results for Deep Inelastic Collisions

5.1) We already showed the trajectories for the case of $^{40}Ar + ^{232}Th$ at $E_{Lab} = 400$ MeV in the introduction. Figs. 5.1 and 5.2 show angular distribution, energy loss and the deflection function $\theta(L)$ for this case.

The cross sections, especially the large spiralling part of it, is nicely reproduced by the calculation. The energy-loss is too small, a fact which appears in all our calculations and may be due to the neglect of the neck formation in the outgoing channels.. The energy distribution will also depend on the details of the transfer mechanism still not included in our model.

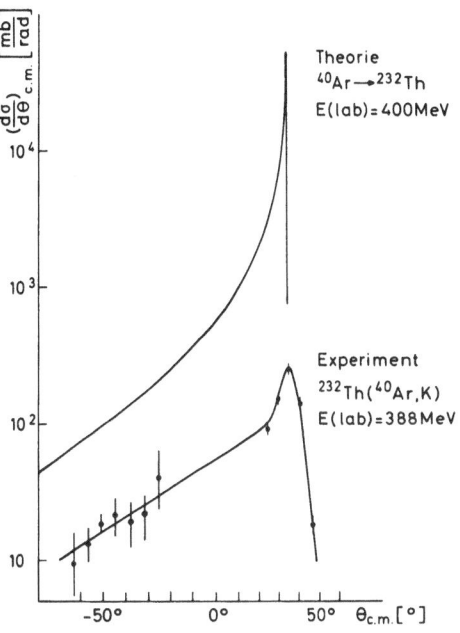

Fig. 5.1: Differential cross-section. The theoretical one in the sum of all transfer channels. $\sigma_{tot}^{th}(\theta < 0^{\circ})$ = 280 mb (exp.: 400±200 mb).

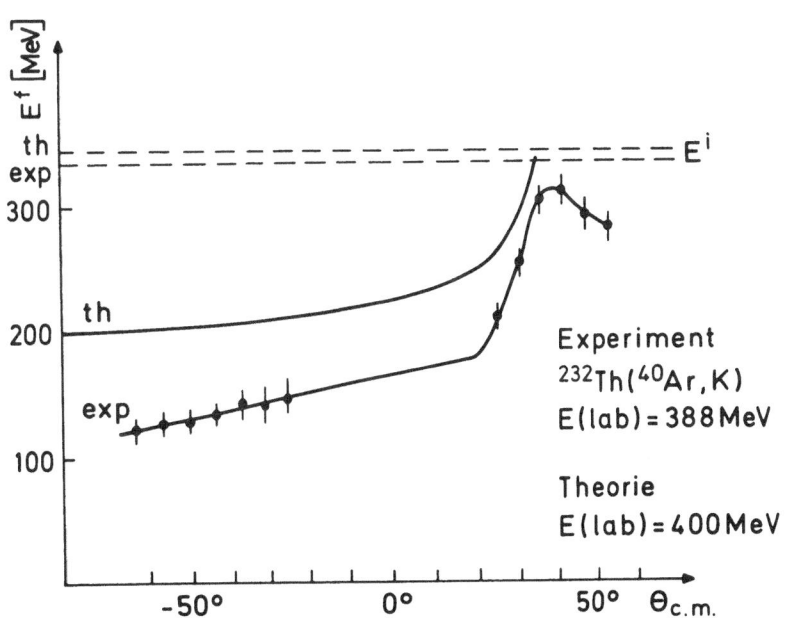

Fig. 5.2a

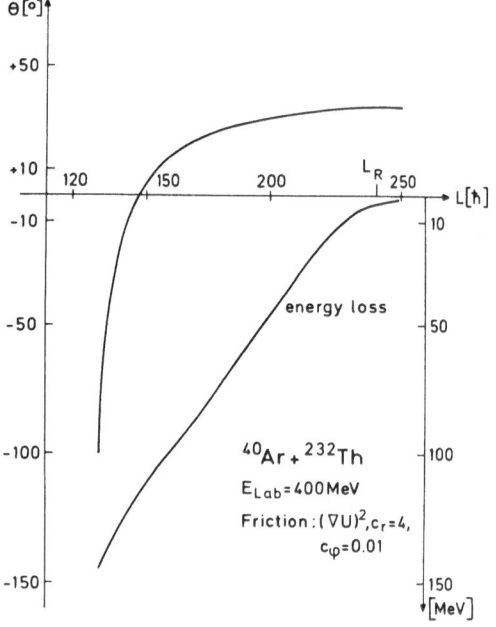

Fig. 5.2b

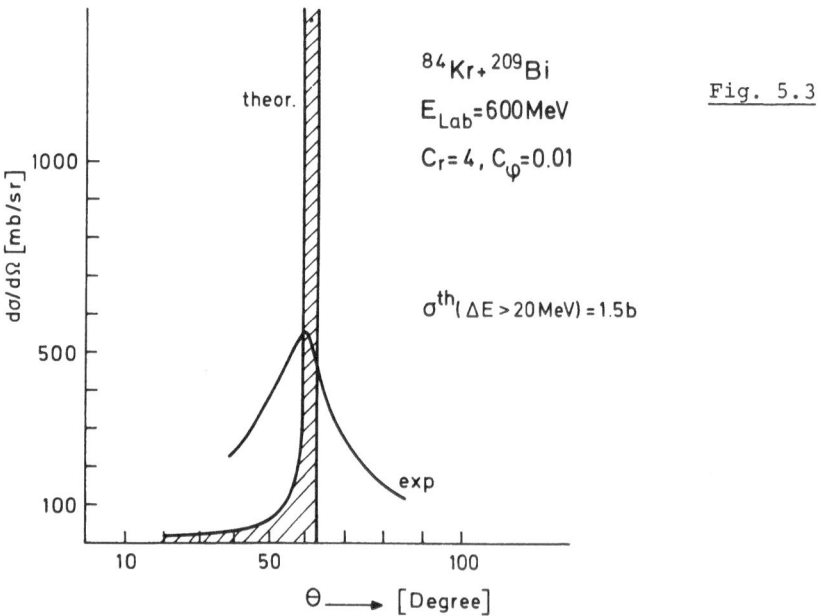

Fig. 5.3

5.2) The results of the experiments of $^{84}Kr+^{209}Bi$ at 525 MeV[13] and 600 MeV[14] were very puzzling. At 600 MeV more than 80 % of the reaction cross-section ($\sigma_R \approx 1.8b$) goes into the deep inelastic collision channels ($\sigma_{d.inel} \approx 1.5$ b). Moreover, nearly all these products come out at an angle of about $60°_{CM}$ (focussing effect). A lot of speculations were made about this fact[14]. Especially this fact together with the small fusion cross section was considered as the main argument against the folding potential (3.2). It was argued that the realistic potential should have pockets for a very small range of angular momenta $0 \le L \le L_{cr}$ only. The present calculation shown in fig. (5.3) prove that this is not necessarily true.

The present model gives a rather simple explanation of the focussing effect: Contrary to the case of $^{40}Ar+^{232}Th$, at 400 MeV the effective barriers are shifted inwards into more overlapping regions for $^{84}Kr+^{209}Bi$ because the Coulomb repulsion is much stronger. Therefore a rainbow trajectory which did not pass much of the friction-region in $^{40}Ar+^{232}Th$, does now penetrate the region where the friction is large. Many orbits with quite different angular momenta are stopped radially near the turning point of the rainbow scattering. Consequently the rainbow, which is quite narrow for elastic (frictionless) scattering is spread over an interval of about 150 $\hbar$ leading to a huge cross-section at $\theta = 63°$ (fig. 5.5, compare also the deflection functions figs. 5.2b and 5.4!)

In addition to this the difference in the mass distribution measured at an angle of about $50°_{CM}$ (less than the grazing angle) and at about $85°_{CM}$ (larger than the grazing angle) (fig. 5.6) can be understood qualitatively by our model.
Projectiles leading to proton pick-up reactions can be deflected to angles $\theta > \Theta_R = 63°$, whereas with no charge transfer or proton stripping, these angles cannot be reached classically. On the contrary to angles $\theta < \Theta_R$ all types of processes can contribute.

5.3) Other deep inelastic collisions are measured also. We have calculated the reaction $^{40}Ca+^{208}Pb$, $E_{Lab} = 288$ MeV measured by Colombani[15] and show comparison between theoretical and experimental angular distribution (here both in the laboratory system) in fig. (5.8).

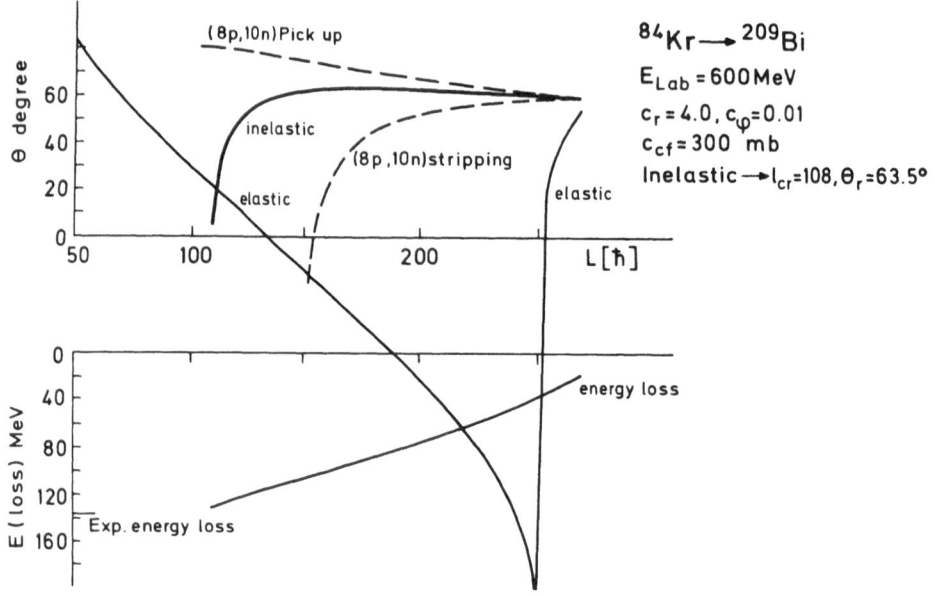

$^{84}Kr \longrightarrow ^{209}Bi$

$E_{Lab} = 600 MeV$

$c_r = 4.0$, $c_\varphi = 0.01$

$c_{cf} = 300$ mb

Inelastic $\rightarrow l_{cr} = 108$, $\theta_r = 63.5°$

Fig. 5.4: (Above). Deflection functions and energy loss v.s. L. The particle transfer was assumed to happen just at the distance of closest approach, there the reduced mass and the Coulomb potential were switched.

Fig. 5.5: Classical trajectories.

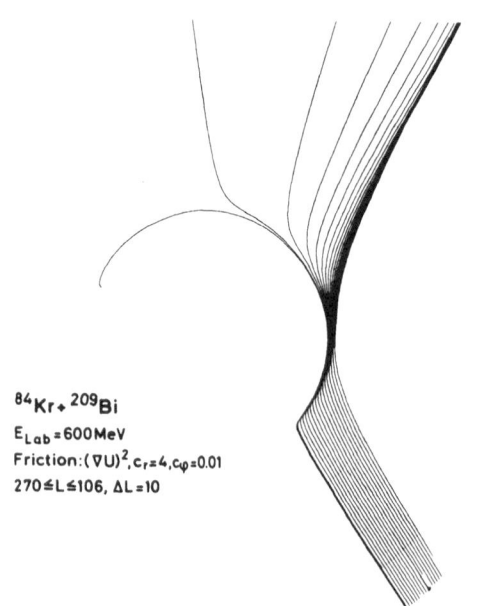

$^{84}Kr + ^{209}Bi$

$E_{Lab} = 600 MeV$

Friction: $(\nabla U)^2$, $c_r = 4$, $c_\varphi = 0.01$

$270 \leq L \leq 106$, $\Delta L = 10$

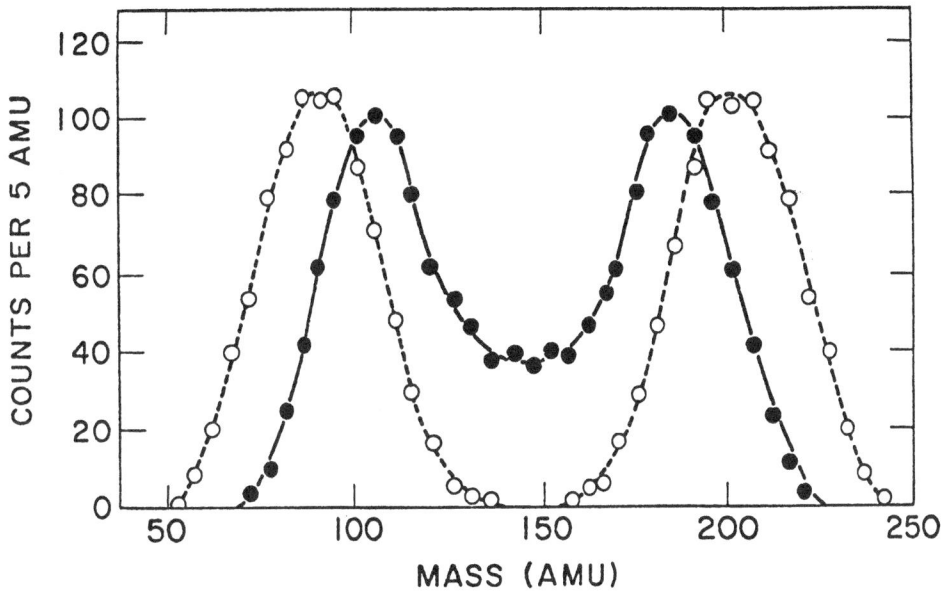

Fig. 5.6: ^{84}Kr+^{209}Bi, 600 MeV, mass distribution, open circles at $\approx 50^{\circ}_{CM}$, full circles at $\approx 85^{\circ}_{CM}$, ref. [14].

In concluding this chapter we should mention that the critical angular momenta L_{cr} for Kr+Bi are still a little bit too large theoretically. This can be changed by a slight modification of the friction parameters: With $c_r = 4.2$, $c_\varphi = 0$ we yield the numbers ^{84}Kr+^{209}Bi:
σ_{CF}(600 MeV) = 229 mb (L_{cr} = 94 ℏ), σ_{CF}(525 MeV) = 28 mb (L_{cr} = 30 ℏ).

The decrease of the tangential friction for heavier projectiles and targets is understandable from the microscopic formula (2.6). The tangential friction is only due to the curvature of the nuclear surfaces. It would vanish for two semi infinite nuclear systems. Thus for two heavy nuclei the tangential friction may be smaller at the same overlap than the one experienced by two light nuclei.

However, the main improvement is due to the slight increase of the radial friction by 5 %. This shows the sensitivity of the value of L_{cr} for very heavy systems at energies near the s-wave barrier on the variation of the radial friction.

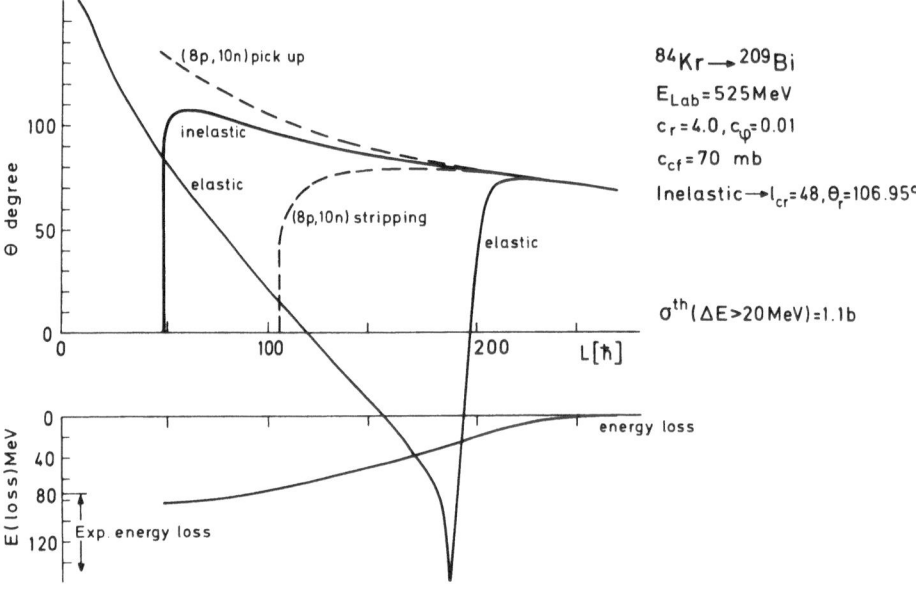

$^{84}Kr \longrightarrow ^{209}Bi$

$E_{Lab} = 525 MeV$

$c_r = 4.0, c_\varphi = 0.01$

$c_{cf} = 70 \ mb$

Inelastic $\rightarrow l_{cr} = 48, \theta_r = 106.95°$

$\sigma^{th}(\Delta E > 20 MeV) = 1.1b$

Fig. 5.7

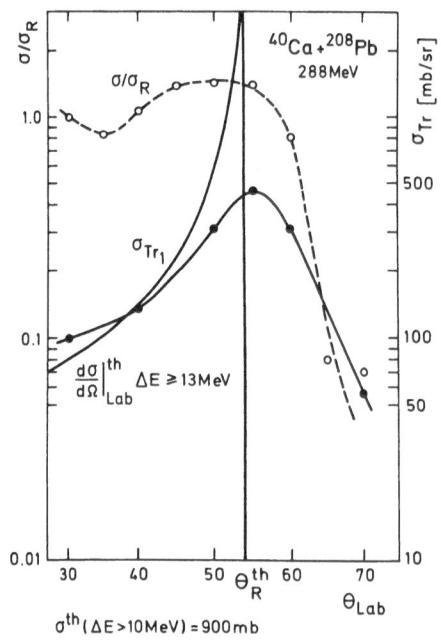

$^{40}Ca + ^{208}Pb$ 288MeV

$\sigma^{th}(\Delta E > 10 MeV) = 900mb$

Fig. 5.8

6) Conclusion

Our calculation has shown that the simple classical model outlined in the introduction gives a fair reproduction of the systematics of heavy ion fusion cross-sections and deep inelastic collisions. Especially the focussing effect observed in $^{84}Kr + ^{209}Bi$ collisions has been explained. Only two parameters had to be adjusted, but were taken as constant for all different systems like $^{12}C + ^{27}Al$ or $^{84}Kr + ^{238}U$ and for all energies. The input quantities, the potential and the friction are quite realistic. An improvement of the fit to the data may even be possible within the model by varying the form factor of the friction somewhat. It should be mentioned that the Fermi-type friction form-factor, which we used in our previous calculation[12] gave an excellent fit for lighter systems like $^{12}C + ^{27}Al$, $^{12}C + ^{58}Ni$, $^{32}S + ^{24}Mg$ and others.

The model explains the main energy loss of deep inelastic collisions but it fails to reproduce it by about 20-60 MeV. Partially, this may be due to the difficulty in measuring the energy loss accurately because the detected ion may have lost an unknown number of nucleons. Perhaps the friction should have in addition a long range Coulomb excitation tail. We have not included the effect of neck formation in our theory, but it might change the energy distribution of the final fragments considerably. Once the neck is formed the barriers in the outgoing channels would be diminished.

The present calculation treats the inelastic process only. Inclusion of the nucleon transfer may lead to an additional energy dissipation. Moreover the energy of the final fragments depends strongly on the charge transfer. Therefore, one has to have a theory giving the charge for each ingoing trajectory and then calculate the new trajectory out. After this procedure one may average over mass- und charge-transfers and obtain the energy loss as function of the scattering angle. It is being planned to extend the present calculation by adding a statistical theory for the nucleon exchange process.

Table 4.1

	R_d [fm]	R_{CB} [fm] ±0.1	$CB(R_{CB})$ [MeV]	E_{lab} [MeV]	σ_{CF}^{exp} [mb]	L_{cr}^{exp} [ℏ]	L_{cr}^{th} [ℏ]	E-loss [MeV]	L-loss [ℏ]	R_{min} [fm]	L_{Max}^{th} [ℏ]	
$^{11}_{5}B \rightarrow {}^{159}_{65}Tb$	8.0	11.6	37.8	115.0	946±150pf	39±4	50-52	33.5	3.1	10.6	64-66	KOZ 74[21]
$^{12}_{6}C \rightarrow {}^{27}_{13}Al$	5.3	9.4	11.0	180.0	834±15%p	35±3	34-36	77.3	3.2	7.5	>50	NAT 72[22]
				97.6	854	26±2	28-30	32.2	2.7	7.7	38-40	"
				96.5	1000	28±3	28-30	30.9	2.5	7.7	38-40	"
				85.7	1110	28±3	28-30	20.8	1.2	7.8	36-38	"
				81.0	1150	28±3	28-30	16.9	0.9	7.8	34-36	"
				63.8	1140	24±2	26-28	8.1	0.4	8.1	28-30	"
				44.3	1040	19±2	22-24	1.5	0.07	8.7	22-24	"
$^{12}_{6}C \rightarrow {}^{48}_{22}Ti$	6.0	9.8	17.8	180.0	1200±15%p	49±4	42-44	90.7	5.5	8.2	70-80	NAT 72[22]
				97.6	938	32±3	36-38	28.7	2.0	8.5	46-48	"
				81.0	943	29±3	34-36	17.6	1.2	8.6	40-42	"
$^{12}_{6}C \rightarrow {}^{58}_{28}Ni$	6.3	10.0	22.4	180.0	1170±15%p	50±4	44-46	94.1	7.6	8.5	>60	NAT 72[22]
				97.6	1240	38±3	36-38	34.9	5.1	8.8	48-50	"
				96.5	1050	35±3	36-38	32.2	3.9	8.8	48-50	"
				81.0	1217	34±3	34-36	19.2	1.9	8.9	42-44	"
				63.8	956	27±3	30-32	9.6	1.0	9.1	34-36	"
$^{12}_{6}C \rightarrow {}^{63}_{29}Cu$	6.4	10.1	22.9	126.0	536±20%p	29±3	42-44	47.9	3.4	8.8	60-62	NAT 70[23]
				97.6	1290±15%p	39±3	38-40	29.7	2.7	8.9	50-52	NAT 72[22]
				96.5	1070	35±3	38-40	28.4	2.4	9.0	50-52	"
				81.0	886	29±3	36-38	16.7	1.2	9.1	42-44	"
				63.8	1110	29±3	32-34	7.3	0.5	9.3	34-36	"
				44.3	882	21±2	24-26	1.0	0.05	10.1	24-26	"
$^{12}_{6}C \rightarrow {}^{107}_{47}Ag$	7.3	10.8	34.9	126.0	906±20%p	40±4	48-50	42.8	3.7	9.8	66-68	NAT 70[23]

Cont. Table 4.1

	R_d [fm]	R_{CB} [fm] ±0.1	$CB(R_{CB})$ [MeV]	E_{lab} [MeV]	σ_{CF}^{exp} [mb]	L_{cr}^{exp} [ℏ]	L_{cr}^{th} [ℏ]	E-loss [MeV]	L-loss [ℏ]	R_{min} [fm]	L_{Max}^{th} [ℏ]	
$^{12}_{6}C \rightarrow ^{158}_{64}Gd$	8.1	11.5	44.9	126.0	1096±163pf	46±4	52-54	37.0	3.6	10.7	68-70	ZEB 73[24]
$^{12}_{6}C \rightarrow ^{197}_{79}Au$	8.6	12.0	53.6	126.0	1403±30%pf	53±8	52-54	34.0	4.0	11.2	68-70	NAT 70[23]
$^{14}_{7}N \rightarrow ^{27}_{13}Al$				262.0	945±140p	47±4	38-40	133.3	9.7	7.5	>60	NAT 74[25]
				157.0	1360±200p	43±4	34-36	62.1	4.6	7.7	54-56	"
$^{14}_{7}N \rightarrow ^{52}_{24}Cr$				262.0	1220±180p	64±5	52-54	140.2	7.6	8.4	>80	NAT 74[25]
				157.0	1220±180p	49±4	46-48	61.8	3.8	8.6	>60	"
$^{14}_{7}N \rightarrow ^{58}_{28}Ni$				262.0	1560±230p	74±6	54-56	139.8	7.5	8.6	>80	NAT 74[25]
				157.0	1650±250p	59±5	46-48	66.2	5.5	8.8	>60	"
$^{14}_{7}N \rightarrow ^{103}_{45}Rh$	7.4	10.8	38.9	121.0	-	52±10	48-50	44.7	9.7	9.9	66-68	GAL 74[26]
$^{16}_{8}O \rightarrow ^{27}_{13}Al$	5.6	9.5	14.6	168.0	860±110p	36±3	38-40	54.9	2.8	7.8	>50	MIL 74[27]
				161.0	500±100p	27±3	26-38	59.2	5.4	7.8	56-58	KOW 68[28]
				126.0	960±120p	33±3	34-36	36.6	3.0	7.9	46-48	MIL 74[27]
				105.0	1040±120p	31±2	32-34	25.9	2.4	8.0	42-44	"
				81.0	1020±140p	27±2	30-32	11.4	0.8	8.2	34-36	"
$^{16}_{8}O \rightarrow ^{59}_{27}Co$	6.6	10.2	28.3	161.0	740±100p	41±3	50-52	62.9	5.7	8.9	74-76	KOW 68[28]
$^{16}_{8}O \rightarrow ^{63}_{29}Cu$	6.7	10.2	30.2	168.0	760±20%p	44±5	52-54	66.1	5.5	9.0	78-80	NAT 70[23]
$^{16}_{8}O \rightarrow ^{107}_{47}Ag$	7.6	10.9	46.0	168.0	710±20%p	46±5	60-62	61.5	5.9	10.0	88-90	NAT 70[23]
$^{16}_{8}O \rightarrow ^{154}_{62}Sm$	8.3	11.6	57.5	137.0	1262±183pf	58±5	58-60	35.0	4.4	10.9	76-78	ZEB 73[24]
$^{16}_{8}O \rightarrow ^{197}_{79}Au$	8.6	12.1	70.8	168.0	1695±30%pf	76±12	64-66	57.1	10.8	11.4	90-100	NAT 70[23]
$^{20}_{10}Ne \rightarrow ^{27}_{13}Al$	5.8	9.6	18.0	210.0	940±150p	44±4	42-44	73.9	7.4	8.0	>60	MIL 74[27]

Cont. Table 4.1

	R_d [fm]	R_{CB} [fm] ±0.1	$CB(R_{CB})$ [MeV]	E_{lab} [MeV]	σ_{CF}^{exp} [mb]	L_{cr}^{exp} [ħ]	L_{cr}^{th} [ħ]	E-loss [MeV]	L-loss [ħ]	R_{min} [fm]	L_{Max}^{th} [ħ]	
$^{20}_{10}$Ne → $^{27}_{13}$Al	5.8	9.6	18.0	200.0	380±100p	27±4	42-44	64.0	4.9	8.0	64-66	KOW 68[28]
				138.0	1170±130p	39±3	38-40	31.0	2.5	8.2	50-52	MIL 74[27]
$^{20}_{10}$Ne → $^{63}_{29}$Cu	7.0	10.4	37.3	210.0	532±20%p	43±5	60-62	89.5	11.6	9.2	>90	NAT 70[23]
$^{20}_{10}$Ne → $^{107}_{47}$Ag	7.9	11.1	56.9	210.0	672±20%p	54±6	70-72	83.2	10.4	10.2	>100	NAT 70[23]
				173.0	868±20%p	56±6	64-66	56.5	8.1	10.3	90-100	"
				163.0	1000 pf	58	62-64	49.7	7.5	10.6	88-90	PLA 74[29]
				129.0	700 pf	43	54-56	25.6	3.7	10.7	70-72	"
				113.0	500 pf	34	48-50	17.2	2.8	10.9	50-60	"
$^{20}_{10}$Ne → $^{209}_{83}$Bi	9.2	12.2	91.3	210.0	1518±30%pf	89±14	76-78	71.7	13.2	11.7	110-120	NAT 70[23]
$^{32}_{16}$S → $^{24}_{12}$Mg	6.2	9.8	26.1	120.0	960±10%p	31±2	32-34	8.6	1.8	9.0	34-36	GUT 73[30]
				110.0	945 "	29±2	30-32	5.4	0.7	9.1	32-34	"
				90.0	669 "	22±2	24-26	2.0	0.2	9.4	24-26	"
				85.0	515 "	19±1	22-24	1.5	0.1	9.5	22-24	"
				80.0	414 "	16±1	20-22	0.8	0.06	9.8	20-22	"
				75.0	269 "	12±1	18-20	0.4	0.02	10.2	18-20	"
				70.0	150 "	9±1	14-16	0.4	0.02	10.1	14-16	"
$^{32}_{16}$S → $^{27}_{13}$Al	6.4	9.9	28.0	336.0	620±80p	45±3	56-58	85.0	6.2	8.6	90-100	MIL 74[27]
				132.5	995±10%p	36±2	38-40	10.3	1.1	9.1	42-44	GUT 73[30]
				120.4	936 "	33±2	36-38	6.5	0.6	9.2	38-40	"
				110.0	914 "	31±2	32-34	7.4	1.9	9.2	34-36	"
				90.0	650 "	23±2	26-28	2.1	0.2	9.6	26-28	"
				87.5	663 "	23±2	26-28	1.1	0.07	9.9	26-28	"

Cont. Table 4.1

	R_d [fm]	R_{CB} [fm] ±0.1	$CB(R_{CB})$ [MeV]	E_{lab} [MeV]	σ_{CF}^{exp} [mb]	L_{cr}^{exp} [ħ]	L_{cr}^{th} [ħ]	E-loss [MeV]	L-loss [ħ]	R_{min} [fm]	L_{Max}^{th} [ħ]	
$^{32}_{16}S \rightarrow ^{27}_{13}Al$	6.4	9.9	28.0	85.0	561±10%p	21±2	24-26	1.4	0.1	9.7	24-26	GUT 73[30]
				80.0	434 "	18±1	22-24	0.7	0.05	10.0	22-24	"
				77.5	336 "	15±1	20-22	0.8	0.06	9.9	20-22	"
				75.0	296 "	14±1	18-20	0.9	0.08	9.8	18-20	"
				73.0	248 "	13±1	16-18	1.1	0.2	9.7	16-18	"
				69.0	139 "	9±1	14-16	0.4	0.02	10.2	14-16	"
				67.0	90 "	7±1	12-14	0.3	0.02	10.2	12-14	"
$^{32}_{16}S \rightarrow ^{40}_{20}Ca$	6.9	10.2	42.0	107.5	670±10%p	32±2	34-36	4.0	0.4	9.9	34-36	GUT 73[30]
				97.5	535 "	27±2	28-30	2.8	0.3	10.0	28-30	"
				90.0	354 "	21±2	22-24	2.3	0.3	10.0	22-24	"
				87.5	300 "	19±1	20-22	1.8	0.2	10.1	20-22	"
				85.0	196 "	15±1	18-20	1.4	0.1	10.2	18-20	"
				82.5	139 "	12±1	14-16	1.6	0.2	10.1	14-16	"
				80.0	65 "	8±1	12-14	0.9	0.04	10.4	12-14	"
$^{32}_{16}S \rightarrow ^{58}_{28}Ni$	7.4	10.5	56.8	120.0	497±10%p	34±2	38-40	8.8	2.2	10.2	42-44	GUT 73[30]
				107.5	350 "	27±2	30-32	4.7	0.7	10.4	30-32	"
				97.5	131 "	15±1	20-22	2.7	0.3	10.5	20-22	"
$^{35}_{17}Cl \rightarrow ^{58}_{28}Ni$	7.5	10.6	59.8	139.75	781	47	46-48	10.6	1.5	10.3	52-54	SCO[31]
				119.75	454	33	34-36	5.7	0.9	10.4	36-38	"
				109.75	271	24	26-28	3.3	0.3	10.6	26-28	"
				104.75	143	17	20-22	2.5	0.2	10.7	20-22	"
				99.75	47	9	10-12	2.0	0.1	10.7	10-12	"

Cont. Table 4.1

	R_d [fm]	R_{CB} [fm] ±0.1	$CB(R_{CB})$ [MeV]	E_{lab} [MeV]	σ_{CF}^{exp} [mb]	L_{cr}^{exp} [ħ]	L_{cr}^{th} [ħ]	E-loss [MeV]	L-loss [ħ]	R_{min} [fm]	L_{Max}^{th} [ħ]	
$^{35}_{17}Cl \rightarrow ^{58}_{28}Ni$	7.5	10.6	59.8	97.75	19	5	<2					SCO [31]
				96.75	6.7	3	<2					"
$^{35}_{17}Cl \rightarrow ^{62}_{28}Ni$	7.6	10.8	59.1	139.75	929	52	50-52	11.1	1.5	10.4	58-60	SCO [31]
				119.75	585	38	38-40	6.5	1.2	10.5	40-42	"
				109.75	401	30	30-32	4.2	0.7	10.6	30-32	"
				104.75	268	24	26-28	2.6	0.2	10.8	26-28	"
				99.75	136	16	18-20	2.4	0.2	10.8	18-20	"
				97.75	89	13	14-16	2.2	0.2	10.8	14-16	"
				96.75	65	11	12-14	2.0	0.1	10.9	12-14	"
				95.75	48	9	8-10	1.9	0.09	10.9	8-10	"
				94.75	37	8	4-6	1.9	0.07	10.8	4-6	"
$^{40}_{18}Ar \rightarrow ^{58}_{28}Ni$	7.7	10.8	62.4	288.0	900±120p	73±5	80-82	70.5	9.8	10.0	120-130	GUT [32]
				197.0	1280±270pf	72±8	64-66	27.6	4.7	10.2	82-84	"
$^{40}_{18}Ar \rightarrow ^{77}_{34}Se$	8.1	11.2	73.2	201.0	-	70±10	72-74	33.9	8.3	10.7	96-98	GAL 74 [26]
				145.0	-	52±10	48-50	9.2	1.7	11.0	54-56	"
$^{40}_{18}Ar \rightarrow ^{109}_{47}Ag$	8.8	11.7	97.3	288.0	1270±190pf	108±9	98-100	78.2	13.7	11.1	>130	GUT [32]
				197.0	920±135pf	76±6	70-72	26.9	5.9	11.3	90-100	"
$^{40}_{18}Ar \rightarrow ^{121}_{51}Sb$	8.9	11.8	104.2	300.0	1820±110pf	136±5	102-104	93.0	23.6	11.3	>150	HAN 73 [33]
				260.0	1500-1600pf	117±2	94-96	59.2	10.8	11.4	140-150	"
				225.0	1340±95pf	101±4	82-84	40.6	9.0	11.5	110-120	"
				200.0	1050±120pf	84±5	72-74	25.7	4.7	11.6	96-98	"
				180.0	587±40pf	59±3	60-62	16.7	2.9	11.7	78-80	"

Cont. Table 4.1

	R_d [fm]	R_{CB} [fm] ±0.1	$CB(R_{CB})$ [MeV]	E_{lab} [MeV]	σ_{CF}^{exp} [mb]	L_{cr}^{exp} [ℏ]	L_{cr}^{th} [ℏ]	E-loss [MeV]	L-loss [ℏ]	R_{min} [fm]	L_{Max}^{th} [ℏ]	
$^{40}_{18}Ar \to ^{121}_{51}Sb$	8.9	11.8	104.2	160.0	240±14pf	35±2	40-42	10.6	2.4	11.8	52-54	HAN 73[33]
$^{40}_{18}Ar \to ^{165}_{67}Ho$	9.6	12.4	131.4	300.0	1430±140f	129±7	104-106	82.6	18.6	12.0	>160	HAN 73[33]
				225.0	860±90f	86±5	74-76	34.4	8.4	12.2	110-120	"
$^{40}_{18}Ar \to ^{238}_{92}U$	10.4	13.1	170.9	400.0	1350±20%f	155±16	128-130	142.5	29.8	12.7	>220	SIK66/68[34]
				300.0	1220±120f	127±7	94-96	72.8	24.1	12.9	>150	HAN 73[33]
$^{84}_{36}Kr \to ^{209}_{83}Bi$	11.2	13.6	297.0	600.0	200±60f	87±13	106-108	130.9	37.6	13.5	>270	WOL 74[35]
				525.0	<40f	<36	46-48	84.2	18.0	13.6	220-230	PET 74[36]
				500.0	<40f	<35	<2					HAN 73[33]
$^{84}_{36}Kr \to ^{238}_{92}U$	11.5	13.9	323.1	500.0	<10f	<18	<2					HAN 73[33]

Tab. 4.1: R_d = distance of touching half density points (eq. 3.4); R_{CB}, CB = position, height (CM) of the Coulomb (s-wave) barrier; $\sigma_{CF} = \pi\lambda^2 (L_{cr} +1)^2$ = compound formation cross-section, p = particle evaporation, f = fission decay, pf = both are measured and added; L_{cr}^{th}: Smaller member = last captured and larger number = first inelastic orbit; E-loss, L-loss, R_{min} = energy-, angular momentum-loss and distance of closest approach of first inelastic orbit. L_{Max}^{th} = angular momentum of first orbit loosing more than 5 MeV.

References

1) D. H. E. Gross, H. Kalinowski and R. Beck, contribution 5.65, "Proceedings of the International Conference on Nuclear Physics", Munich, August 27-September 1,1973
2) R. Bass, Phys. Lett. 47B (1973) 139, preprint Frankfurt 1974
3) M. Lefort, Y. Le Beyec, J. Peter, invited paper given at the Conference on Reactions between Complex Nuclei, Nashville, June 10-14, 1974 and: J. Galin, D.Guerreau, M. Lefort and X. Tarrago:"On the limitation to complete fusion during the collision between two complex nuclei", Orsay preprint IPNO-RC-73-09 (1973), Institute de Physique Nucléaire, Université Paris-Sud, B.P. no. 1, 91406-Orsay, France
4) J. Wilczynski, Phys. Lett. 47B (1973) 484
5) D. Glass and U. Mosel, Phys. Lett. 49B (1974) 301
6) R. Beck and D. H. E. Gross, Phys. Lett. 47B (1973) 143
7) L. Wilets: Theory of nuclear fission, Clarendon Press Oxford (1964)
8) D. H. E. Gross (1974) submitted to Nucl. Phys.
9) H. J. Krappe,invited talk, this conference
10) J. P. Bondorf. Invited talk presented at the International Conference on Reactions between Complex Nuclei, Nashville, Tenn.,USA, June 10-14 (1974)
11) A. Bohr and B. R. Mottelson, Nuclear Structure, Vol. 1, Benjamin Inc. (1969)
12) D. H. E. Gross and H. Kalinowski, Phys. Lett. 48B (1974) 302
13) F. Hanappe, M. Lefort, C. Ngô, J. Péter, B. Tamain, Phys. Rev. Lett. 32 (1974) 738
14) K. L. Wolf, J. P. Unik, J. R. Huizenga, V. E. Viola, D. Birkelund and H. Freiesleben, August 1974, submitted to Phys. Rev. Lett.
15) P. Colombani:"Interactions entre ions tres lourds", Orsay preprint IPNO-Ph-N-74-09
16) Y. Le Beyec, M. Lefort, M. Sarda, Nucl. Phys. A192 (1972) 405
17) H. H. Gutbrod, W. G. Winn, M. Blann, Phys. Rev. Lett. 30 (1973)1259
18) Y. Le Beyec, M. Lefort, A. Vigny, Phys. Rev. C3 (1971) 1268
19) H. Gauvin, Y. Le Beyec, M. Lefort, C. Deprun, Phys. Rev. Lett. 28 (1972) 697
20) R. Bimbot, H. Gauvin, Y. Lebeyec, M. Lefort, N. T. Porile, B. Tamain, Nucl. Phys. A189 (1972) 539
21) R. L. Kozub, J. M. Miller, L. Kowalski, D. Logau, N. H. Lu, contribution to the International Conference on "Reactions between Complex Nuclei", Nashville, Tenn., USA, June 1974
22) J. B. Natowitz, E. T. Chulick, M. N. Namboodiri, Phys. Rev. C6 (1972) 2133
23) J. B. Natowitz, Phys. Rev. C1 (1970) 623
24) A. M. Zebelman, J. M. Miller, Phys. Rev. Lett. 30 (1973) 27
25) M. N. Namboodiri, E. T. Chulick, J. B. Natowitz, R. A. Kenefick, Texas-preprint 1974
26) J. Galin, B. Fatty, D. Guerreau, C. Rousset, U. C. Schlotthauer-Voos, X. Tarrago, Phys. Rev. C9 (1974) 1126
27) R. L. Kozub, N. H. Lu, J. M. Miller, D. Logan, T. W. Debiak, L. Kowalski, preprint, Columbia, New York 1974
28) L. Kowalski, J. C. Jodogne, J. M. Miller, Phys. Rev. 169(1968)894
29) F. Plasil, private communication
30) H. H. Gutbrod, W. A. Winn, M. Blann, Nucl. Phys. A213(1973) 267
31) W. Scobel, private communication
32) H. H. Gutbrod, F. Plasil, H. C. Britt, B. H. Erkila, R. H. Stokes, M. Blann, proceedings to 3. IAEA, Symposium on Physics and Chemistry of Fission, Rochester 1973, Vol. II, 309
33) J. Péter, F. Hanappe, C. H. Ngô B. Tamain, IPNO-RC-73-07,Orsay preprint 1973
34) T. Sikkeland, Phys. Rev. Lett. 27B (1968)277 and Arkiv f.Fys. 36 (1966) 539
35) K. L. Wolf, private communication
36) J. Péter, private communication

FUSION AND FISSION INDUCED BY HEAVY IONS

Yu.Ts.Oganessian

Joint Institute for Nuclear Research

Dubna, USSR

Abstract

Some problems associated with the mechanism of compound nucleus formation in reactions induced by ions with mass $\geqslant$ 40 amu are considered. Experiments to synthesize neutron-deficient nuclei with Z=100, 104 and 106 by bombarding Pb isotopes with 40-Ar, 50-Ti and 54-Cr ions are described. Some data on delayed fission of neutron-deficient Am isotopes in the Pb + Al reaction are given.

The properties of heavy nuclei are discussed in terms of the present-day fission theory. The possibilities of synthesizing elements with Z > 106 are investigated.

1. INTRODUCTION.

The mechanism of the fusion of complex nuclei is an interesting problem of heavy ion physics. The experimental results available in this field are closely related to the possibilities of heavy ion accelerators of the first generation. It is therefore not surprising that the major part of these data have been obtained from reactions involving relatively light ions from B to Ne, for which the fusion process has been investigated in detail. The analysis of the results of numerous experiments permits a number of general conclusions about the formation probability for compound nuclei and the characteristic features of their decay.

Nuclear fusion is a dominating process in the vicinity of the Coulomb barrier. As the projectile energy increases, the compound nucleus formation process begins to compete with those of the direct emission of individual nucleons and even complexes of different nucleons such as deuterons, alpha particles and others. This fact imposes some limitations on the compound nucleus formation probability, which are generally expressed through the critical angular momentum l_{cr}. The value of l_{cr} is very difficult to calculate and is essentially deduced from experimental data. One can indicate only the upper limit on l_{cr} based on the simple logical conclusion that the rotational energy cannot exceed the total excitation energy of the nucleus [1], i.e.

$$\mathcal{E}_{rot} = \frac{\hbar^2 l^2}{2\mathcal{J}} \leqslant E^* \qquad\qquad l^2 \leqslant \frac{E_I - Q}{\hbar^2} 2\mathcal{J} ,$$

where E_I and $\mathcal{J}$ are the ion energy and the moment of inertia of the compound nucleus, respectively.

At the same time, in reactions induced by ^{12}C, ^{16}O and ^{20}Ne ions, compound nuclei produced may have excitation energies up to 150 MeV [2] (Fig. 1). Different decay characteristics of excited nuclei are rather well described by the thermodynamic approach, in which nuclear properties are defined by macroscopic parameters such as entropy, level density, the moment of inertia, etc. In fact, experimental data in the region of light and medium-mass nuclei (e.g. partial cross sections, energy spectra and angular distributions of the particles emitted, the characteristics of gamma radiation, and others) are generally in satisfactory agreement with the calculations based on this approach.

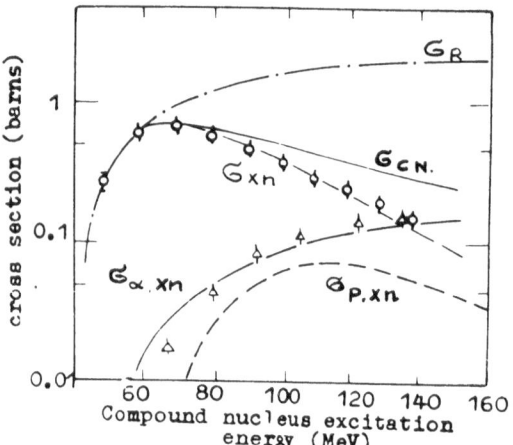

Fig. 1. Cross section for compound nucleus formation in the reaction $^{130}Te + ^{12}C$ (ref.[2]).

A statistical approach proves applicable also to the region of heavy nuclei, which decay mainly by fission [3]. From analysing the mass and charge distributions and the kinetic energy spectra of fission fragments, one has obtained information both about the formation probability and the fission mechanism for highly excited heavy nuclei.

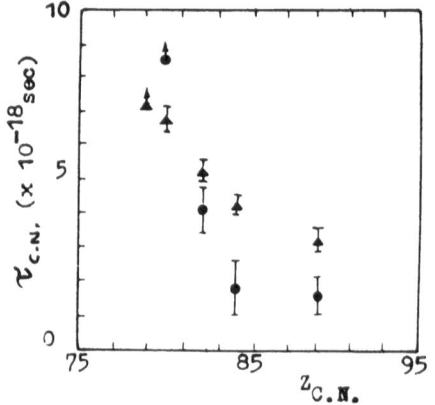

Fig. 2. Compound nucleus half-lives measured in the reactions (HI,f) at an excitation energy of 60 MeV using the blocking effect technique.

We note that theoretical models are based on the classical concept of a compound nucleus as an intermediate state whose lifetime considerably exceeds the characteristic time of nucleon motion in the nuclear volume ($\frac{\tau_{c.N.}}{\tau_n} \geqslant 10^3$). This assumption has recently been substantiated by the direct lifetime measurements for compound nuclei with Z=75-95, formed in reactions induced by ^{12}C, ^{16}O and ^{22}Ne ions in a wide range of excitation energies [4] (Fig. 2).

All this applies to reactions induced by ions with mass $\leqslant 20$ and an energy range up to 10-12 MeV/nucl.

It is not excluded that at higher energies the nature of the interaction may change, and it will be more justified to employ the hydrodynamical description of the process, as considered in the subsequent theoretical papers [5,6].

However, the second generation of heavy ion facilities presently under construction in Darmstadt, Berkeley and Dubna, as well as the projects for future machines in Oak Ridge and in France indicate that heavy ion physics is developing in the direction of increasing projectile masses rather than increasing energies.

The hypothesis about the possible existence of the island of stability in the region of superheavy elements seems to me to have played an important role in this development.

It is known that all of the heavy elements with Z=102-105 have been produced by heavy ion reactions involving compound nucleus formation followed by neutron evaporation. It was natural to make an attempt to synthesize also superheavy elements by using ions with mass $\geqslant$ 40.

However, a large number of experiments carried out at different laboratories have yielded only the upper limits on the production cross sections for superheavy nuclei (table 1).

Table 1.

Reaction	Compound nucleus		$T_{1/2}$	Cross section
	Z	N		upper limit(cm^2)
^{248}Cm + ^{40}Ar	114	174	10^{-9} sec	10^{-32}
^{232}Th + ^{84}Kr	126	190	10^{-7} sec	5×10^{-30}
^{238}U + ^{84}Kr	128	194	10^{-7} sec	5×10^{-30}
^{238}U + ^{68}Zn	122	184	10^{-9} sec	5×10^{-32}
^{243}Am + ^{68}Zn	125	186	10^{-9} sec	5×10^{-32}
^{232}Th + ^{76}Ge	122	186	5×10^{-3} sec	10^{-34}
^{238}U + ^{76}Ge	124	190	5×10^{-3} sec	2×10^{-34}
^{238}U + ^{65}Cu	121	182	3 days	10^{-34}

It is not excluded that the absence of any effect in these experiments is accounted for by the properties of the nuclei being synthesized, since in all cases isotopes far from the most stable nucleus with Z=114 and N=184 might be produced. However the results of the subsequent experiments suggest that the processes associated with the mechanism of the interaction between nuclei as complex as these may play an important role here.

We shall discuss this point in more detail. With an increase in the incident ion mass, one can expect a monotonic decrease in the

probability of compound nucleus formation because of limitations im-
posed by the value of the critical angular momentum. On the other hand,
this decrease in production cross section must not be dramatic, as confirmed by the experimental data obtained in reactions with ^{40}Ar, ^{70}Zn,
^{74}Ge and ^{84}Kr ions for compound nuclei with $A \leqslant 220$ (refs.[12-16])
(see Fig. 3).

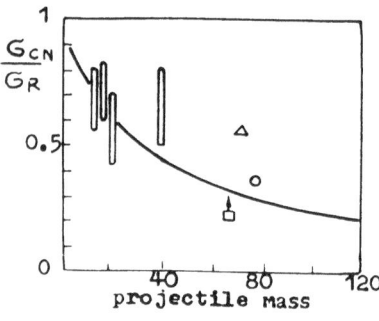

Fig. 3. Cross section for
the formation of nuclei
with $A \leqslant 220$, obtained in
reactions induced by ions
with different masses.

Therefore, in accordance
with the hypothesis ad-
vanced by Swiatecki[18],
the probability of the
formation of a heavy com-
pound nucleus will have
a second restriction
associated with the cri-
tical shape or critical
deformation β_{crit} of the
nucleus. At first approximation, this critical de-
formation is determined by
the position of the fission
barrier vertex. Thus, in
the classical representa-
tion, the compound nucleus
formation probability is
directly related to the
probability of its production with deformation
$$\beta < \beta_{crit}.$$

The situation in the region of heavy
nuclei is more complicated. In accordance with the present-day concepts, the
fission barrier of transuranic elements
has a complex structure, which results
from the strong influence of shell effects
on the deformation energy of the nucleus[17].
In an extreme case, the fission barrier
may entirely be determined by the shell
correction, and its structure will then
be substantially different from that observed for masses $A_c \leqslant 220$ (Fig. 4).

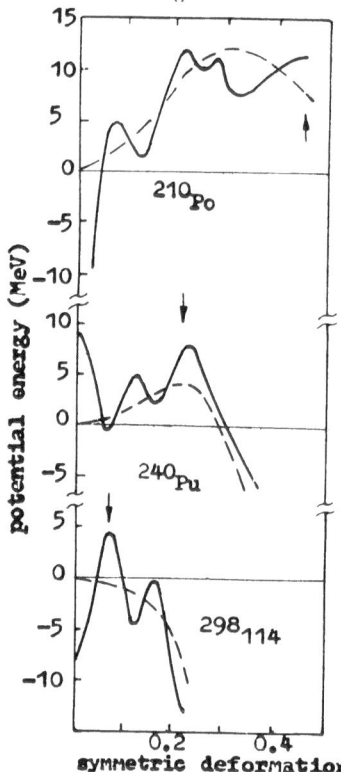

Fig. 4. Variations in nuclear potential
energy with increasing deformation, according
to Nix[22]. Arrows show fission barrier ver-
tices, the dashed line is the liquid drop
energy.

This, however, may be hindered by some factors associated with the properties and nature of the interaction potential of the two colliding nuclei [19].

It is natural that nuclei with $\beta > \beta_{crit}$ will be quite unstable against disintegration into two parts. In contrast to compound nucleus fission, this process is conventionally called "quasifission" [20], implying that the characteristic features of fission should be substantially different from those of "quasifission".

Indeed, if the original configuration of the compound system determines the nature of the disintegration of the nucleus, the mass distribution of the fragments will then be analogous to that observed in deep inelastic processes involved in multinucleon transfer reactions [21]. On the contrary, if this kind of correlation does not exist and by the moment of scission the system becomes statically balanced with respect to the fission degrees of freedom, the characteristics of the fragments will slightly differ from those expected for fission of a compound nucleus [3].

In the experiments carried out by Lefort et al. [13], in which ^{84}Kr was used to irradiate ^{186}W and ^{209}Bi ions, the reaction products appeared to be grouped in the vicinity of the masses of the interacting nuclei and no symmetric fission was observed. Unfortunately the sensitivity of these experiments was rather poor.

In bombarding ^{181}Ta with ^{84}Kr and ^{136}Xe ions, in addition to the products of multinucleon transfer reactions, we observed the formation of fission fragments with symmetric mass and charge distributions, as expected for the decay of a compound nucleus [11]. In the subsequent experiments of G.Seaborg et al. [23], involving the bombardment of ^{238}U with ^{40}Ar and ^{84}Kr ions, one also observed fission fragments with a wide range of both Z and A, although the use of a ^{238}U target may seem to create additional difficulties during the data analysis due to the presence of fission products of nuclei adjacent to uranium.

Thus, the problem of the formation of compound nuclei with Z > 110 remains open.

In reverting to the synthesis of superheavy elements, one should emphasize the necessity of producing a compound nucleus with a relatively low excitation energy. At the same time, in moving from light ions to those of Ge, Kr and Xe, one observes a decrease in the interaction effective radius. This in turn leads to an additional enhancement in the reaction Coulomb barrier. In the bombardments of ^{232}Th with ^{76}Ge ions, the Coulomb barrier for the fission reaction has turned out to be 30 MeV higher than that corresponding to few-nucleon transfer reactions [11]. The enhancement of the Coulomb barrier in the ^{84}Kr bombardments of ^{74}Ge and ^{238}U is about 50 MeV according to refs.[15,24]. If so, one can show that the minimum excitation energy of compound nuclei is about 40-50 MeV practically in all the experiments on the synthesis of superheavy elements. At this excitation energy, the fission barrier is practically equal to zero as a consequence of vanishing shell effects with the increasing temperature of the nucleus. This is expected to lead to a sharp decrease in the cross section for the ground state production of heavy and superheavy nuclei [25].

This is the possible reason why the unjustified optimism in attempts to land on the "island of stability" has presently given up place to pessimism.

I think, however, that the available information is insufficient to make definite conclusions about the complicated problem of the synthesis of superheavy elements.

Following the 1973 Munich Conference, we have carried out a number of experiments to synthesize different isotopes of elements with atomic numbers 100, 104 and 106 using reactions induced by Ar, Ti and Cr ions. Late in 1974 we hope to start experiments with iron ions.

The main point of the experiments performed has been as follows.

If in going from ions with $A_I \leqslant 20$ to those with $A_I \sim 80$, the mechanism of the nuclear interaction suffers essential changes, the same tendency should also manifest itself in the intermediate region of ion masses, $A_I \sim 40$-50. These sufficiently heavy ions enable one to investigate the (HI,xn) reactions, which lead to the formation of very fissile neutron-deficient isotopes of fermium and, possibly, still heavier elements. This offers the possibility of studying the mechanism of compound nucleus formation directly, and, at the same time, eliminates numerous uncertainties involved in the identification of the final product.

The description of these experiments is given below.

2. EXPERIMENTS WITH ^{40}Ar IONS. FORMATION OF Fm COMPOUND NUCLEI (Z=100)

An analysis of the limited number of target-projectile combinations leads to the choice of the ^{206,207,208}Pb(^{40}Ar,xn)^{244,246}Fm reactions. The isotope ^{244}Fm is one of the known strongly fissionable nuclei. It suffers spontaneous fission with a 100% probability and has a half-life of about 3.5 msec. The properties of ^{244}Fm have been established by Nurmia et al.[26], who synthesized it by the reaction ^{233}U(^{16}O,5n)^{244}Fm.

The choice of a lead target has a certain sense. Since Pb isotopes are "magic" nuclei and the nucleus being synthesized is a deformed one, there is a large gain in the reaction Q value, and the compound nucleus may have a small excitation energy. This is exemplified by Fig. 5 where the calculated values of the minimum excitation energy E_{min}^{*} of the compound nuclei ^{248}Fm, ^{258}Ku and ^{262}106 are presented as a function of the incident ion mass A_I. It is seen that with increasing ion mass up to $A_I \sim 20$-30 E_{min}^{*} reaches a value of 40-50 MeV and then decreases to a minimum value at $A_I \sim 40$-50. The decrease in the excitation energy of the compound nucleus is expected to lead to a decrease in the number of neutrons emitted, which in turn may result in an enhancement of cross sections for nuclear production in the ground state. The dependence presented in Fig. 5 is valid provided that the fusion barrier is determined by the classical expression

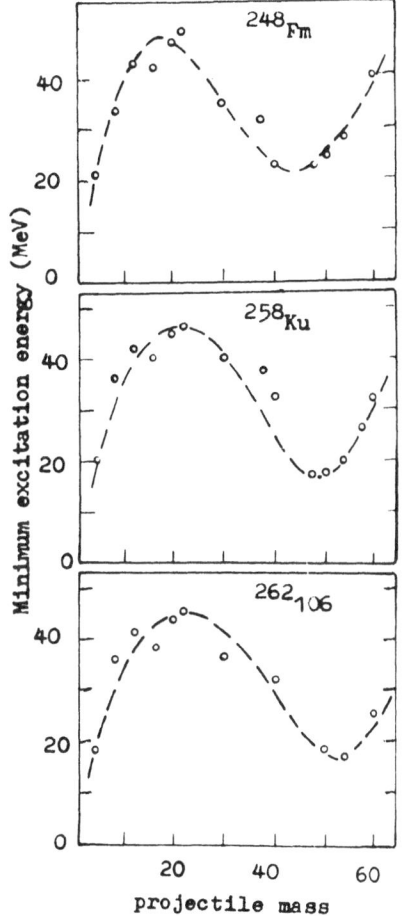

Fig. 5. Minimum excitation energy of the compound nucleus formed in reactions induced by ions with different masses.

$$B_{int} = Z_1 Z_2 e^2 / r_e (A_1^{1/3} + A_2^{1/3}) \quad (2.1)$$

with a constant value of the interaction radius $r_e = 1.45 \times 10^{-13}$ cm^2 both for light and heavy nuclei.

The B_{int} values have been determined in separate experiments involving the measurements of fission cross sections as a function of the ion energy in the bombardments of ^{208}Pb with ^{40}Ar and ^{54}Cr ions [27]. In the energy range investigated, this dependence can be presented by the following classical relation

$$\sigma_{C.N.} = \pi r^2 (A_1^{1/3} + A_2^{1/3})^2 (1 - \frac{B_{int}}{E_I}) \quad (2.2)$$

and the B_{int} value can be determined from the linear extrapolation of $\sigma(I/E_I)$ at $\sigma_{c.N.} \rightarrow 0$. The experimental data presented in Fig. 6 suggest that $r_e = 1.44 \pm 0.02$ f both for ^{40}Ar and ^{54}Cr ions.

Thus, in the combination of the Pb target with ions of mass up to 54, no enhancement in the Coulomb barrier has been observed, which contradicts the theoretical predictions [28] (Fig. 7).

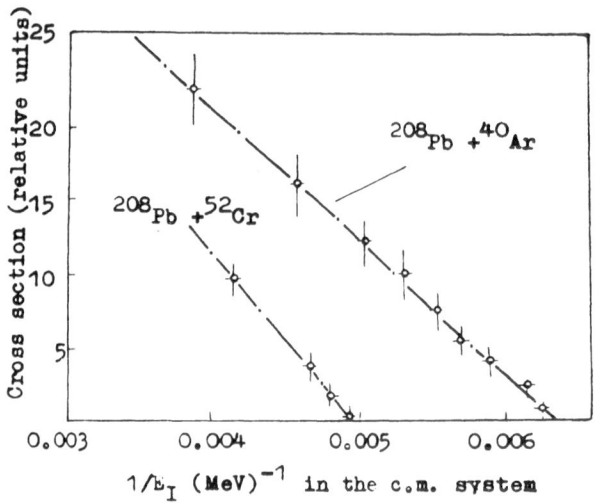

Fig. 6. The fission cross section as a function of the energy of Ar and Cr projectiles. Curves refer to calculations using eq. (2.2) at r_e = 1.44 f.

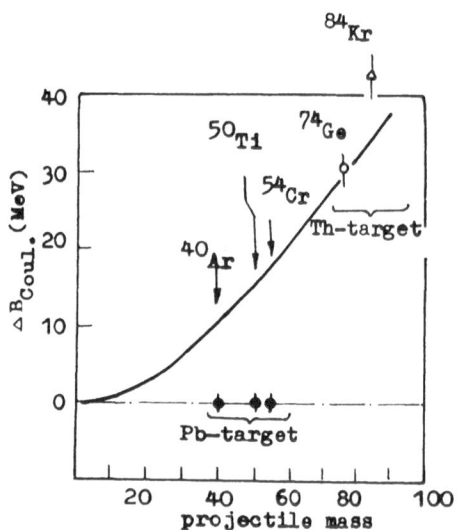

Fig. 7. Enhancement of the reaction Coulomb barrier over the calculated value (at r_e = 1.45 f), as a function of the projectile mass. Experimental points show the results obtained using Ar, Ti and Cr ions, and the data from refs. 11,24) obtained using Ge and Kr ions.

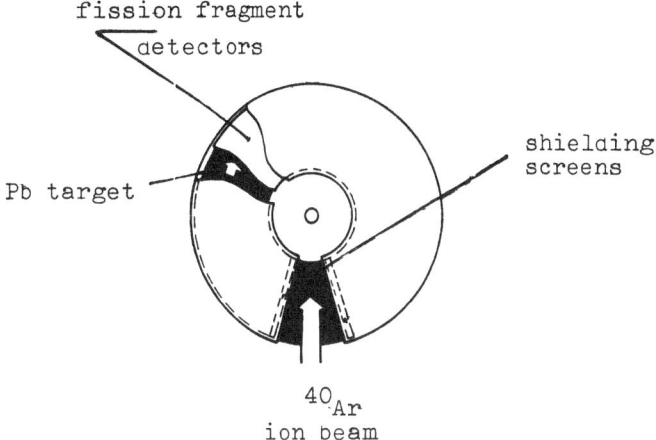

Fig. 8. Schematic view of the device for detecting
short-lived spontaneous fission nuclei.

In order to detect the spontaneously fissioning nuclei of ^{244}Fm,
^{246}Fm, we used the device shown schematically in Fig. 8.

The beam of ^{40}Ar ions with an energy of 225 MeV and intensity
up to 10 μA passed through diaphragms to strike a water-cooled Dural
disk serving as a target and having the shape of a truncated cone with
a base angle of 20° and maximum diameter of 250 mm. By vacuum vapouri-
zation, a lead layer was deposited onto the lateral face of the disk,
the thickness of the layer being from 2 to 5 mg/cm² in different experiments. The rotation velocity of the disk varied in the range of 3 to
3000 rev/min. The disk temperature was controlled during the experiments
using special detectors and did not exceed 60°C. At a distance of 3 mm,
the target was surrounded by dielectric detectors for fission fragments,
which were mica plates with the uranium and thorium content of less than
10^{-7} g/g.

In the experiment described, the lead layer served as both a thick
target in which the excitation functions of the Pb(Ar,xn) reaction were
integrated, and a recoil catcher foil. At the Ar ion energy of 225 MeV,
recoil nuclei stopped at a depth of 1 to 3 mg/cm² in the direction perpendicular to the plane of the detectors. The detection efficiency under
these experimental conditions was determined to be about 50%.

The separated isotopes ^{206}Pb, ^{207}Pb and ^{208}Pb with high enrich-
ment, as well as ^{209}Bi were used as targets. Under these experimental
conditions the background due to both spontaneous fission of heavy elements and spontaneous fission isomers was practically eliminated.

The sensitivity of the experimental device permitted the detection

of one spontaneous fission track after a 10 hour bombardment with the
^{40}Ar beam of nearly 5 μA intensity, which corresponds to the production cross section of 2 x 10^{-36} cm^2.

The experimental values of the cross sections for the reaction
(^{40}Ar,xn) at x = 1, 2, 3 and 4 as well as some data on the cross section for the formation of ^{244}Fm in the reaction ^{209}Bi(^{37}Cl,2n) are
presented in Table 2.

Table 2.

Reaction	E_I(lab) (MeV)	B_{Coul} (lab,MeV)	Total flux x10^{16}	Number of tracks	$T_{1/2}$	Cross section (nb)
^{208}Pb(^{40}Ar,4n)^{244}Fm	220		6	214	(4$\pm$0.5)ms	1.5
^{207}Pb(^{40}Ar,3n)^{244}Fm		187$\pm$1.5	2	111	- " -	5
^{208}Pb(^{40}Ar,2n)^{246}Fm			10	70	(1$\pm$0.2)s	7
^{207}Pb(^{40}Ar,1n)^{246}Fm			6	1	- " -	<0.1
^{206}Pb(^{40}Ar,2n)^{244}Fm	220		1	35	4 ms	3
^{209}Bi(^{37}Cl,2n)^{244}Fm	240	179	5	18	- " -	0.1

To analyze the results obtained, we carried out calculations for
the excitation functions of xn-reactions assuming the mechanism of nuclear fusion to remain unchanged in going from relatively light projectiles to ions such as ^{37}Cl and ^{40}Ar (ref.[29]).

As a result, the cross section of the (HI,xn) reaction can be
written in the following form

$$\sigma_x (E_I) = \left\{ \prod_{i=1}^{x} [\Gamma_n/(\Gamma_n + \Gamma_f)]_i \right\} \sum_{l=0}^{l_{cr}} \sigma_l(E_I) P_{x,l}(E^*) \qquad , \quad (2.3)$$

where E_I is the ion energy, σ_l is the partial cross section,
$P_{x,l}(E^*)$ is the probability for the emission of x neutrons from the
compound nucleus with excitation energy E^* and angular momentum l.

The l_{crit} value was determined from the empirical relation [30)]

$$\sum_{l=0}^{l_{cr}} \sigma_l \Big/ \sum_{l=0}^{\infty} \sigma_l = (1 + 0.03 A_I)^{-1} \qquad , \quad (2.4)$$

where

$$\sigma_l = \pi \lambda^2 (2l+1) T_l \qquad , \quad (2.5)$$

T_1 is the coefficient of transmission of the l-th partial wave through
the interaction potential of the type [31)]

$$V(r) = \frac{Z_1 Z_2 e^2}{r} + \frac{\hbar^2 l(l+1)}{2 \mu r^2} + V_0 \exp \frac{r_0(A_1^{1/3} + A_2^{1/3}) - r}{d} \cdot \quad (2.6)$$

The following potential parameters were used: $V_0 = -70$ MeV, $r_0 = 1.25 \times 10^{-13}$ cm, $d = 0.44 \times 10^{-13}$ cm and 0.34×10^{-13} cm(ref.[32]). The transmission coefficients T_ℓ have been calculated in the inverted parabola approximation.

The quantity $P_{x,\ell}(E^*)$ was calculated by a formula given in ref.[29]. The values of the parameters incorporating this formula were obtained from the condition of the best agreement with the experimental data for the shape of the excitation functions of the reactions (HI,xn) (refs.[29,30]).

In calculating the ratio of the partial widths of neutron emission and fission, Γ_n/Γ_f, the empirical relation of Sikkeland [33] was used.

The results of the calculations are presented in Fig. 9, while a comparison with experimental data is given in Figs. 10 and 11.

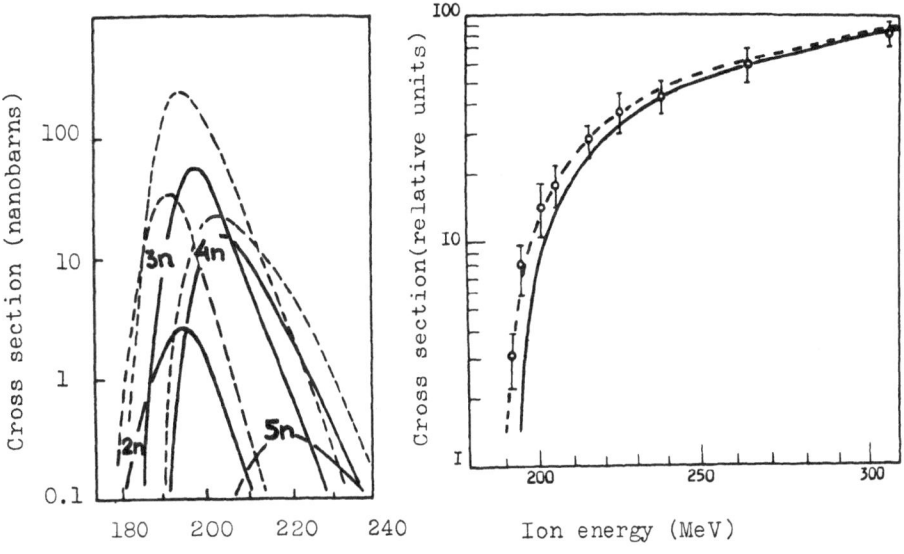

Fig.9. Excitation functions of the reactions $^{208}Pb(^{40}Ar,xn)^{248-x}Fm$, calculated for the two diffuseness parameters d = 0.44 f (dashed line) and d = 0.34 f (solid line).

Fig. 10. Cross section for the formation of the compound nucleus ^{248}Fm as a function of the ^{40}Ar ion energy. Experimental points refer to the fission data, the solid line corresponds to the best agreement with the data on the reaction $(^{40}Ar,xn)$.

The magnitudes of the xn-reactions cross sections and their qualitative agreement with the calculated values indicate that the interaction of ^{40}Ar ions with a heavy nucleus leads, with a fairly high probability, to compound nucleus formation. A more accurate quantitative analysis

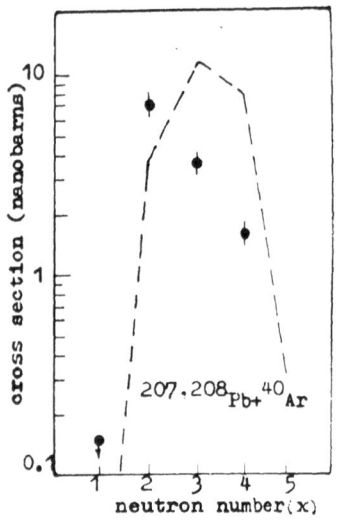

Fig. 11. Experimental
and calculated cross
sections in the maxima
of excitation functions
for the Pb(Ar,xn)Fm
reactions.

of the data is impossible since the production cross sections measured for highly fissionable nuclei are mainly 10^{-7}-10^{-10} th of
the total fusion cross section. Such an analysis necessitates the precise determination
of the fraction of nuclei that did not undergo fission, which requires more detailed
measurements of the ratio Γ_n/Γ_f as a
function of the excitation energy and the
angular momentum of the nucleus.

At the same time, after considering
the Pb+Ar reaction separately, one can note
an interesting feature, namely that the cross
sections of reactions with two and three
neutrons emitted are comparable with and even
exceed the cross section for the reaction
involving 4 neutrons evaporated. This fact
seems to be rather important for both the
production of weakly excited compound nuclei
and the mechanism of their production. Therefore, we shall consider this problem in more detail.

As indicated previously, the probability for the nucleus to be
produced in the ground state is principally dependent on its excitation
energy. The minimum value of the compound nucleus excitation energy
in turn is determined by the interaction barrier and the reaction Q-
value, i.e.

$$E^*_{min} = B_{int} + Q, \qquad (2.7)$$

At the same time, from Fig. 5 it follows that E^*_{min} is strongly
dependent on the projectile mass. This accounts for the well known
experimental fact that, in contrast to the $^{208}Pb(^{40}Ar,xn)$ reaction,
in the synthesis of heavy elements with Z=100-105 using ^{12}C, ^{14}N, ^{16}O
and ^{22}Ne ions no reactions with the emission of a small number of neutrons were observed.

In view of the fact that the $(^{40}Ar;2-3n)$ reactions occur at projectile energies close to the interaction barrier, their cross sections
are extremely sensitive to the absolute barrier value. As seen from
Fig. 9, a 4 MeV change in the B_{int} value changes the cross section for
the reaction with the emission of 2 neutrons by over a factor of 10.
This makes it possible to determine the B_{int} value from the experimental cross sections for reactions with $x \leqslant 3$ exactly enough. The value
obtained is in good agreement with the previous measurements(Fig.10).

On the other hand, variations in the masses of the interacting nuclei cause corresponding irregular changes in the Q value. Therefore, for close target-projectile combinations, several MeV deviations from the smooth E^*_{min} dependence shown in Fig. 5 can be observed. This should also lead to a noticeable difference in the cross sections of reactions with $x < 3$. As an example, one can mention the $^{209}Bi(^{37}Cl,2n)^{244}Fm$ and $^{206}Pb(^{40}Ar,2n)^{244}Fm$ reactions, in which the E^*_{min} quantity takes on the corresponding values equal to 30 MeV and 33 MeV, respectively. Both the experimental and calculated cross sections for these reactions differ over a factor of 10. This is a consequence of the small difference in the E^*_{min} values.

Thus, the intrinsic self-consistency of the data for the formation of compound nuclei in the Ar+Pb and Cl+Bi reactions permits the use of this method of calculation to analyse the possibilities of producing heavier compound nuclei. Therefore, further experiments were aimed at the synthesis of the isotopes of element 104, kurchatovium.

3. EXPERIMENTS WITH ^{50}Ti IONS. PRODUCTION OF COMPOUND KURCHATOVIUM NUCLEI (Z=100).

At present, the properties of 5 isotopes of element 104 (2 even-even nuclei) and 3 isotopes of element 105 are known. The spontaneously fissioning Ku isotope with Z=104, mass number 260 and half-life of one tenth sec was first synthesized by the $^{242}Pu(^{22}Ne,4n)$ nuclear reaction at Dubna [34]. Subsequently, the chemical properties of kurchatovium[35-37] were investigated. Later, the alpha decay of the odd isotopes of this element with mass numbers 257, 259 and 261 was studied at Berkeley[38-40]. In 1970, we found [41] that ^{259}Ku undergoes spontaneous fission in 15-20% of the cases (the partial SF half-life is about 20-30 sec), whereas the American authors [42] determined the spontaneous fission half-lives for the isotopes ^{258}Ku and ^{261}Ku to be equal to 10 msec and $\geqslant$ 10 min, respectively.

From comparing these data with the systematics of Ghiorso et al. [38,43] obtained by extrapolating the properties of elements 100 and 102, it follows that spontaneous fission is strongly forbidden for the odd isotopes of element 104 and the half-life of the even isotope ^{260}Ku is 10^6-10^7 times as large as the predicted one (Fig. 12). This long half-life of ^{260}Ku was the reason of prolonged discussions between the LBL at Berkeley and the JINR Laboratory of Nuclear Reactions at Dubna. However, the experiments on the synthesis and investigation of the properties of this isotope, repeated in 1969 using an improved technique, confirmed the original data.

The problem of the properties of heavy nuclei is in principle

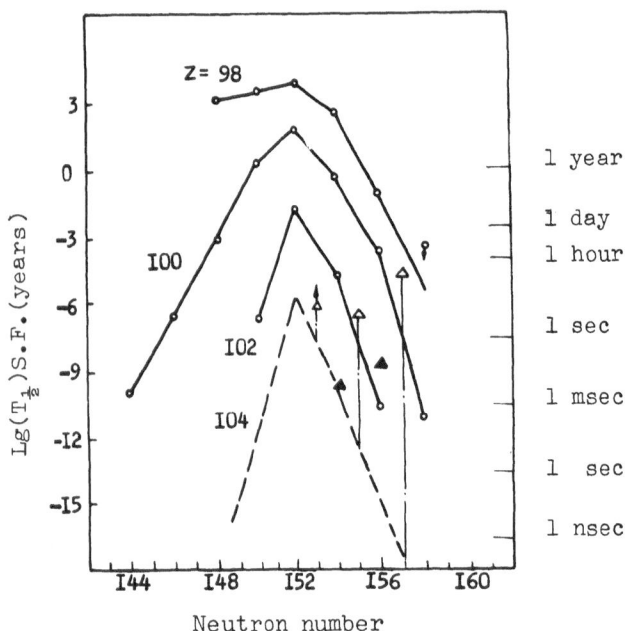

Fig. 12. Spontaneous fission half-lives of even-even nuclei with Z=98-104. The dashed line refers to the systematics of Ghiorso et al.(refs. 38,43) for isotopes with Z=104.

very important not only in terms of the discovery of a new element and investigation of its physical and chemical properties, but also in order to establish direct relationship between the spontaneous fission half-lives of heavy elements and the existence of the region of stability of superheavy nuclei.

With this in mind, it is necessary to investigate the properties of nuclei in a wider range of mass numbers, in particular, by determining the spontaneous fission half-lives of isotopes with N ≤ 152 for nuclei with Z=104.

However, this kind of investigations involve certain difficulties. Heavy nuclei occurring in fusion reactions have a high excitation energy, and only $10^{-8}-10^{-10}$th of them can de-excite by emitting neutrons and gamma rays. As a result, the production cross sections for heavy nuclei appear to be equal to $10^{-34}-10^{-32}$ cm^2. On the other hand, as the atomic number increases, the spontaneous fission half-life decreases, especially in the region of neutron-deficient isotopes. This leads to a considerable increase in the background due to short-lived spontaneous fission isomers which are formed with a large cross section in multinucleon transfer reactions.

However, the situation changes substantially if, instead of the heavy Pu, Cm and Cf isotopes, one uses the targets of the stable Pb nuclei and bombards them with the beam of ^{50}Ti ions.

Fig. 13 exemplifies this by showing the excitation functions for the ^{207}Pb(^{50}Ti,xn)$^{257-x}$Ku reaction. The maximum cross section corresponds to the emission of two or three neutrons from the compound nucleus ^{257}Ku and sharply decreases at x ≥ 3. It has been shown previously[44] that reactions involving the emission of two neutrons are very sensitive to the value of the minimum excitation energy of the compound

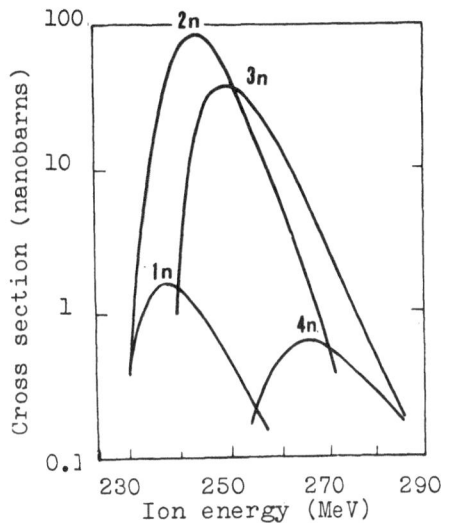

Fig. 13. Calculated excitation function for the reaction $^{207}Pb(^{50}Ti,xn)^{257-x}Ku$.

nucleus. Our estimates suggest that as one goes from ^{48}Ti ions to those of ^{50}Ti, the cross sections for the 2n-reactions increase tens of times. Therefore, in our experiments we made use of a ^{50}Ti ion beam.

3.1. ACCELERATION OF Ti IONS.

The calculated value of the interaction barrier in the reaction $^{208}Pb+^{50}T$ is equal to 230-235 MeV in the lab. system. The maximum beam energy on the 310 cm cyclotron of the JINR Laboratory of Nuclear Reactions is determined by the ratio $E_{max} = 250 \ Z_i^2/A_I$ and equals 245 MeV and 320 MeV for $^{50}Ti^{+7}$ and $^{50}Ti^{+8}$, respectively.

In order to facilitate the acceleration of these highly charged ^{50}Ti ions, a special-type ion source was developed which used the enriched metal isotope ^{50}Ti as a working material. The intensity of the external $^{50}Ti^{+7}$ ion beam was 5×10^{10}-10^{11} particles per second, whereas that of the internal $^{50}Ti^{+8}$ ion beam was 2×10^{11} particles/sec. The Ti^{+7} ion energy was 10-15 MeV higher than the interaction barrier. Therefore the external beam was used to investigate reactions with the emission of two neutrons, while the majour part of the experiments were carried out at the internal beam.

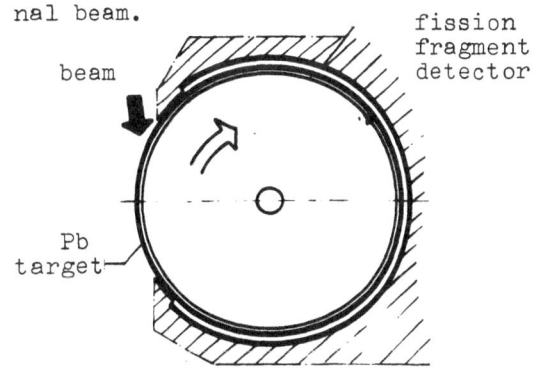

Fig. 14. Experimental device for detecting spontaneous fission nuclei at the internal beam.

The experimental device intended for the detection of short-lived spontaneously fissioning nuclei is shown schematically in Fig. 14. The ^{50}Ti beam is incident tangentially upon the lateral face of a hollow cylinder placed vertically, which is rotated with a maximum angular velocity of 3000 rev/min. The cylinder lateral face is covered by a Pb layer of about 2 mg/cm² deposited by vacuum vaporization. This layer serves as both target and recoil catcher foil. As the angle between the beam direction and the cylinder surface is very small, the Pb layer is an "infinitely thick" target, in which the excitation functions are integrated from the Coulomb barrier to E_{max}.

Spontaneous fission fragments were detected using mica detectors with small admixtures of U and Th. This technique was tested in previous experiments on producing ^{244}Fm in the ^{206,207,208}Pb$(^{40}$Ar,xn$)^{244}$Fm reactions. These experiments have shown that it can be used successfully for the detection of short-lived spontaneously fissioning activities provided that their half-lives exceed 3 msec and the production cross section is larger than 10^{-35} cm^2.

3.2. SYNTHESIS OF NEW KURCHATOVIUM ISOTOPES

The isotope ^{256}Ku can be produced by the ^{208}Pb$(^{50}$Ti,2n$)^{256}$Ku reaction. It follows from the systematics (see Fig. 12) that its alpha-decay half-life lies within the limits of 50 to 200 msec, whereas its partial spontaneous fission half-life is considerably longer because of the stabilizing effect of the subshell N=152. Therefore, by using the ^{208}Pb+^{50}Ti reaction one could hopefully detect spontaneous fission of the isotope ^{252}102 ($T_{\frac{1}{2}} \sim 2$ sec, 30% S.F.), which is formed as a result of the alpha-decay of the nucleus ^{256}Ku.

In the initial experiments, the rotation velocity of the cylinder was chosen to be equal to 8 rev/min, which permitted the detection of spontaneous fission with $T_{\frac{1}{2}} \geqslant 0.5$ sec. In the bombardment of ^{208}Pb with the integral flux of 8 x 10^{15} ^{50}Ti ions, 12 spontaneous fission tracks were observed. This number is substantially smaller than that expected. This may indicate either that the mechanism of the fusion reaction changes essentially as one goes from ^{40}Ar ions to those of ^{50}Ti, or the properties of the isotope ^{256}Ku considerably differ from those predicted.

In order to verify the second assumption, we repeated the experiments at the rotation velocity of 1500 rev/min. With an integral flux of about 10^{16} ions, 70 tracks of fission fragments were recorded. The time distibution of these tracks is shown in Fig. 15.

The subsequent experiments were carried out using a ^{207}Pb target at the cylinder rotation velocity of 1500 rev/min. With an integral flux of about 1.2 x 10^{16} ions, 53 spontaneous fission fragments were detected, whose time distribution is presented in Fig. 16a. It is seen from this Figure that the half-life considerably exceeds the time interval chosen. Therefore, the experiment was repeated at a rotation velocity of 3.6 rev/min, the interval being 1.5 to 13 sec . The time distribution of these tracks is presented in Fig. 16b.

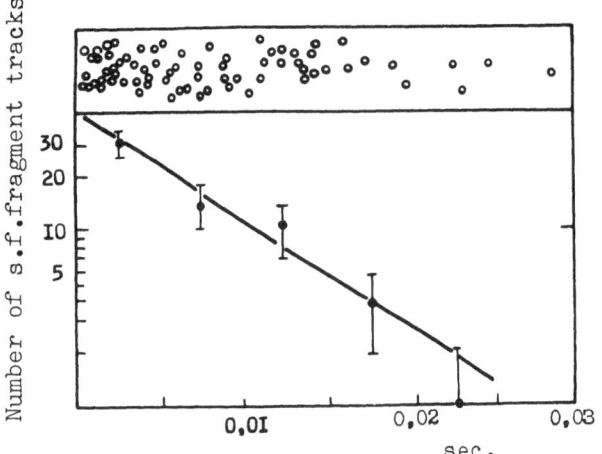

Fig. 15. Time distribution of spontaneous fission tracks in the reaction $^{208}Pb + ^{50}Ti$.

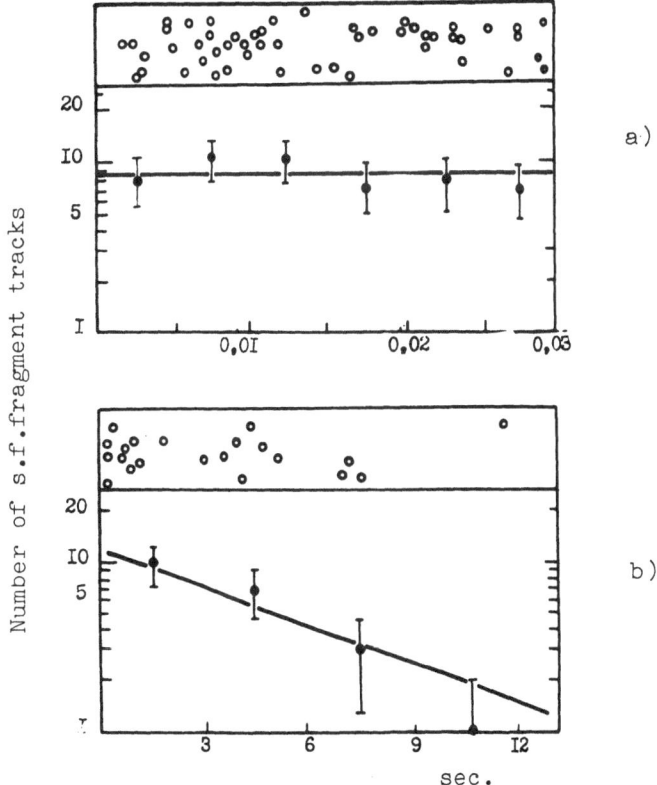

Fig. 16. Time distribution of fission tracks in the reaction $^{207}Pb + ^{50}Ti$ a) at a rotation velocity of 1500 rev/min; b) at a rotation velocity of 3.6 rev/min.

The experiments carried out suggest that at the bombardment of ^{207}Pb with ^{50}Ti ions a spontaneous fission activity with a half-life of several seconds has been formed.

Finally, the last series of experiments involved the bombardment of a ^{206}Pb target. With an integral flux of about 0.4×10^{16} ions, only two tracks were recorded. In other words, in the system ^{206}Pb+^{50}Ti, only the upper limit on the cross sections for the formation of spontaneous fission nuclei can be determined. All the experimental results obtained are presented in Table 3.

Table 3.

Reaction	E_I lab. (MeV)	$B_{Coul.}$ lab. (MeV)	Total flux x 10^{16}	Number of tracks	$T_{\frac{1}{2}}$	Cross section (nb)
^{208}Pb + ^{50}Ti	260	233	1	70	~5 ms	6
			1	16	>1 s	1.5
^{207}Pb + ^{50}Ti			2	90	~4 s	3
			1.2	–	~5 ms	<0.3
^{206}Pb + ^{50}Ti			0.4	2	≥3 ms	<0.3
^{206}Pb(^{40}Ar,2n)^{244}Fm	220	187	1	35	4 ms	3
^{205}Tl(^{45}Sc;α,2n)^{244}Fm	240	219	0.6	0	≥1 ms	<0.1
^{203}Tl(^{45}Sc;α,0n)^{244}Fm	240		0.6	0	≥1 ms	<0.1

The absolute error in the cross section determination is estimated to be no more than a factor of 2, the relative inaccuracy being about 30%.

Thus, at the bombardment of the targets from separated Pb isotopes with ^{50}Ti ions, one has observed the formation of two spontaneous fission activities with strongly different half-lives, about 5 msec and several seconds.

The 5 msec activity observed in the ^{208}Pb+^{50}Ti reaction cannot be attributed to nuclei with atomic number Z< 104, since all the isotopes of element 102 through N=148 are known, and they do not possess such properties [45], while spontaneous fission is strongly forbidden for the odd isotopes of element 103 (ref. [46]). The maximum yield of this activity has been observed in the reaction on ^{208}Pb. The yield decreases over a factor of 10 in the case of a ^{207}Pb target and is practically absent for ^{206}Pb.

On the basis of the experimental cross sections of the reactions
and the properties of the known Ku isotopes and those of the lighter
elements, one may assume that the activity observed is due to the decay
of the isotope ^{256}Ku, which is formed in the ^{208}Pb(^{50}Ti,2n)^{256}Ku reaction.

As it follows from the next sections, the problem of the properties of the Ku isotope with N=152 is in principle very significant.
Therefore, the exact identification of the mass number of the $\sim$5 msec
activity is of importance. As spontaneous fission is strongly forbidden for odd isotopes, the possible candidates are the even nuclei
^{256}Ku and ^{258}Ku, which are formed in reactions with the emission of
2 and 4 neutrons, respectively.

In spite of the fact that the calculated cross sections for reactions with the emission of 2 and 4 neutrons considerably differ (see
Fig. 13), we have carried out separate experiments to measure the integral excitation function for the activity with $T_{\frac{1}{2}}\sim$5 msec. Fig. 17 shows

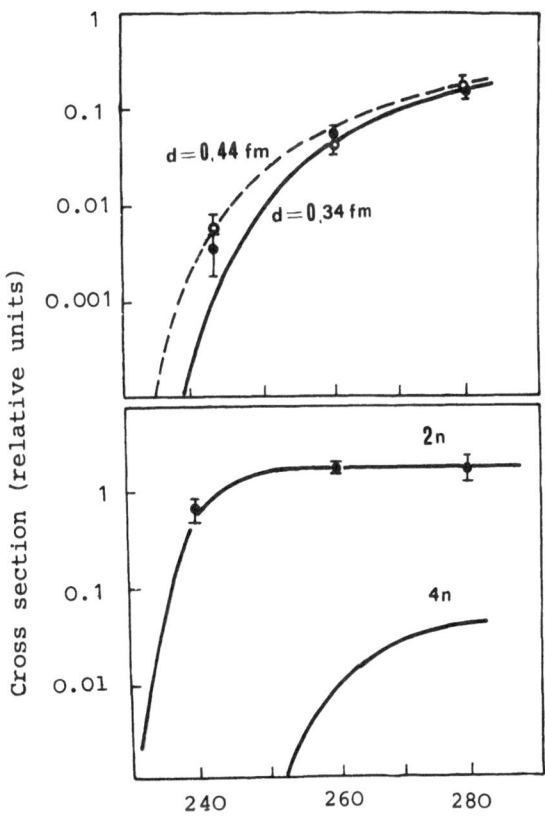

Fig. 17 Integral excitation
functions of the ^{151}Pm nuclei
(fission fragments), ^{204m}Pb
(transfer reaction products)
(upper part) and of the isotope
^{256}Ku (lower part). Solid
curves show calculated data.

the experimental and calculated values of the integral cross sections for the formation of the 5 msec activity and of the nuclei ^{151}Pm (the fragments of fission of a compound nucleus) and ^{204m}Pb (transfer reaction products) at the bombardment of ^{208}Pb with ^{50}Ti ions. The variations presented in Fig. 17 provide additional evidence for the formation of the short-lived activity in the reaction with two neutrons emitted and its being the isotope ^{256}Ku.

From the direct comparison of the yields of the short-lived and long-lived activities in the ^{208}Pb+^{50}Ti reaction, one can conclude that the isotope ^{256}Ku undergoes spontaneous fission in most cases.

It should be noted that the experimental cross sections of the Pb(^{50}Ti,2n)Ku reaction turned out to be tens of times less than the theoretical ones (see Figs. 13 and 18). However, in the σ (xn) calculation for the partial width ratio Γ_n/Γ_f the semiempirical dependence of Sikkeland et al. [33] was used assuming the influence of the subshell N=152 on the value of Γ_n/Γ_f. On the other hand, it will be shown later that this assumption is not well-grounded for nuclei with Z=104. The exclusion of the influence of the subshell N=152 on the Γ_n/Γ_f value leads to a substantial decrease in the calculated cross sections. This practically removes the discrepancy mentioned above(Fig.18).

The long-lived activity with a half-life of about 4 sec is most likely to be the isotope ^{255}Ku, which is formed with a maximum cross section in the reaction ^{207}Pb(^{50}Ti,2n), with lower probability in the ^{208}Pb(^{50}Ti,3n) reaction, and it is absent in the reaction ^{206}Pb(^{50}Ti,1n).

It is however noteworthy that the half-life of this activity turned out to be close to that of the known isotope

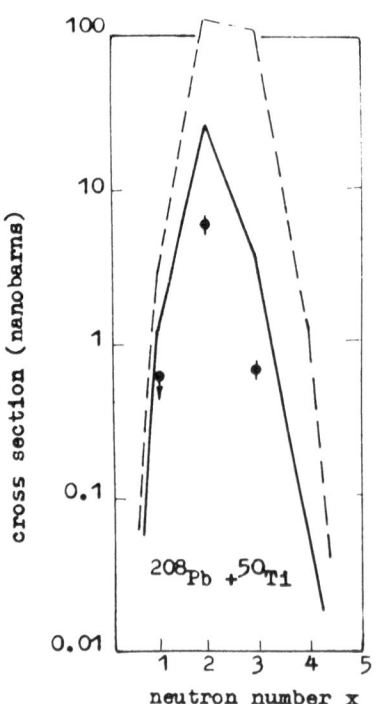

Fig. 18. Experimental and calculated cross sections of the reactions ^{208}Pb(^{50}Ti,xn)$^{258-x}$Ku. The calculation employed the ratios Γ_n/Γ_f shown in Fig. 22 by dashed and solid curves.

$^{252}102$, which might be produced in the reaction of the type $^{207}Pb(^{50}Ti; \alpha, 1n)^{252}102$.

We shall consider the probability of this kind of process for reactions with ions such as ^{40}Ar and ^{50}Ti. At the maximum ^{50}Ti projectile energy of 260 MeV, the maximum excitation energy of the compound nucleus ^{257}Ku is about 40 MeV. Estimates show that the threshold of the reaction with the evaporation of an alpha particle and one neutron is about 260 MeV. In addition, in this nuclear region the partial width ratio Γ_α/Γ_n as well as Γ_p/Γ_n is rather small, $\leq 10^{-2}$. Therefore, a contribution from the processes passing through the evaporation of charged particles from the compound nucleus is negligibly small in our case.

At the same time, similarly to the case of reactions with the light ions of C, N and O, the process of the direct emission of an alpha particle with the subsequent evaporation of a neutron is possible energetically. However, in refs. [47-49] it was shown that the mechanism of such reactions considerably depends on the mass of the bombarding ion. As one goes from light ions to particles such as ^{40}Ar and ^{50}Ti, the direct emission probability for alpha particles decreases tens and hundreds of times.

We carried out control experiments in which the spontaneously fissioning nuclei of ^{244}Fm were produced in the bombardment of ^{203}Tl and ^{205}Tl with ^{45}Sc ions and ^{206}Pb with ^{40}Ar ions. As seen from Table 3, the cross sections ratio $\sigma(\alpha, 0-2n)/\sigma(2n) \leq 0.03$. Hence one can conclude that the formation probability for the isotope $^{252}102$ is negligibly small at the bombardment of ^{207}Pb with ^{50}Ti ions.

From comparing the formation cross sections for the isotopes ^{255}Ku and ^{256}Ku, it follows that the odd isotope ^{255}Ku has comparable values of the alpha-decay and spontaneous-fission half-lives.

Finally, the absence of the effect in the $^{206}Pb + ^{50}Ti$ reaction may indicate that the lifetime of the even isotope ^{254}Ku is < 3 msec.*)

3.3. SPONTANEOUS-FISSION HALF-LIVES OF HEAVY NUCLEI.

Let us see in what way the properties of the neutron-deficient Ku isotopes synthesized agree with the known data on the stability of heavy nuclei with respect to spontaneous fission.

---

*) At the same time, by using a more rapid technique we have observed, in the $^{206}Pb + ^{50}Ti$ reaction, spontaneously fissioning nuclei with $T_{\frac{1}{2}} \sim 1.5$ msec, which may supposedly be assigned to the decay of the isotope ^{254}Ku.

For the isotope ^{256}Ku with N=152, the experimental half-life turned out to be a factor of 10^3-10^4 less than the expected one, tens or hundreds of seconds. The properties of this isotope essentially change the understanding of the stabilizing effect of the subshell N=152 in moving from Z=102 to Z=104. The dependence of $T_{\frac{1}{2}}^{SF}$ on neutron number N, shown in Fig. 19, indicates the monotonic increase in the lifetimes of even-even nuclei with increasing mass without considerable variations in the region of N=152.

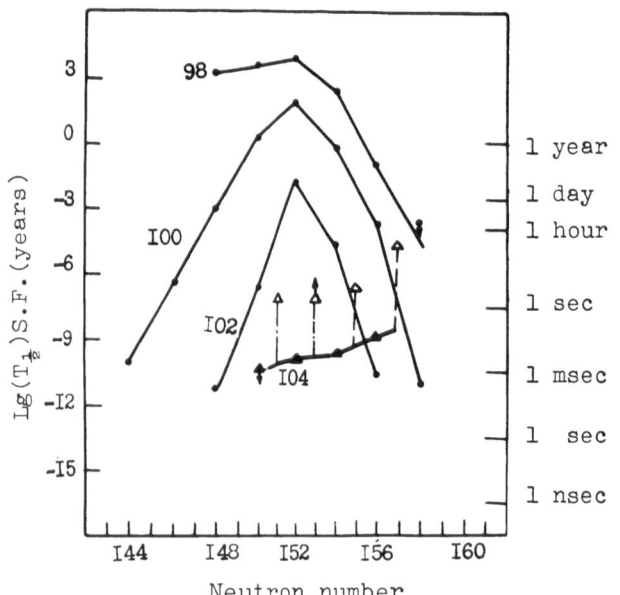

Fig. 19. Systematics of spontaneous-fission half-lives for even-even nuclei taking into account the latest results for isotopes with Z=104. The plot incorporates data for the isotope ^{250}102($T_{\frac{1}{2}} \sim$ 0.2 msec) synthesized in the reaction ^{233}U(^{22}Ne,5n)^{250}102.

As regards the odd nucleus ^{255}Ku, its lifetime is determined by the hindrance factor for spontaneous fission, which is about 10^3 in this case.

Now it is not surprising that the isotope ^{260}Ku has $T_{\frac{1}{2}}^{SF} \sim$ 0.1 sec, and there is no necessity in assuming hindrance factors as large as 10^7-10^{12} for other odd Ku isotopes with mass numbers 257 and 259, as was done in the papers by Ghiorso et al.[38,43] It is seen from the systematics given in Fig. 19 that the hindrance factor for all odd Ku isotopes reduces the spontaneous fission rate by a factor of 10^3-10^4.

As in going from the isotope ^{254}102 to ^{256}Ku, the spontaneous fission half-life changes very strongly (more than a factor of 10^8), it is natural to try to explain the data obtained at least qualitatively.

In accordance with the present-day concepts, in the region of transfermium elements the fission barrier has a complex structure and nuclear stability against spontaneous fission is mainly determined by a contribution from the shell effect correction to the total deformation energy of the nucleus.

From this point of view, we shall consider the Fm isotopes for which the subshell N=152 was shown to strongly influence the half-life value. For the isotope ^{252}Fm, the fission barrier calculated using Strutinsky's method [17] has the shape of a two-humped curve with two minima that correspond to the ground and isomeric states of the nucleus. The lifetime of this nucleus for spontaneous fission is determined by its transition from the ground state to the point behind the second barrier (see Fig. 20).

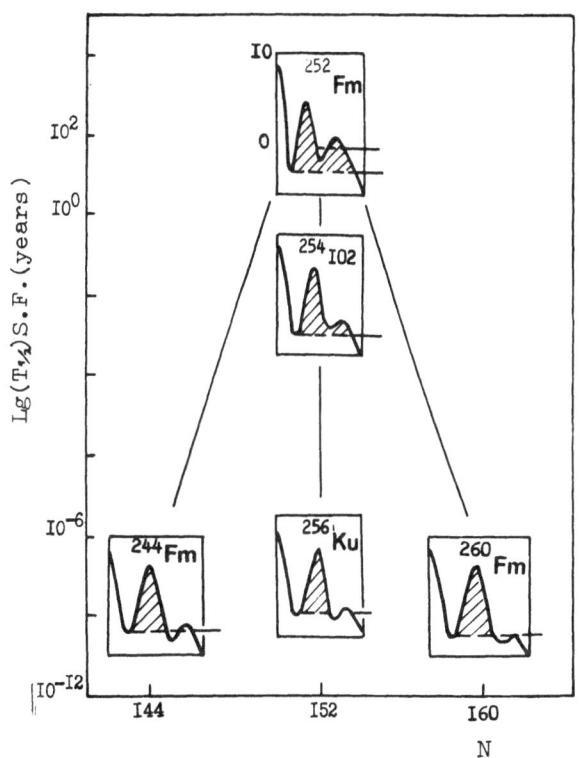

Fig. 20. Illustrative representation of fission barrier structures for even-even nuclei with Z ⩾ 100.

As the neutron number decreases in moving to the lighter Fm isotopes, the second barrier height will be reduced as a result of decreasing liquid-drop deformation energy. This should lead to a decrease in spontaneous fission half-lives. In an extreme case where the ground state appears to lie above the second maximum, the value of $T_{\frac{1}{2}}^{SF}$ will be determined by penetrability through the first barrier. One can assume that this circumstance will lead to a decrease in $T_{\frac{1}{2}}^{SF}$ by a factor of 10^{12} in going from ^{252}Fm to ^{244}Fm.

At the same time, it was shown in a paper by Randrup et al. [50] that this situation may occur in moving from N=152 to the heavier Fm isotopes. In this case the superposition of the shell correction and the liquid-drop deformation energy may also lead to a considerable decrease in the second maximum height and, consequently, a sharp decrease in the spontaneous fission half-lives of heavy isotopes. Indeed, as one moves from ^{252}Fm to ^{258}Fm, the $T_{\frac{1}{2}}^{SF}$ value changes by nearly a factor of 10^{13}.

Now we shall vary the number of protons in the nucleus at the fixed N=152. As Z increases, the liquid-drop part of the deformation

energy will decrease; this will lead to the same consequences as those
that took place in the above considered case of the neutron-deficient
Fm isotopes. One can assume that the maximum change in the $T_{\frac{1}{2}}^{SF}$ value
will occur at the moment when nuclear lifetime is mainly determined by
penetrability through the first barrier. It is not excluded that the
same situation occurs as one goes from ^{252}Fm to ^{256}Ku. In this case
the spontaneous fission half-life also changes by nearly a factor of
10^{12}.

If this assumption is valid, neither further penetration into
the region of large Z values nor the variations of N at given Z should
lead to very strong variations in $T_{\frac{1}{2}}$, since the first barrier height,
according to the calculation of ref. [17], is comparatively insensitive
to the nucleon content of the nucleus. In our opinion, this point of
view receives qualitative confirmation from the recent calculations of
Pauli, Ledergerber [51], Möller and Nix [52,53]. However, quantitative
conclusions will require a more detailed theoretical analysis taking
into account the latest data on the properties of Ku isotopes and the
heavier elements.

Such an interpretation of the experimental data indicates that
the spontaneous fission half-life of even-even Ku isotopes is almost
entirely determined by the shell effects.

At the same time, the experimental data suggest that the reactions Pb(^{40}Ar,2n)Fm and Pb(^{50}Ti,2n)Ku have nearly equal cross sections,
as predicted theoretically. This allowed one to hope that this type of
reaction can be successfully used for the production of nuclei with
Z>104. Therefore, in the subsequent experiments an attempt was made
tp synthesize the new element with Z=106.

4. EXPERIMENTS WITH ^{54}Cr IONS. FORMATION OF COMPOUND NUCLEI WITH Z=106.

Calculations based on the experimental data obtained have shown
that the combination Pb+^{54}Cr is one of the most suitable reactions for
the synthesis of isotopes with Z=106.

Some improvements in the ion source have made it possible to
produce a ^{54}Cr^{+8} ion beam with an intensity of $\sim 2 \times 10^{11}$ part/sec
at the heavy ion cyclotron. The experiments were carried out using
the technique which was employed in the synthesis of kurchatovium isotopes. The separated isotopes ^{206}Pb, ^{207}Pb and ^{208}Pb were used as targets which were bombarded with ^{54}Cr ions with a maximum energy of
280 MeV.

In the ^{207,208}Pb+^{54}Cr reactions, we observed the formation of
spontaneously fissioning nuclei with a half-life of several milliseconds

(see Fig. 21). The yield of this activity decreases by more than a factor of 5 as one uses the ^{206}Pb target. In addition, a large number of control experiments were carried out, in which different Pb isotopes and ^{209}Bi were bombarded with ^{51}V, ^{52}Cr and ^{55}Mn ions. In these experiments, one succeeded only in determining the upper limit on the production cross section for spontaneously fissioning nuclei.

The results of the experiments are presented in Table 4.

As has been shown previously, the contribution from the (p,xn) and (α,xn) reactions which may in this case lead to neutron-deficient isotopes with Z=104 and 105 in the energy range indicated, is negligible. Therefore, we are apt to think that the 50 events observed are due to spontaneous fission of nuclei with Z=106.

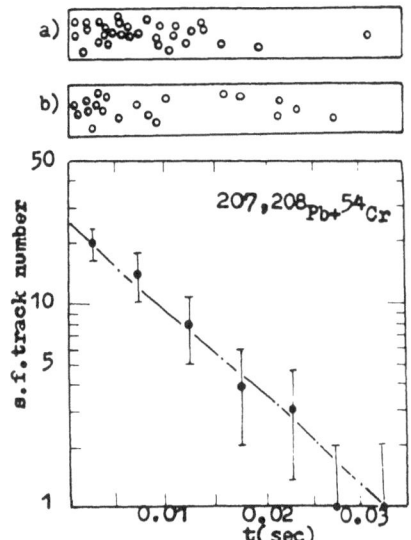

Fig. 21. Time distribution of spontaneous fission tracks in the reactions
a) ^{207}Pb+^{54}Cr, and
b) ^{208}Pb+^{54}Cr.

Table 4

Reaction	E_I,lab. (MeV)	$B_{Coul.}$ (MeV)	Total flux x 10^{16}	Number of tracks	$T_{\frac{1}{2}}$	Cross section (nb)
^{208}Pb +^{54}Cr	280		2	20	∼ 7 ms	1
^{207}Pb +^{54}Cr		254±1.5	3	31	"	1
^{206}Pb +^{54}Cr			4	-		⩽0.2
^{208}Pb +^{52}Cr	280		1	1	⩾ 2 ms	<0.1
^{209}Bi +^{54}Cr	280	258	1.5	1	"	<0.1
^{209}Bi +^{51}V	270	246	1.5	0	"	<0.1
206 207 Pb +^{55}Mn 208	290	267	1	1	"	<0.1

On the basis of the yields of this activity in experiments with different Pb isotopes, and the new understanding of the spontaneous fission systematics, one may assume that the activity observed is due to the odd isotope $^{259}106$ formed in the $^{207,208}Pb(^{54}Cr;2n,3n)^{259}106$ reactions.

In the future, we hope to investigate this region of nuclei more thoroughly and try to produce other isotopes with Z=106 using the isotopes ^{204}Pb and ^{210}Pb as targets. On the other hand, it seems fairly interesting to study the properties of the heavier isotopes (N=157) since in this region one can expect a considerable increase in nuclear lifetimes. Unfortunately, with a Pb target used, these isotopes cannot be produced and one should revert to the conventional method of using heavy targets. At our Laboratory, we employ the $^{246}Cm(^{22}Ne,5n)^{263}106$ reaction to synthesize the isotope $^{263}106$, while at Berkeley, as far as I know, they use the combination $^{249}Cf(^{18}O,4n)^{263}106$.

The situation, however, changes in the case of nuclei with Z>106. For instance, one of the most promising combinations for the synthesis of element 108 is $^{208}Pb(^{58}Fe,xn)^{266-x}108$. If the nature of the nuclear interaction does not change in moving from ^{54}Cr toward ^{58}Fe, one may then hope for the production of heavy isotopes of this element with N=156,157.

In order to obtain information about the production cross section of these nuclei, we made an attempt to estimate the Γ_n/Γ_f value from the experimental data on the production of the neutron-deficient isotopes of Fm, Ku and element 106 by reactions induced by ^{40}Ar, ^{50}Ti and ^{54}Cr. For this purpose, one had to compare the calculated cross sections with the experimental ones (see eq. (2.3)). Fig. 22 shows the values of Γ_n/Γ_f for the wide range of isotopes produced in heavy ion reactions. It is interesting to note that the influence of the sub-shell N=152 on the Γ_n/Γ_f value, which has been observed for isotopes with Z=98,100 and 102, does not manifest itself in the case of nuclei with Z=104, although the isotopes ^{256}Ku and ^{255}Ku had been produced in reactions with the emission of 2 neutrons only.

At the same time, with these Γ_n/Γ_f values, the cross section for the Pb(HI,2n) reactions induced by ^{40}Ar, ^{50}Ti and ^{54}Cr ions vary as little as several times. Therefore, by extrapolating the available data on the Γ_n/Γ_f values to the region of the heavy isotopes of nuclei with Z~106, one can assume that, in this case, the cross section for reactions with the emission of two neutrons will also be about 10^{-33} cm^2.

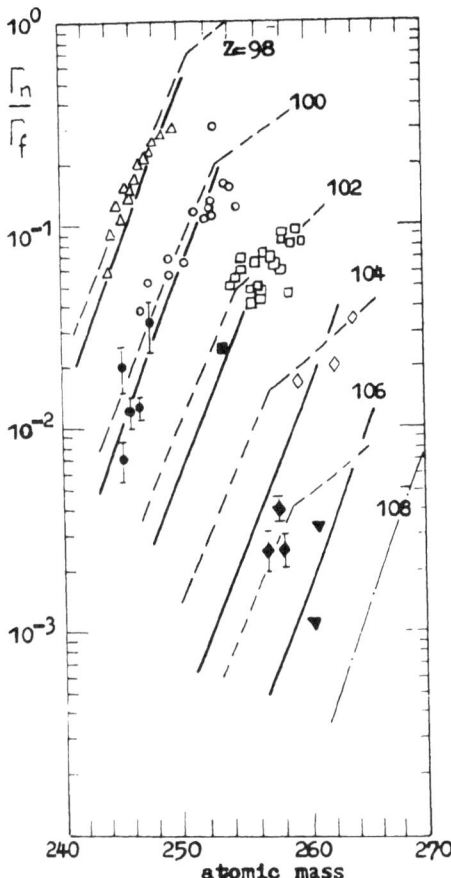

Fig. 22. The Γ_n/Γ_f values, obtained in heavy ion reactions, for compound nuclei with Z from 98 to 106. Black points refer to the results of the present paper. The dashed curve is taken from a paper by Sikkeland (ref. 33).

5. DELAYED FISSION EXPERIMENTS. PROPERTIES OF NEUTRON-DEFICIENT PLUTONIUM ISOTOPES.

In 1966-1967, G.N.Flerov, V.I.Kuznetsov and N.K.Skobelev observed spontaneously fissioning nuclei with half-lives of 1.4 min and 2.6 min and production cross sections of about 5 x 10^{-34} cm^2 by bombarding ^{230}Th with ^{10}B and ^{11}B. These activities were identified as the americium isotopes; ^{232}Am and ^{234}Am (ref.[54]).

Neutron-deficient Am isotopes can also be produced by another reaction induced by ^{27}Al ions on Pb. This reaction has been investigated using the technique developed earlier for experiments with ^{50}Ti and ^{54}Cr ions (see Fig. 14). In bombarding ^{206}Pb with ^{27}Al we observed spontaneous fission fragments, the time distribution of which is presented in Fig. 23.

Similar experiments have been carried out with ^{207}Pb and ^{208}Pb targets. In these experiments, spontaneously fissioning nuclei with $T_{\frac{1}{2}} \sim 1$ min were also observed. On the basis of the calculated values given in Fig. 24, one can assume that the reactions ^{206,207,208}Pb+^{27}Al yield two isotopes, ^{230}Am and ^{232}Am, with close values of half-lives.

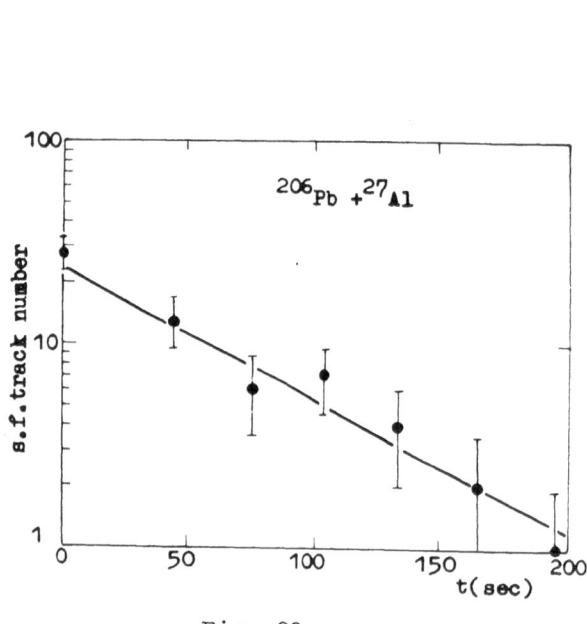

Fig. 23.

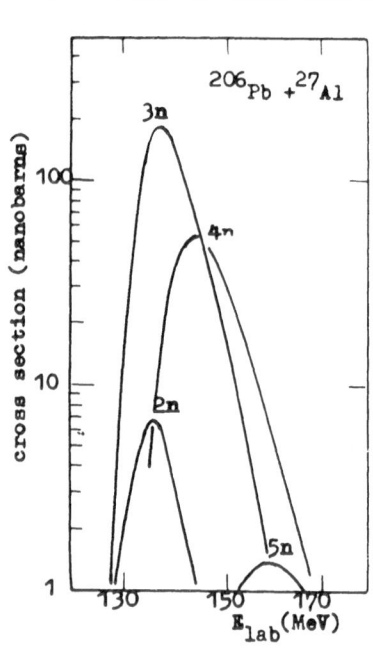

Fig. 24. Calculated excitation functions of the reaction ^{206}Pb(^{27}Al,xn)$^{233-x}$Am.

Since the half-lives for spontaneous fission from the ground state for Am isotopes are 10^{12} years according to the systematics, and those for fission from an isomeric state is about $10^{-5} - 10^{-3}$ sec, it should be assumed that the half-life observed is determined by electron capture followed by fission of the nuclei ^{230}Pu and ^{232}Pu.

The problem is from which energy level fission occurs. Here, the following factor is worth noting.

According to the calculations of V.V.Pashkevich, neutron-deficient Pu nuclei may have the considerably different structure of the fission barrier. If in the region of heavy nuclei with $Z \geqslant 104$ variations in the fission barrier structure are associated with changes in the height of the second maximum in the nuclear deformation energy, then for Pu isotopes, decreasing neutron number considerably reduces the first barrier height (Fig. 25).

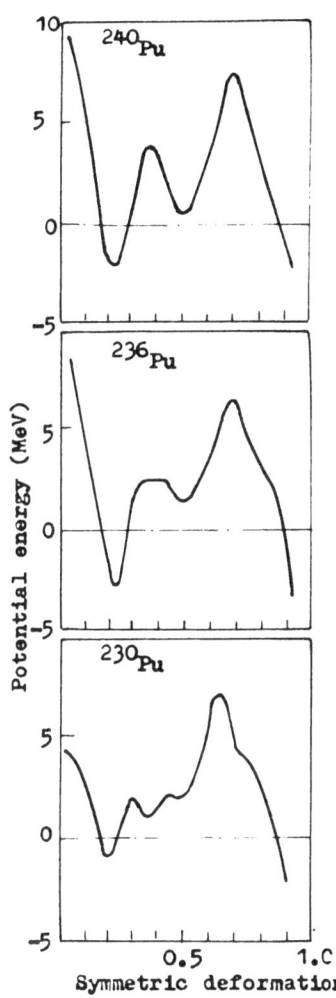

Fig. 25. Nuclear poten-
tial energy as a func-
tion of deformation
for plutonium isotopes.

The position of the second maximum for the isotope ^{230}Pu corresponds to ∽ 8 MeV. The Q value for the transition ^{230}Am → ^{230}Pu is about 5.5 MeV. Therefore, one can expect the probability for these nuclei to undergo subbarrier fission to be considerably higher than that in the case of spontaneous fission isomers in the known region of the heavier Pu isotopes.

This offers new possibilities for investigating the behaviour of highly deformed nuclei, in particular, the structure of the second barrier by using different techniques based on the (β-f) or (X,f) coincidences.

Such investigations can be conducted for a wide range of neutron-deficient nuclei using reactions with compound nucleus formation at the bombardment of different Hg, Tl, Pb and Bi isotopes with the ^{27}Al, ^{28}Si, ^{31}P and ^{32}S ions.

6. CONCLUDING REMARKS. FUTURE PERSPECTIVES.

The investigation of the mechanism of interactions between complex nuclei, in particular, the fusion process, plays an important role for the future studies of interactions between complex nuclei. Of course, the advent of new accelerators capable of producing the beams of ions with $A_I \gtrsim 100$ and intensity up to 10^{13} part/sec will offer wide possibilities for investigating the problems mentioned. I think that in going this way we shall encounter a lot of interesting and unexpected phenomena. The currently available information is related only to the first steps in this direction.

From the experimental data on fission of heavy nuclei produced by reactions with ions heavier than ^{40}Ar, it is regrettably difficult to draw unambiguous conclusions about the mechanism of compound nucleus formation. Different theoretical hypotheses about a "composite system", "the quasifission mechanism" and others, each of which implies many assumptions and requires by itself experimental verification, interpret the fission data from different points of view. More definite conclusions can be drawn from the experimental studies of the (HI,xn) reactions in the region of highly fissionable heavy nuclei with Z > 100.

The experimental investigations described in this paper are, in fact, an attempt to advance in this direction. By using Pb isotopes as targets and bombarding them with ions ranging from ^{40}Ar to ^{54}Cr, one has succeeded in progressing from Z=100 to Z=106. Estimates suggest that one can go farther this way using the heavier ions, ^{58}Fe or ^{64}Ni, unless new phenomena prevent compound nucleus formation.

Despite the fact that experiments on the synthesis of new elements are very difficult (their production cross section is expressed in nanobarns), they yield important information about the properties of heavy nuclei. The substantial changes in the systematics of spontaneous fission half-lives for even-even isotopes with Z=104 are very likely to be associated with the structure of the fission barriers of these nuclei. On the other hand, delayed fission will enable one to investigate the neutron-deficient isotopes of plutonium and its neighbours. In this region, one can also expect considerable changes in the fission barrier shape. Therefore, it is necessary to carry out a detailed theoretical analysis of the data obtained in the framework of the modern nuclear fission theory. This, we think, will make it possible to predict the properties of heavy and superheavy elements more reliably. The problem of the synthesis of superheavy nuclei occupies a special place.

The (HI,xn) reactions induced by ^{76}Ge and ^{84}Kr ions have not resulted in the production of superheavy elements. However, it has become clear that due to increasing Coulomb barrier the minimum excitation energy of a compound nucleus (if formed) is, in these cases, > 50 MeV. It is not excluded that another combination using Pb targets or the combination ^{248}Cm+^{48}Ca may prove more advantageous.

Another method of using the combination ^{238}U+^{136}Xe (ref.[55]) is not, strictly speaking, associated with the formation of a classical compound nucleus. The wide distribution of the reaction products over Z and A allows one to hope that this reaction may be used for the production of very long-lived nuclei. Indeed, the formation of superheavy elements as fission fragments is a rather intricate process, which gives rise to many questions. However, in this particular reaction one has succeeded in observing the formation of spontaneously fissioning nuclei with a half-life of about 150 days.

In the subsequent experiments with the ^{136}Xe ions we shall make an attempt to identify this activity. It is however evident that all the experiments on the synthesis of superheavy elements require enhanced sensitivity. My optimism is based on the fact that this possibility will occur in the nearest future after the advent of the new heavy ion accelerators.

ACKNOWLEDGMENTS.

It is a pleasure for me to express my thanks to Academician G.N.Flerov for his great attention and support, as well as valuable critical comments on the work at all stages. I am deeply thankful to my colleagues A.G.Demin, A.S.Iljinov, A.A.Pleve, S.P.Tretyakova, M.I.Ivanov, Yu.P.Tretyakov and others, in collaboration with whom this work has been performed. Thanks are also due to V.V.Pashkevich for carrying out some calculations, and L.V.Pashkevich for translating the paper into English.

REFERENCES

1) J.R.Grover, Phys.Rev. 157(1967)832
2) Yu.Ts.Oganessian, Yu.E.Penionzhkevich, A.Shamsutdinov and Nguyen Tac Anh, JINR Preprint P7-5912, Dubna, 1971
3) Yu.Oganessian, Nuclear Structure, IAEA, Vienna, 1968, p.489; S.A.Karamyan, Yu.Ts.Oganessian, B.I.Pustylnik and G.N.Flerov, Proc. Second Symposium on Physics and Chemistry of Fission, IAEA-SM 112/130 p. 759, Vienna, 1969
4) V.V.Kamanin, S.A.Karamyan, F.Normuratov and S.P.Tretyakova, JINR Preprint P7-6302, Dubna, 1972
5) C.Y.Wong and T.A.Welton, Phys.Lett. 49B(1974)243
6) W.Scheid, H.Müller and W.Greiner, Phys.Rev.Lett. 32(1974)741
7) G.N.Flerov and I.Zvara, JINR Communication D7-6013, Dubna, 1971
8) S.G.Nilsson, S.G.Thompson and C.F.Tzang, Phys.Lett. 28B(1969)458
9) A.A.Pleve, A.G.Demin, V.Kush, M.B.Miller and N.A.Danilov, JINR Preprint P7-7279, Dubna, 1973
10) R.Bimbot, C.Deprun, D.Gardes, H.Gauvin, Y.Le Beyec, M.Lefort and J.Peter, Nature 234(1971)215; P.Colombani, B.Gatty, J.C.Jacmart, M.Lefort, J.Peter, M.Riou, C.Stephan and X.Tarrago, Europ. Conf. on Nuclear Physics, Aix-en-Provence, v.II p.91, 1972
11) Yu.Oganessian, Proc. Intern. Conf. on Nuclear Physics, Munich, 1973 v.2, p.351
12) T.Sikkeland, R.J.Silva, A.Ghiorso and M.J.Nurmia, Phys.Rev. 4C(1970)1564; H.H.Gutbrod et al., Proc. Third Symposium on Physics and Chemistry of Fission, Rochester, 1973, IAEA, Vienna, 1974, paper IAEA SM 174/59
13) J.Peter, F.Hanappe, C.Ngô and B.Tamain, Preprint IPNO-RC-73-07, Orsay, 1973
14) F.Plasil, Preprint ORNL-TM-4599, Oak Ridge, 1974
15) H.Gauvin, R.L.Hahn, Y.Le Beyec and M.Lefort, Preprint IPNO-RC-74-03, Orsay, 1974
16) G.N.Flerov, S.A.Karamyan, Yu.E.Penionzhkevich, S.P.Tretyakova and I.A.Shelayev, JINR Preprint P7-6262, Dubna, 1972
17) M.Brack, J.Damgaard, A.S.Jensen, H.C.Pauli, V.M.Strutinsky and S.V.Wong, Rev.Mod.Phys. 44(1972)320
18) W.J.Swiatecki, Communications of Europ. Conf. on Nuclear Physics, Aix-en-Provence, 1972, Preprint LBL-972, Berkeley, 1972

19) W.J.Swiatecki and S.Bjornholm, Phys.Reports 4C(1972)325;
    A.J.Sierk and J.R.Nix, Proc. Third Symposium on Physics and
    Chemistry of Fission, Rochester, 1973 (IAEA, Vienna 1974)
    paper IAEA-SM-174/74;
    J.Wilczynski, ibid, paper IAEA-SM-174/208;
    J.Galin et al. Phys.Rev. C9(1974)1018
20) F.Hanappe, M.Lefort, C.Ngô, J.Peter and B.Tamain, Phys.Rev.Lett.
    32(1974)738
21) A.G.Artukh, G.F.Gridnev, V.L.Mikheev, V.V.Volkov and J.Wilczynski,
    JINR Preprint E7-6970, Dubna, 1973
22) M.Bolsterli et al. Phys.Rev. 5C(1972)1050
23) G.Seaborg, Paper presented at the Nobel Symposium on Superheavy
    Elements, Ronneby, Sweden, June 11, 1974
24) M.Lefort, Y.Le Beyec and J.Peter, invited Paper at the Conference
    on Reactions Between Complex Nuclei, Nashville, Tenn., 1974
25) L.G.Moretto, Nucl.Phys. A180(1972)337
26) M.Nurmia, T.Sikkeland, R.Silva and A.Ghiorso, Phys.Lett.26B(1967)78
27) Yu.Ts.Oganessian, Yu.E.Penionzhkevich, K.A.Gavrilov and Kim De En,
    JINR Preprint P7-7863, Dubna, 1974
28) R.Bass, Phys.Lett. 47B(1973)139
29) A.S.Iljinov, JINR Communication P7-7108, Dubna, 1973
30) T.Sikkeland and V.E.Viola, Proc. Third Conf. on Reactions Between
    Complex Nuclei, Asilomar, USA, 1963
31) G.Igo, Phys.Rev. 115(1959)1665
32) V.E.Viola and T.Sikkeland, Phys.Rev. 128(1962)767
33) T.Sikkeland, A.Ghiorso and M.J.Nurmia, Phys.Rev. 172(1968)1232
34) G.N.Flerov et al. Atom.Energ. 17(1964)310; Phys.Lett. 13(1964)163
35) I.Zvara, Yu.T.Chuburkov, R.Zaletka and M.R.Shalayevsky,
    JINR Communication P7-3783, Dubna, 1968; Radiochimiya 11(1969)163
36) I.Zvara et al. JINR Communication D7-4542, Dubna, 1969
37) I.Zvara et al. JINR Communication D12-5845, Dubna, 1971
38) A.Ghiorso, Proc. R.A.Welch Found.Conf. on Chemical Research, XIII.
    The Transuranium Elements-The Mendeleev Centennial, Nov. 17-19,1969,
    Houston, Texas, p.107
39) A.Ghiorso, M.Nurmia, J.Harris, K.Eskola and P.Eskola,
    Phys.Rev.Lett. 22(1969)1317
40) A.Ghiorso, M.Nurmia, K.Eskola and P.Eskola, Phys.Lett.32B(1970)95
41) Yu.Ts.Oganessian et al. At.Energ. 28(1970)393
42) M.J.Nurmia, Nucl.Chem. Annual Report, LBL-666, p.42, Berkeley,1971
43) A.Ghiorso and T.Sikkeland, Physics Today, 20, no.9(1967)25
44) Yu.Ts.Oganessian, A.S.Iljinov, A.G.Demin and S.P.Tretyakova,
    JINR Preprint D7-8194, Dubna, 1974
45) G.N.Flerov, At.Energ. 24(1968)5
46) G.N.Flerov, Yu.Ts.Oganessian, Yu.V.Lobanov, Yu.A.Lazarev and
    S.P.Tretyakova, JINR Communication P7-4932, Dubna, 1970
47) J.Galin et al. Preprint IPNO-RC-7303, Orsay, 1973
48) L.G.Moretto et al. Preprint LBL-1966, Berkeley, 1973
49) A.G.Artukh et al. JINR Communication P7-7189, Dubna, 1973
50) J.Randrup et al. Nucl.Phys. A217(1973)221
51) H.C.Pauli and T.Ledergerber, Proc. Third Symposium on Physics
    and Chemistry of Fission, Rochester, 1973(IAEA, Vienna) 1974,
    paper IAEA-SM-174/206
52) P.Möller and J.R.Nix, Proc.Third Symposium on Physics and Chemistry
    of Fission, Rochester, 1973(IAEA, Vienna, 1974), paper IAEA-SM-
    174/202
53) P.Möller and J.R.Nix, Preprint LA-UR-74-417, Los Alamos, 1974
54) V.I.Kuznetsov, N.K.Skobelev and G.N.Flerov, Yad.Fiz. 4(1966)279;
                                                  Yad.Fiz. 5(1967)271,1136
55) G.N.Flerov, in Comptes Rendus du Congrès Intern. de Physique
    Nucléaire, 1964, Paris, vol.1 p. 373.

DEEP INELASTIC TRANSFER REACTIONS
AND THE FORMATION OF A DOUBLE NUCLEAR SYSTEM

V.V.Volkov
Joint Institute for Nuclear Research, Dubna, USSR

Abstract

Experimental data on multi-nucleon transfer reactions, obtained
mainly by bombarding ^{232}Th with the ^{16}O, ^{22}Ne and ^{40}Ar ions with an
energy of 7-9 MeV/nucleon are discussed. The energy spectra of light
reaction products and isotopic production cross sections show some
features which are difficult to interpret in the framework of conventional direct processes. The concept of a double nuclear system is
proposed, which allows one to explain these features easily.

1. INTRODUCTION.

Heavy ion transfer reactions are usually considered as quasielastic direct processes occurring in the grazing collisions of two
nuclei. However, lately experimental data have been obtained, especially for multi-nucleon transfer reactions, which cannot be described in
this traditional framework. Transfer reactions in deep inelastic collisions of two nuclei are meant. Deep inelastic transfer reactions (DITR)
have been observed for the first time by the authors of refs.[1-3].
However, the peculiarity of the mechanism of these reactions and their
close relation to the interaction of two nuclei have become evident
only recently. Now the experiments aimed at studying DITR are under
way in Dubna [4-10], Orsay [11-15] and Berkeley [16,17]. The theoretical
analysis of DITR is being carried out elsewhere [18-35].

There are some reasons why these reactions attract the attention
of physicists. DITR allow one to obtain unique information on the interactions of nuclei. Indeed, the compound nucleus forgets the histo-
ry of its formation. Quasi-elastic processes, both elastic and inelastic scattering, give information on nuclear interaction in peripheral
collisions when the nuclear surfaces overlap inconsiderably. On the
contrary, DITR provide information on nuclear interaction just with
considerable overlapping of nuclear surfaces, when collisions turn out

to be close to the head-on ones. DITR allow deep reconstruction of interacting nuclei and the obtaining of such exotic isotopes as ^{8}He, ^{11}Li, ^{14}Be, ^{15}B, ^{20}C, ^{21}N, and ^{24}O. The transfer reactions of this type may prove useful for superheavy element synthesis as well. Similarly to amphoteric elements in chemistry, the DITR mechanism itself is combining the properties of two opposite processes: direct reactions and compound nucleus formation.

Here the peculiarities of DITR are considered, basing mainly on experimental data obtained in Dubna [5-10] *). An attempt has been made to interpret these peculiarities qualitatively on the basis of the concept of a double nuclear system (DNS) formed in deep inelastic collisions of two nuclei. In DNS the nuclear surfaces considerably overlap , while the velocity of their relative motion is small. As a DNS has a considerable angular momentum, it rotates as a whole and evolves in time, coming from one state to another. The DNS lifetime turns out to be much larger than the characteristic nuclear time (10^{-22} sec).

2. EXPERIMENTAL DATA

2.1. Energy spectra of transfer reactions and nuclear viscosity.

Figs. 1-3 show the elemental energy spectra of the light products of transfer reactions in the ^{232}Th+^{40}Ar system at 388 MeV [9]. The spectra are given in the lab. system. The $\Delta E, E$ method has been used for measurements without separating isotopes. The spectra have revealed the following peculiarities:

- a broad range spectrum extends to 200 MeV and that means the transfer of considerable amount of kinetic energy to excitation,
- the spectra are divided into two parts in the case of few-nucleon transfer reactions and at large emission angles,
- the spectra maxima are reduced with decreasing emission angle for few-nucleon transfer reactions,
- the spectra shape becomes symmetric with   increasing number of stripped nuclei mainly due to the damping of the high energy part,
- the spectra maxima of multi-nucleon transfer reactions turn out to be close to the exit Coulomb barrier. A noticeable amount of products have energies much lower than the exit Coulomb barrier.

Fig. 4 shows a comparison of the energy spectra of S, P, Si and Al at three emission angles and at two ^{40}Ar projectile energies, 297

---

*)DITR have been studied experimentally by a team of Dubna physicists, namely, A.G.Artukh, G.F.Gridnev, V.L.Mikheev, V.V.Volkov and J. Wilczynski of the Krakow University, Poland.

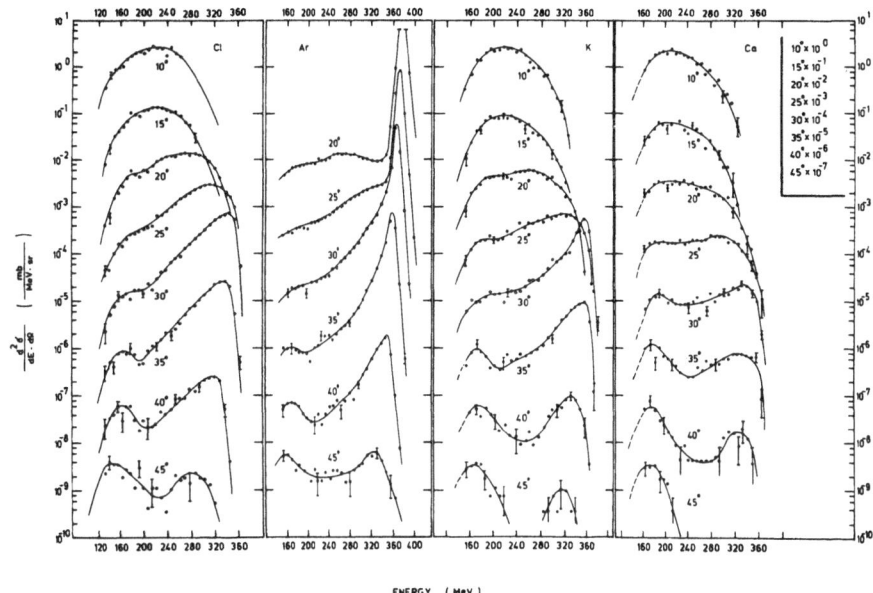

Fig. 1. Energy spectra of Ca, K, Ar and Cl transfer reaction products obtained by bombarding ^{232}Th with ^{40}Ar of 388 MeV (lab. system).

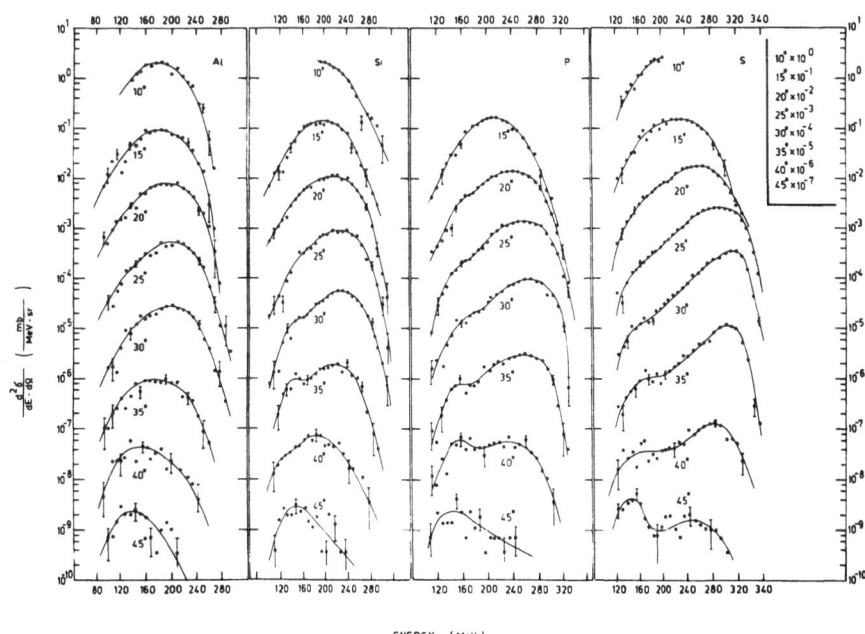

Fig. 2. Energy spectra of S, P, Si and Al transfer reaction products obtained by bombarding ^{232}Th with ^{40}Ar of 388 MeV(lab. system).

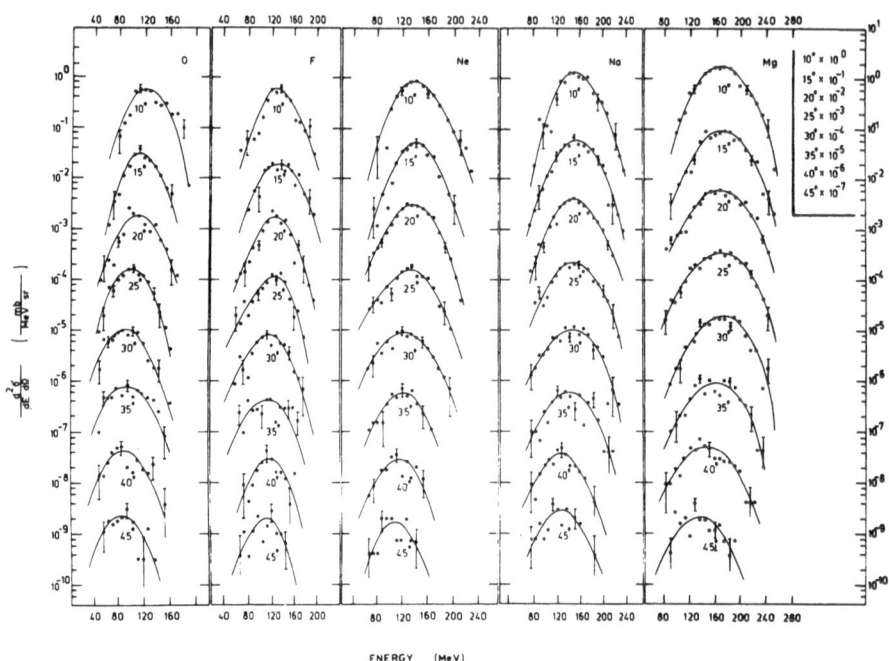

Fig. 3. Energy spectra of Mg, Na, Ne, F and O transfer reaction products obtained by bombarding ^{232}Th with ^{40}Ar of 388 MeV (lab.system).

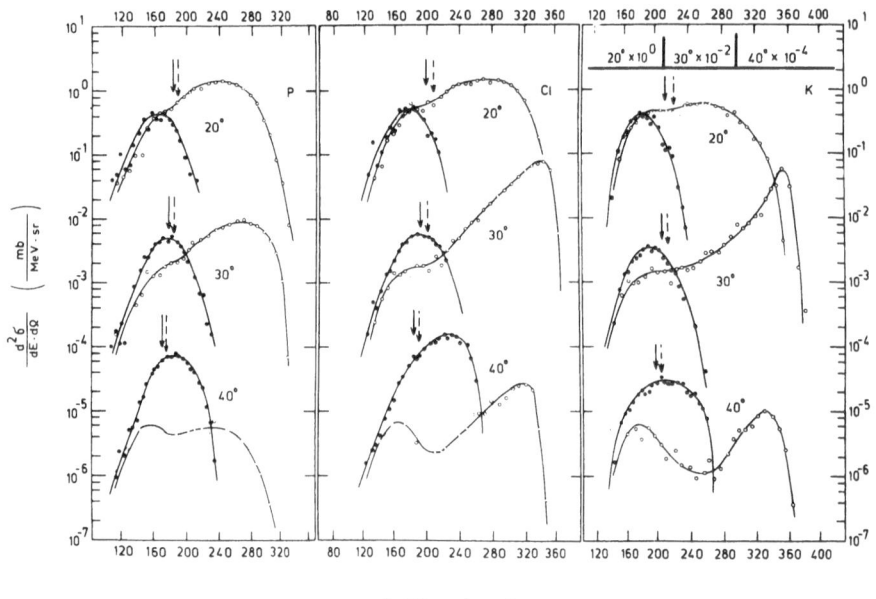

Fig. 4. Comparison of the energy spectra of P, Cl and K obtained by bombarding ^{232}Th with ^{40}Ar of 279 and 388 MeV.

and 388 MeV. Arrows show the exit Coulomb barrier: the solid and dashed
arrows correspond to 297 and 388 MeV, respectively. The spectrum width
greatly increases with increasing projectile energy. However, the low
energy parts of the spectra, especially at small angles, coincide. For
an angle of 20° the greater part of the energy spectrum at 288 MeV is
lower than the exit Coulomb barrier.

Fig. 5 shows the Q_m value of reactions for the maxima of the
energy spectra with respect to Z and the emission angle of the reaction
light product, Θ. They have been calculated on the basis of the kinetic
energy of light reaction products under the assumption of a two-body re-
actionmechanism.Some facts give evidence for the two-body character of
transfer reactions. The most important of them is that the peculiarities of the energy spectra and angular distributions observed in nucleon
stripping are conserved in pick-up reactions when the dissociation of
the bombarding nucleus in collisions is cancelled out. Potassium and
chlorine nuclei produced in proton pick-up reactions and proton stripping may be considered as the "labeled" argon since the particle charge
and mass are varied in this case slightly. The deviation to smaller
angles with energy loss may be due to only nuclear attraction. The larger the nuclear surface overlapping in collisions, the greater is the nuclear
attraction effect. The Q_m behaviour for Cl and K allows one to draw the
conclusion that kinetic energy losses greatly increase with increasing
overlap between the nuclei. The losses amount to about 150 MeV at a narrow angular interval of about 20° (at 379 MeV). This implies that nuclear behaviour in collisions is that of quite viscous objects.

Fig. 6 shows a comparison of kinetic energies of two final nuclei
and their exit Coulomb barriers. The abscissa is the emission angle for a
light product in the   c.m.s. The ordinate is the decay kinetic energy at
the maximum of the energy spectrum reduced by the height of the exit
Coulomb barrier. The mark "O" at the ordinate axis implies in this case
a situation when reaction products gain all their kinetic energy due to
Coulomb repulsion. As it follows from data of Fig. 6, essential nuclear
viscosity may result in total dissipation of kinetic energy in collisions
and the reduction of the relative velocity of two nuclei to small values. Low energy maxima of few-nucleon transfer reactions and the maxima of multi-nucleon transfer reactions lie dozens of MeV lower than
the exit Coulomb barrier. If one makes a mirror reflection of low energy maxima of few-nucleon transfer reactions onto the negative angle
region, one obtains a good conjugation with their high energy maxima.

2.2. Angular Distributions, Element Production Cross Sections.
Fig. 7 shows the differential cross section for the production

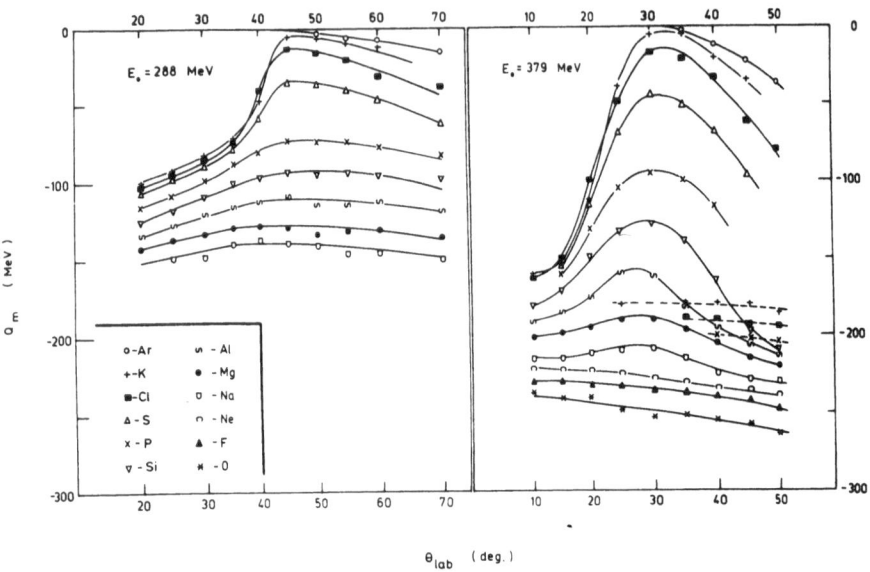

Fig. 5. Dependence of $Q_m(Z,Q)$ upon Z and Θ of reaction light products (lab. system).

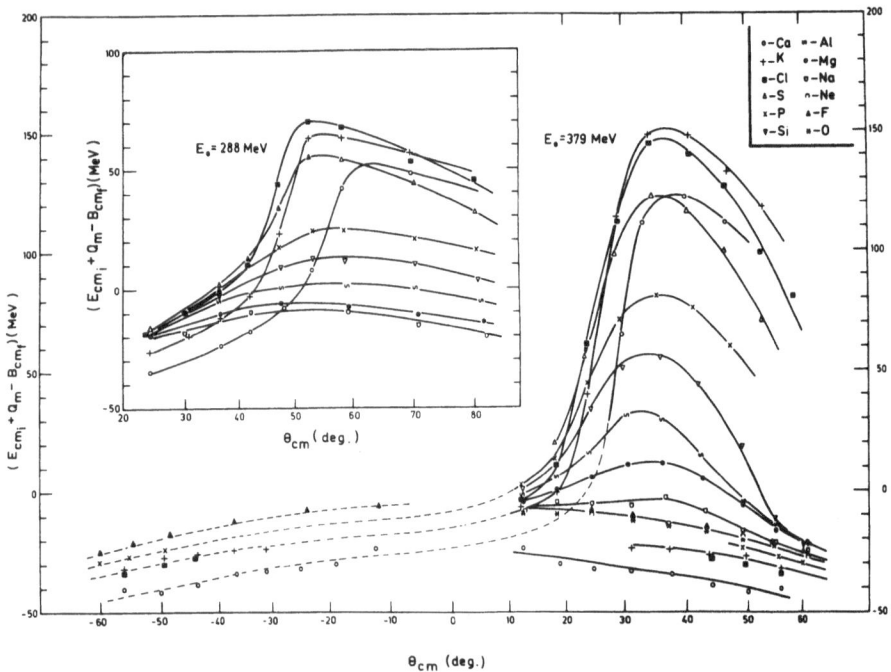

Fig. 6. Relation of the total kinetic energy of reaction products to the exit Coulomb barrier with respect to Z and the emission angle of the light product.

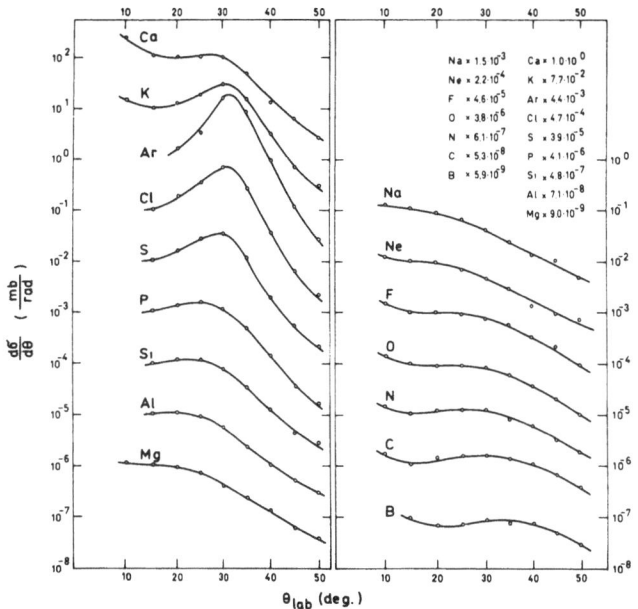

Fig. 7. Differential cross sections $d\sigma/d\theta$ of transfer reaction light products obtained by bombarding ^{232}Th with ^{40}Ar ions of 379 MeV (lab. system).

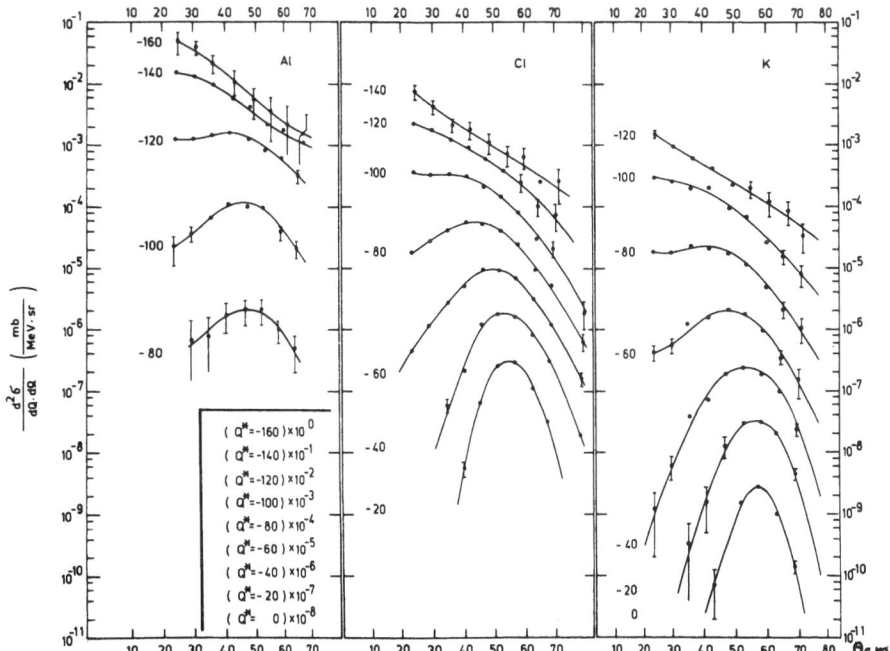

Fig. 8. Partial angular distributions of Al, Cl and K for various $Q^x \equiv Q$ values obtained by bombarding ^{232}Th with ^{40}Ar of 288 MeV.

of light elements - transfer reaction products in the bombardment of
^{232}Th with 379 MeV ^{40}Ar ions. The shape of the angular distributions
varies regularly with the number of nucleons transferred. The peculiar
maximum at the grazing angle is quite pronounced only for few-nucleon
transfer reactions. With an increase in the number of nucleons transferred, its half-width increases and the maximum is shifted towards
smaller angles. For multi-nucleon transfer reactions $d\sigma$ / $d\Omega$ increases
at first with decreasing emission angle. However, when the largest number of nucleons are transferred, there appears a broad maximum at an
angle of $30°-35°$. Fig. 8 shows the partial angular distributions
$d^2\sigma$ / $d\Omega$ dQ for Al, Cl and K for various values of $Q^x = Q$. The data
have been calculated for the system ^{232}Th$+^{40}$Ar at 288 MeV in the c.m.
system. The shape of partial angular distributions extremely changes
with increasing inelasticity of interaction. It is worth noting that
the maximum at a "grazing" angle for quasi-elastic transfer reactions
rapidly decreases towards smaller and larger angles. Thus, at small
and large angles the main contribution to the cross section comes from
inelastic processes.

Fig. 9 shows the cross sections of light element production in
bombarding ^{232}Th with ^{40}Ar and ^{22}Ne ions. The abscissa is the number
of protons stripped from the incident nucleus or picked up by it. The
data obtained for isotopic yields show that nearly the same number of
neutrons is transferred together with protons. It is seen that at first
the reaction cross section rather quickly decreases with an increase
in the number of protons transferred. However, in the case of ^{40}Ar at
379 MeV this reduction will be moderated (from O to B). A similar behaviour of the element cross section for the ^{232}Th$+^{40}$Ar system (379 MeV)
is also observed in pick-up reactions [10]. For instance, the production cross section for Fe exceeds 10 mb. On summing up the cross sections for direct processes, it is 800 mb for the ^{232}Th$+^{22}$Ne system at
σ_R = 1960 mb, 1.7 b for the ^{232}Th$+^{40}$Ar system at 297 MeV and 2.4 b
at 388 MeV. The σ_R values are equal to 1.9 and 2.9 b, respectively.
With increasing projectile energy and mass, transfer reactions become
a dominating process in the inelastic interaction of nuclei.

2.3. Isotope Production Cross Sections.

The cross sections of isotope production in transfer reactions
vary from dozens of millibarns to fractions of a microbarn. For multi-nucleon transfers involving proton stripping from the bombarding nucleus,
the authors of ref.[5] have found systematics for the cross sections.
For the ^{232}Th$+^{16}$O system at 137 MeV it is shown in Fig. 10. The abscis-

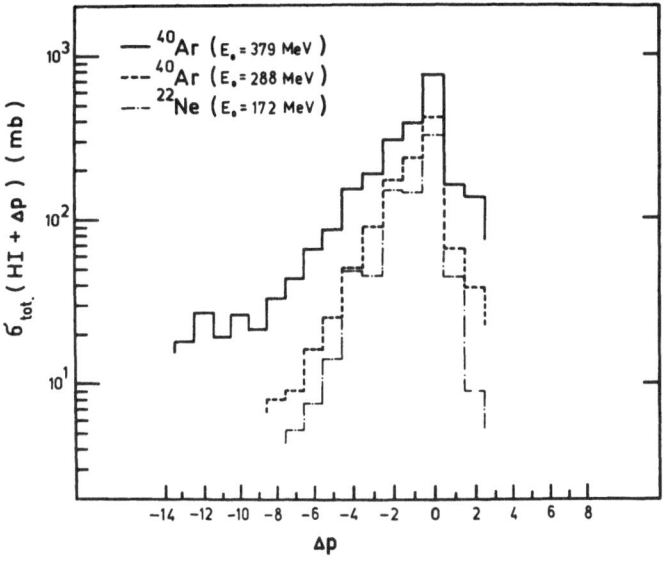

Fig. 9. Cross section of light element production by bombarding ^{232}Th with ^{22}Ne (174 MeV) and ^{40}Ar ions (288 MeV and 379 MeV). The abscissa is the number of stripped or picked-up protons.

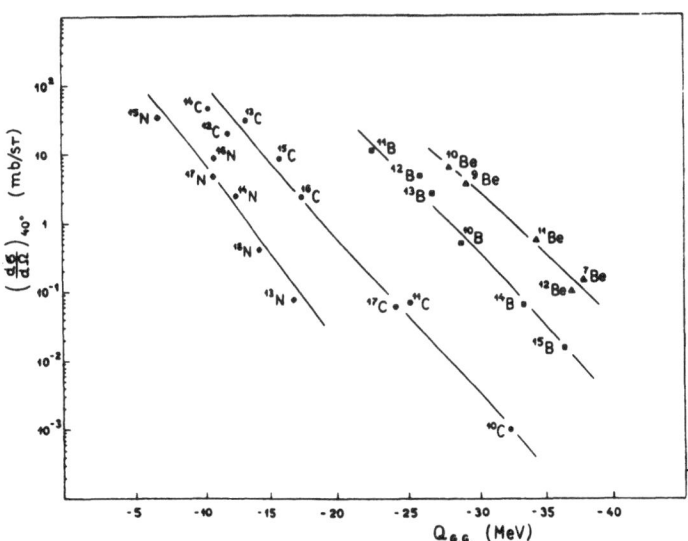

Fig. 10. Systematics of the differential cross section for isotope production for 40° obtained by bombarding ^{232}Th with ^{40}Ar at 137 MeV (lab.system). The target is 20 mg/cm^2 thick.

sa is Q_{gg}, $Q_{gg} = (M_1 + M_2) - (M_3 + M_4)$ being the energy required for accomplishing the reaction when the final nuclei turn out to be in the ground state. The ordinate is $d\sigma / d\Omega$. It is seen that the production cross sections of the isotopes of each element lie fairly well on the straight elemental lines. The systematics has been obtained for some combinations of targets and projectiles and emission angles of $40°$ and $12°$. Figs. 11-13 show the systematics for the ^{232}Th+^{15}N ($40°$)(ref.[6]), ^{232}Th+^{22}Ne ($12°$)(ref.[10]) and ^{94}Zr+^{22}Ne ($12°$) (ref.[10]) systems. It is seen that the systematics is fairly good and is fulfilled the better, the deeper the incident nucleus reconstruction.

The systematics is characterized by the exponential dependence $d\sigma / d\Omega$ upon Q_{gg} and the displacement of elemental lines towards the Q_{gg} axis with an increase in the number of stripped protons. The slope of elemental lines varies with Z of reaction products.

2.4. Difficulties in Interpreting Experimental Data within the Traditional Scheme of Direct Reactions

In classical direct reactions the energy of light products in the lab. system increases with decreasing emission angle as a result of a more favourable summation over the vectors of transfer velocity and that in the c.m.s. The spectra shown in Figs. 1-3 reveal the reverse dependence. The spectra are divided into two parts and the appearance of particles of energy lower than the Coulomb barrier is also difficult to explain. In transfer reactions the centrifugal barrier amounts to several dozens of MeV, while in the potential of two nuclei interaction, $V(R)$, the potential well is either absent or not deep. Therefore, it is difficult to explain the appearance of a low energy particle by the penetrability of the potential barrier.

The cross section of direct reactions is determined by the properties of the quantum states between which nucleon transfer occurs, and by the appropriate values of Q. The multi-nucleon transfer reactions are accompanied by considerable excitation of the final nuclei, and the probability of their production in the ground state is extremely small. From this point of view the interpretation of the Q_{gg} systematics involves serious difficulties. It has first been discussed in ref. [20]. It has also been pointed out there that the Q_{gg} dependence of isotope production cross section shows that in two-body collisions the conditions are fulfilled which are close to the statistical equilibrium. The authors of ref.[20] have proposed the concept of partial statistical equilibrium for the states in which nucleons most easily pass from nucleus to nucleus through the contact region.

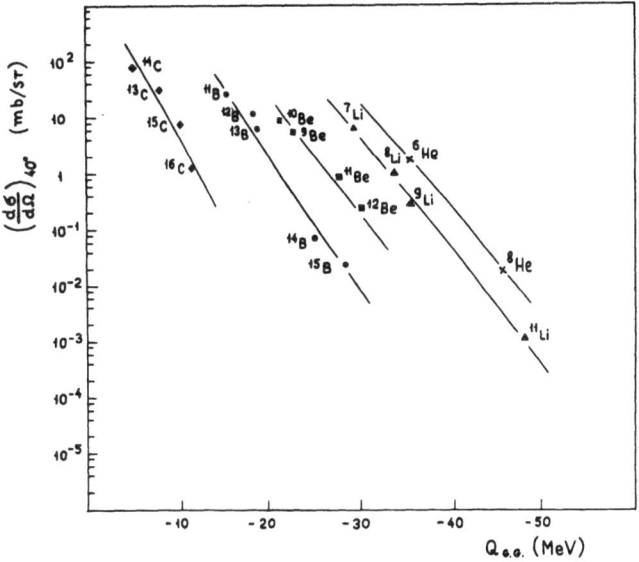

Fig. 11. Systematics of the differential cross section of isotope production for 40° obtained by bombarding [232]Th with ^{15}N at 145 MeV (lab. system). The target is 20 mg/cm[2] thick.

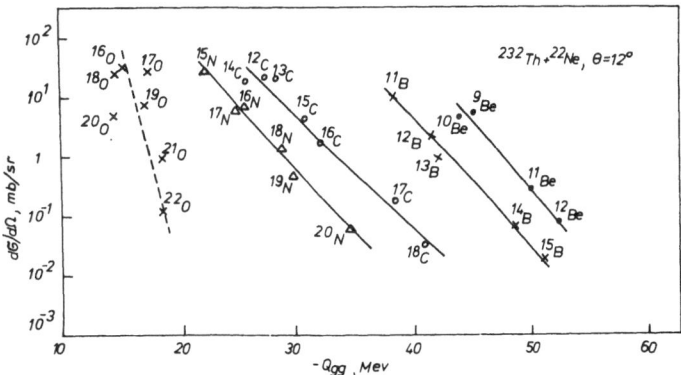

Fig. 12. Systematics of the differential cross sections of isotope production for 12° obtained by bombarding [232]Th with [22]Ne at 174 MeV (lab. system). The target is 2.5 mg/cm[2] thick.

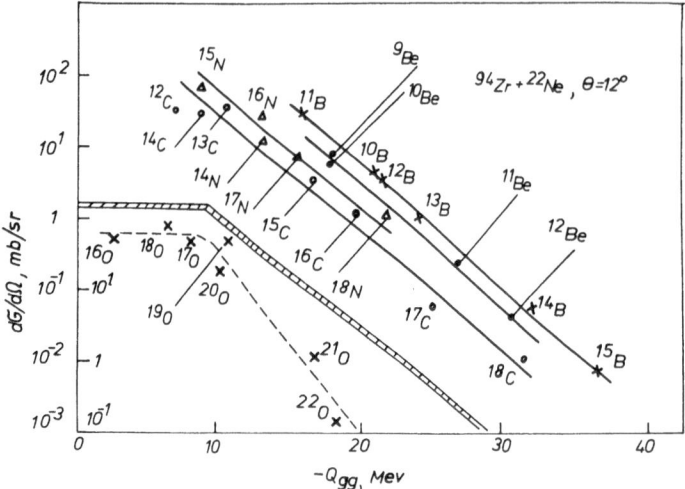

Fig. 13. Systematics of the differential cross sections for isotope production for 12° obtained by bombarding ^{94}Zr with ^{22}Ne of 174 MeV (lab. system). The target is 2.5 mg/cm^2 thick.

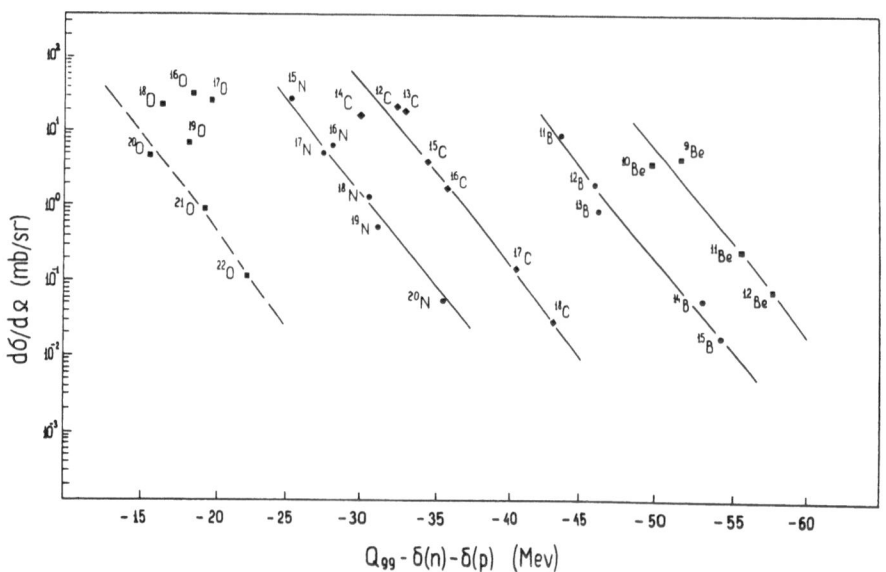

Fig. 14. The Q_{gg} systematics for the ^{232}Th + ^{22}Ne system, E = 174 MeV, θ = 12° (lab. system) after introducing corrections taking into account the effect of neutron and proton pairing.

3. POSSIBLE INTERPRETATION OF DEEP INELASTIC TRANSFER REACTIONS IN TERMS OF THE DOUBLE NUCLEAR SYSTEM CONCEPT

3.1. Formation, Evolution and Disintegration of the Double Nuclear System (DNS)

The formation of a double nuclear system is a natural consequence of strong nuclear viscosity and strong incompressibility of nuclear matter. The analysis of the interaction potential of two nuclei

$$V(R) = V_n(R) + V_c(R) + V_{rot}(R),$$

where $V_n(R)$, $V_c(R)$ and $V_{rot}(R)$ are the nuclear, Coulomb and centrifugal potentials, respectively, has been performed in refs.[24,25,36]. It has been shown that at small radii a rise is observed in the $V_n(R)$ potential which shows that repulsive forces, instead of attractive ones, become active. The appearance of repulsive forces has been interpreted as a result of the Pauli principle effect and strong incompressibility of saturated nuclear matter.

We consider the collision of two nuclei (Z_1, A_1) and (Z_2, A_2) (c.m.s.) at energies exceeding the Coulomb barrier, $E_o > B_o$, the orbital momentum $\hbar \ell$ being close to the critical angular momentum ℓ_{cr}. The spins of both nuclei are assumed to be equal to zero. At the moment of contact between the nuclear surfaces the kinetic energy is $E_o - B_o$. We divide it into two parts: the radial part E_R and the tangential one E_t. They are as follows

$$E_t = \frac{\hbar^2 \ell(\ell+1)}{2\mu R^2} \qquad \text{and } E_R = E_o - B_o - \frac{\hbar^2 \ell(\ell+1)}{2\mu R^2} .$$

Here μ is the reduced mass, $R = r_o(A_1^{1/3} + A_2^{1/3})$ and $r_o = 1.5$ fm. Due to the radial portion of the kinetic energy and nuclear attraction, the nuclei start to penetrate into each other energetically. It can be expected that under conditions of strong nuclear viscosity the tangential motion practically damps near the trajectory turning point, where the radial velocity is reduced to zero, and the overlapping of nuclear surfaces is maximally large. E_t is partly converted into excitation energy, partly to the rotation of the system of two tightly bound nuclei. The ratio of these two parts depends upon the variation of the moment of inertia of the system. At the first stage of the process $\mathfrak{J} = \mu R^2$. After the damping of the relative motion $\mathfrak{J} = \mathfrak{J}_1 + \mathfrak{J}_2 + \mu R_*^2$, where $\mathfrak{J}_1$ and $\mathfrak{J}_2$ are the intrinsic moments of inertia of the nuclei, and R_* is the distance between the centres of the nuclei at the turning point. The intrinsic moments of inertia $\mathfrak{J}_1$ and $\mathfrak{J}_2$ are close to those of the rigid body ones, since their excitation energy is about several tens of an MeV.

The radial part of the kinetic energy E_R also greatly dissipates. It may be assumed that a larger portion of E_R is converted into nuclear excitation. After the turning point the light nucleus possessing some reverse potential energy starts "to slide down" along the radius following the V(R) potential. This resembles the pendulum motion in a viscous liquid. The light nucleus may be slightly accelerated under the effect of the Coulomb field when passing nuclear periphery where the stopping effect of nuclear viscosity weakens due to smaller density of nuclear matter.

A majour part of collision kinetic energy dissipates within the period from the moment of contact to the turning point. At the collision energy of several MeV per nucleon, this time is about $1-2\times10^{-22}$ sec. This time is too short for the nuclei to become strongly deformed and change their structure to a great extent.

Thus, as a result of strong nuclear viscosity and repulsion at small radii at the beginning of collision, a double nuclear system is formed at the turning point. The nuclear surfaces of DNS are considerably overlapping and their relative velocity is small. The DNS has an angular momentum $\hbar\ell$ and the moment of inertia corresponding to the common rotation of two tightly bound nuclei. The DNS has its rotation energy $E_{rot} = \dfrac{\hbar^2\ell(\ell + 1)}{2\mathfrak{J}}$ and the excitation energy $E^* = E_0 - V^*(R_*)$, where V^* is the potential of the nuclear interaction, in which the centrifugal potential $V_{rot}(R)$ is taken to be that for a rotating DNS.

What happens to DNS later? If $\ell < \ell_{cr}$, the interaction potential $V^*(R)$ has a minimum to which the system slides down along R. However, this is a local minimum since the shape of DNS does not correspond to that of the potential energy. Therefore, the system starts evolving by exchanging nucleons and changing its shape for the one corresponding to the potential energy minimum. If the Z and A values of the DNS are not very large, the excited compound nucleus of equilibrium deformation comes as a final result.

With $\ell > \ell_{cr}$, the Coulomb and centrifugal forces exceed nuclear attraction (the interaction potential $V^*(R)$ has no minimum) and the DNS starts disintegrating. However, since with $\ell = \ell_{cr}$ the three forces are balanced, the domination of repulsion over attraction may be assumed to be small if ℓ is somewhat larger than ℓ_{cr}. As a result, the disintegration of the DNS under the conditions of high nuclear viscosity will proceed slowly compared with the characteristic nuclear time. In the course of disintegration the system may turn by a considerable angle, and the emission of light products will take place in the region of negative angles. The comparatively long lifetime of the DNS provides the possibility of transferring a large number of nucleons from one nucleus to the other. During this period of time the nuclei incorporating the DNS may undergo noticeable deformation.

3.2. Interpretation of Experimental Data.

The concept of a double nuclear system permits a natural interpretation of the peculiarities of transfer reactions. The double peak in the energy spectra of few-nucleon transfers results from the fact that a transfer of a small number of nucleons takes place in two different processes, namely in the quasielastic and deep inelastic ones. In the former case the ion is scattered mainly in the Coulomb field with the emission into the region of positive angles, and most of the kinetic energy is conserved. In the latter case the turning double system passes through 0° and disintegrates by emitting a light product with an energy close to the Coulomb barrier in the region of negative angles.

Multi-nucleon transfer reactions mainly occur in deep inelastic processes, since the transfer of a considerable number of nucleons requires a longer contact time and higher excitation of the nuclei. The development of deformations in the double nuclear system reduces the exit Coulomb barrier and makes possible the emission of particles with energies substantially lower than the Coulomb barrier for non-deformed nuclei. The shape and width of the energy spectra of deep inelastic transfers can be interpreted in terms of the spread of the lifetime of the double nuclear system. The distribution of specific lifetimes t over an average time τ may be assumed to be characterized by the Gaussian $t \sim \exp - (\frac{t-\tau}{\sigma})^2$. The deformation of the double nuclear system will increase proportionally to time at first approximation. Correspondingly, the Coulomb barrier will decrease. As a result, the shape of the energy spectrum may be described by the Gaussian, while its width will be proportional to $Z_3 Z_4$, the atomic numbers of the reaction products. The possible dependence of τ on the Z_3/Z_4 ratio can also affect the spectrum width.

An analysis of the angular distribution was carried out in the quasielastic approach, according to which all partial waves corresponding to surface collisions take part in the interference [37]. However, under the real conditions the inelasticity of the interaction increases rapidly as ℓ goes from ℓ_s to ℓ_{cr}. This leads to the destruction of the coherence of partial waves if one takes the entire interval $\ell_{cr} - \ell_s$. In the exit channels only the groups of partial waves close in the ℓ values interfere. The width of the packet $\Delta \ell$ decreases with the increasing gradient of energy changes $dE/d\ell$. The elemental angular distributions are a complicated superposition of partial angular distributions differing both in shape and intensity. The analysis made in ref. [9] has shown that the quasielastic part of the angular distributions of few-nucleon transfers (the Q_* values are small) is well described by the Strutinsky model [37] provided use is made of the Gaussian distribution of partial wave amplitudes. However, this model is incapable of describing the angular distributions at larger energy losses. It has recently been shown in ref. [34] that some progress can be made by using a modified version of the Strutinsky model and introducing a more complicated ℓ dependence of the amplitudes and phases of partial waves.

The angular distributions of deep inelastic transfers are to be analysed separately since the methods developed for direct processes are hardly applicable here. The disintegration of the double nuclear system resembles the fission of a compound nucleus with a large angular momentum, but the basic difference between them is that the disintegration of the double system takes a shorter time than required for one revolution. Nucleon transfer in the double nuclear system can be regarded as one of the aspects of its evolution in time. In this case there should exist a relationship between the lifetime of the double nuclear system and the number of nucleons transferred. As this system rotates, this correlation should manifest itself in the fact that the average emission angle of a light product (in the region of negative angles) will somehow increase with the number of transferred nucleons.

The disintegration of the double nuclear system is accompanied by the emission of a light product in the region of negative angles. This implies that the average turning angle of the system prior to its disintegration is no less than $\pi/2$. The estimates of the double nuclear system lifetime for $\ell \sim \ell_{cr}$ and the rigid body moment of inertia give a value of about 2-3×10^{-21} sec, which much exceeds the characteristic nuclear time, $\sim 10^{-22}$ sec. During the turning time of the system, conditions close to the statistical equilibrium with respect to energy and nucleon exchange between the nuclei can be established[20].

The excitation energy of the DNS at the turning point is equal to $U_i(R_*) = E_o - V^*(R_*) - \Delta(p,n)$, where $\Delta(p,n)$ is an energy consumed to break proton and neutron pairs in both nuclei. The DNS can disintegrate in many directions, since the exchange of energy and nucleons between the nuclei is a process of statistical nature.

We assume the probability for the disintegration of the DNS with the production of two nuclei (3 and 4) in the exit channel to be proportional to the product of the level densities in these nuclei,

i.e. $\rho_3 \cdot \rho_4$. The level densities are determined by the final excitation energy U_f and its distribution between the nuclei. Since the
Q_{gg} systematics describes the cross sections for the formation of light
reaction products in bound states, the ρ value for a light product
should be understood to be the number of bound states.

We shall express U_f in a form, which clearly reflects the influence of the nucleon transfer process on the value of the final excitation energy

$$U_f = U_i + Q_{gg} + \Delta E_c + \Delta E_{rot} - \delta(n) - \delta(p) .$$

Here ΔE_c is a change in the Coulomb energy of the interaction in the
exit channel as compared with the entrance one, ΔE_{rot} is a change in
the rotational energy of the double nuclear system at nucleon transfer,
which is due to a change of the moment of inertia, and $\delta(n)$ and $\delta(p)$
are corrections taking account of the effect of neutron and proton
pairing at nucleon transfer.

The Q_{gg} factor automatically includes the energy consumption on
the breaking of pairs in the donor-nucleus for those nucleons that were
transferred to the acceptor-nucleus. The transferred nucleons occupy,
as a rule, the excited levels in the acceptor-nucleus and turn out to
be unpaired. However, the Q_{gg} factor, which characterizes the energy
consumption for the transfer of nucleons from the ground state of the
donor-nucleus to the ground state of the acceptor-nucleus, neglects
this fact. As a result, the excitation energy U_f appears to be overestimated. The over-all pairing correction for a given reaction channel (for a definite isotope) is equal to the sum of the pairing energies
in the acceptor-nucleus of the additional nucleon pairs formed in the
acceptor-nucleus as a result of nucleon transfer. The exception of this
rule are the channels, in which extremely neutron-rich light nuclei
with an even number of neutrons are formed. Such nuclei may not have
excited bound states corresponding to the breaking of a neutron pair.
If such nuclei are formed as a result of a pick-up of two neutrons,
these neutrons may be transferred only in the form of a pair, for these
two neutrons no pairing correction is introduced.

In accordance with the hypothesis about the partial statistical
equilibrium with respect to energy and particle exchange, we shall
assume the final excitation energy U_f to be distributed between both
nuclei proportionally to their level densities. In order to describe
the level density ρ , we make use of the expression for ρ at
a constant temperature

$$\rho \sim \exp \frac{U}{T} .$$

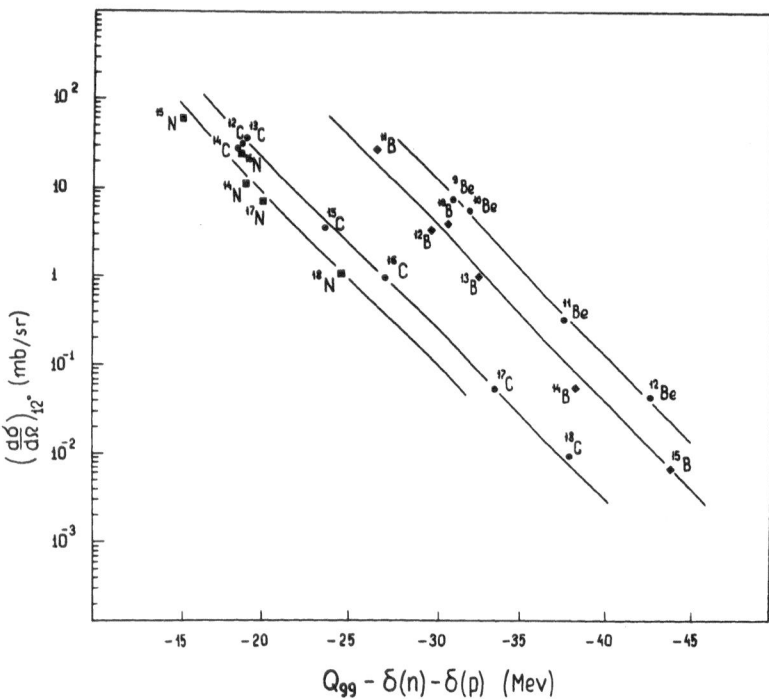

Fig. 15. The Q_{gg} systematics for the ^{94}Zr + ^{22}Ne system, E = 174 MeV, Θ = 12° (lab. system) after introducing corrections taking into account the effect of neutron and proton pairing.

It should, however, be noted that T can noticeably differ from the temperature of the corresponding compound nucleus, since the double nuclear system does not reach the state of complete statistical equilibrium. This quantity can be regarded as the temperature of a partial statistical equilibrium or simply as a parameter. Generally, the T values are the same for both nuclei forming the DNS. However, if one of the reaction products is a light nucleus with large neutron excess and a small number of bound states, practically the entire excitation energy will then be concentrated in the heavy nucleus. The light nucleus will remain unexcited. For the general case, the product of the level densities of the final nuclei $\rho_3 \cdot \rho_4$ can be written as follows:

$$\rho_3 \cdot \rho_4 \sim \exp \frac{U}{T} .$$

The energy spectra of the isotopes produced by deep inelastic transfers differ inconsiderably (for a given Z_f). The range of angular momenta ℓ_{cr}, ℓ_{cr} + $\Delta \ell$ for deep inelastic transfers is comparatively narrow, and in calculating U_f one can use some average value

of ℓ . Thus, the isotopic production cross section is mainly determined by the final excitation energy U_f, i.e.

$$\sigma \sim \exp \frac{U_f}{T} \quad .$$

On heavy nuclei, the main contribution to the changes in excita-
tion energy U is made by Q_{gg} and ΔE_c , the Q_{gg} value reaching tens
of MeV and ΔE_c being equal to 5-10 MeV per transferred proton. The
value of ΔE_{rot} does not exceed several hundred keV per transferred
nucleon, and the pairing energy in heavy nuclei is about 1 MeV. Therefore, in the first approximation one can leave only the main terms in
the expression for U_f. This will give the following expression for the
cross section:

$$\sigma \sim \exp \left(\frac{Q_{gg} + \Delta E_c}{T} \right) \quad .$$

This is just the expression that describes the empirical Q_{gg} systematics of isotopic production cross sections in multi-nucleon transfer
reactions.

In the empirical Q_{gg} systematics, the slope of the isotopic lines
changes to some extent as one progresses from one element to another.
The introduction of corrections for neutron pairing (in this case the
abscissa corresponds to $Q_{gg} - \delta(n)$) results in the identity of the
slopes. The role of the pairing corrections manifests itself most vi-
vidly in the case of the bombardments of ^{94}Zr with ^{22}Ne. In ^{94}Zr, the
pairing energy increases and the Coulomb energy decreases compared
with ^{232}Th. As a result, nitrogen isotopes appear to be located to the
right of the carbon isotopes, while boron isotopes are to the right of
berillium ones. Corrections for the pairing of neutrons and protons
permit the reestablishment of the normal sequence of the isotopic lines
in accordance with the Z value (Fig. 15).

4. CONCLUSION

The experimental information presented in this paper and its
analysis allow to conclude that transfer reactions in the deep inelastic collisions of two complex nuclei are characterized by a new, in
principle, mechanism of nuclear reactions, which is unknown for light
bombarding particles. The characteristic feature of this mechanism is
the production of a double nuclear system, in which, in spite of the
strong interaction, the nuclei retain most of their individual properties. The originality of these reactions consists in the fact that they
combine the features typical both for classical direct processes and
a compound nucleus. The light products "forget" neither the initial
direction of the projectile motion, nor its characteristics; in other

words, their angular distribution is directed forward and the maximum
yield corresponds to isotopes close to the initial nucleus both in A
and Z. At the same time, the production cross sections for some isotopes are determined by the statistical equilibrium conditions characteristic of a compound nucleus. These conditions allow the transfer
of a considerable number of nucleons from one nucleus to the other.
Under equilibrium conditions, the heavy nucleus possessing a high
level density absorbs practically the entire excitation energy. This
fact provides the possibility of producing neutron-excessive isotopes
of light elements with an extremely low neutron (or neutron pair)
binding energy in transfer reactions.

The experimental investigations described in this paper were
possible due to the high parameters of heavy ion beams from the U-300
cyclotron of the JINR Laboratory of Nuclear Reactions. The author is
deeply thankful to Academician G.N.Flerov, Director of the Laboratory
of Nuclear Reactions, for his every support and valuable advice, which
largely contributed to the success of the work. The author is grateful
to his colleagues A.G.Artukh, G.F.Gridnev and V.L.Mikheev for fruiful
discussions.

REFERENCES

1) J.Wilczynski, V.V.Volkov, P.Dezowski. Yad.Fiz. 5(1967)942
2) G.F.Gridnev, V.V.Volkov and J.Wilczynski. Nucl.Phys. A142(1970)385
3) J.Galin, D.Guerreay, M.Lefort, J.Peter, X.Tarrago and R.Basile.
   Nucl.Phys. A159(1970)461
4) Yu.Ts.Oganessian, Yu.E.Penionzhkevich and A.O.Shamsutdinov.
   Yad.Fiz. 14(1971)54
5) A.G.Artukh, V.V.Avdeichikov, J.Ero, G.F.Gridnev, V.L.Mikheev,
   V.V.Volkov and J.Wilczynski.Nucl.Phys. A160(1971)511
6) A.G.Artukh, V.V.Avdeichikov, G.F.Gridnev, V.L.Mikheev, V.V.Volkov,
   and J.Wilczynski. Nucl.Phys. A168(1971)321
7) A.G.Artukh, G.F.Gridnev, V.L.Mikheev, V.V.Volkov and J.Wilczynski.
   Nucl.Phys. A215(1973)91
8) A.G.Artukh, J.Wilczynski, V.V.Volkov, G,F.Gridnev and V.L.Mikheev.
   Yad.Fiz. 17(1973)1126
9) A.G.Artukh, G.F.Gridnev, V.L.Mikheev, V.V.Volkov and J.Wilczynski.
   Nucl.Phys. A215(1973)91
   Proc. of the Intern. Conf. on Nuclear Physics, Munich, 1973,
   vol. 1 Contributed Papers. 5.164, 5.1C5, 5.166, 5.229.
10) A.G.Artukh, G.F.Gridnev, V.L.Mikheev and V.V.Volkov. Intern. Conf.
    on Reactions Between Complex Nuclei, Nashville, 1974
    vol. 1 Contributed Papers pp.72, 86, 87.
11) J.Peter, F.Hanappe, C.Ngo and B.Tamain. Proc. of the Intern. Conf.
    on Nuclear Physics, Munich, 1973, vol. 1 Contributed Papers 5.282
12) F.Hanappe, M.Lefort, C.Ngo, J.Peter and B.Tamain. Phys,Rev.Lett.
    32(1974)738
    Intern. Conf. on Reactions Between Complex Nuclei, Nashville, 1974,
    vol. 1 Contributed Papers p.116
13) B.Gatty, D.Guerreau, M.Lefort, J.Pouthas, X.Tarrago, J.Galin,
    B.Gauvin, J.Girardo and H.Nifenecker. Report IPNO-RC-74-07
14) J.C.Jacmart, P.Colombani, H.Doubre, N.Frascaria, N.Poffe, M.Riou,
    J.C.Roynette, C.Stephan and A.Weidinger.Report IPNO-Ph.N.-74-10

15) R.Bimbot, D.Gardes, R.L.Hahn, Y.de Moras and M.F.Rivet.
Report IPNO-RC-74-01(1974). Intern. Conf. on Reactions Between
Complex Nuclei, Nashville, 1974, vol 1 Contributed Papers, p.78
16) L.G.Moretto, D.Hennemann, R.G.Jared, R.C.Gatti and S.G.Thompson.
LBL-1966, July 1973, Third Sysmposium on Physics and Chemistry of
Fission, Rochester, New York (1973)
L.G.Moretto. Personal Communication.
17) S.G.Thompson. Review paper presented at the Nobel Symposium on
Superheavy Elements, June 11-15, 1974. Ronneby, Sweden, submitted
to Physica Scripta
18) A.Y.Abul-Magd and M.El-Nadi. Phys.Rev. C3(1971)1645
19) C.Toepffer. J.de Phys. 32(1971)C6-291
20) J.P.Bondorf, F.Dickman, D.H.E.Gross and P.J.Siemens.
J.de Physique 32(1971)C6-145
21) P.J.Siemens, J.P.Bondorf, D.H.E.Gross and F.Dickman, Phys,Lett.
36B(1971)24
22) A.Y.Abul-Magd. Phys.Rev. C5(1972)559
23) A.Y.Abul-Magd, E.I.El-Abed and M.El-Nadi. Phys.Lett. 39B(1972)166
24) R.Basile, J.Galin, D.Guerreau, M.Lefort and X.Tarrago.
J.de Phys. 33(1972)9
25) R.Basile, D.Mas, J.Galin, D.Guerreau, M.Lefort and X.Tarrago.
Phys.Lett. 44B(1973)245
26) V.M.Strutinsky. Phys.Lett. 44B(1973)245
27) J.P.Bondorf and W.Nörenberg. Phys.Lett. 44B(1973)487
28) A.Y.Abul-Magd. ICTP, Trieste, Internal Report IC/73/19
29) J.Wilczynski. Phys.Lett. 47B(1973)124
30) R.Bass. Phys.Lett. 47B(1973)139
31) R.Beck and D.H.E.Gross. Phys,Lett. 47B(1973)143
32) J.Wilczynski. Phys.Lett. 47B(1973)484
33) D.H.E.Gross and H.Kalinowski. Phys.Lett. 48B(1974)302
34) V.P.Aleshin. Program and Abstracts of the XXIVth Meeting on
Nuclear Spectroscopy and Nuclear Structure, Kharkov January 29-
February 1, 1974, p. 299
35) K.K.Gudima, A.S.Iljinov and V.D.Toneev. Intern. Conf. on Reactions
Between Complex Nuclei, Nashville, 1974. Contributed Papers
Received after the Absolute Deadline, p. 13; JINR Preprint P7-7915
36) H.Muller, W.Scheid and W.Greiner. Europ. Conf. on Nuclear Physics,
Aix-en-Provence, 1972, Communications p.48.
37) V.M.Strutinsky. JETP 46(1964)2078.

CRITICAL DISCUSSION OF THE CONCEPT OF CRITICAL ANGULAR
MOMENTUM IN HEAVY ION REACTIONS

Marc Lefort
Chimie Nucléaire, Institut Physique Nucléaire, Orsay
B.P. N° 1, 91406 Orsay - France

Abstract

After a presentation of the concept of critical angular momentum, $\ell_{cr}\hbar$, and its relation to the compound nucleus cross section, the experimental methods used to determine ℓ_{cr} are discussed. Criteria for testing the compound nucleus formation are presented. Then a number of experimental results is given, mainly on argon induced reactions which show how $\ell_{cr}\hbar$ strongly depends on the entrance channel and particularly on the bombarding energy. A <u>critical distance</u> of approach, R_{cr}, is extracted as the relevant parameter limiting complete fusion of two heavy ions. It corresponds to an approximate formulation for the compound nucleus cross section $\sigma_{CN} = \pi R_{cr}^2 (1 - V_{cr}/E)$ where V_{cr} is the Coulomb + nuclear potential at the distance R_{cr} between the centers. The fission barrier for a rotating nucleus is discussed, as well as the separation between evaporation residues and fission. Finally, the possibility of existence of an inferior critical angular momentum for very heavy projectiles is presented.

1. Introduction. Compound nucleus cross section and partial wave summation.

It has been argued many years ago, that heavy ions, because of their short wave length and very low transparency, should be absorbed by target nuclei with a large probability. Only very few ℓ-waves corresponding to the diffuse edges of the nuclei should lead to direct reactions. That was observed indeed for a number of heavy ion induced reactions at energies not too high above the barrier, where σ_{CN} was found close to σ_R. A typical example is given by the study of sulfur and chlorine induced reactions by Gutbrod et al.[1]. When σ_{CN} was plotted versus $1/E$, a linear relationship was observed and the slope of the line allows one to deduce the fusion reactions and the intercept with the abscissa gives the interaction barrier, V_{IB}. Experimental measurements of σ_{CN} could be reproduced by a simple relationship like $\sigma_{CN} = \pi R_B^2 (1 - \frac{V_{IB}}{E})$ where the parameters R_B and V_{IB} fitted well with the assumption that

$\sigma_{CN} \approx \sigma_R$.

However, for some heavy ions, even at low energies, the particular structure of the projectile strongly enhances direct reaction cross sections. That was observed of course with lithium ions, but also with nitrogen and carbon ions. At higher energies, a number of general limitations occur for the compound nucleus formation. They have been experimentally observed and predicted on various theoretical bases. An obvious limitation was pointed out by Cohen, Plasil and Swiatecki[2]: since large angular momenta are brought into the composite system by heavy projectiles, the rotating energy may prevent the system to fuse entirely into a spherical compound nucleus. Therefore the idea was raised that a critical value of the angular momentum could be reached over which the possibility of formation of a compound nucleus would disappear. Later on, experimental measurements of the compound nucleus cross sections were made[3,4], and the resulting data were found to be much lower than σ_R, especially for bombarding energies around 10 MeV per amu. It was assumed that the missing part $\sigma_R - \sigma_{CN}$, corresponding to incomplete fusion processes, was occurring with the projectiles of higher impact parameters (grazing trajectories). The simplest formulation was to make a classical summation of partial waves:

$$\sigma_{CN} = \pi \lambdabar^2 \sum_0^\infty (2\ell+1) \, P_\ell \ , \tag{1}$$

where P_ℓ is the product of the transmission coefficient T_ℓ with the probability $P_\ell(CN)$ for the ℓ-wave to end up in a compound system. For the highest ℓ-values, where T_ℓ is smaller than 1, $P_\ell(CN) = 0$, so we can write:

$$\sigma_{CN} = \pi \lambdabar^2 \sum_0^{\ell_{cr}} (2\ell+1) \, P_\ell(CN) \ , \tag{2}$$

where $\ell_{cr} \hbar$ is the limit above which $P_\ell(CN) = 0$. A simplification is to assume that the change from $P_\ell(CN) = 1$ to $P_\ell(CN) = 0$ is sharply dependent on $\ell \hbar$ and the limit is formulated by the well known sharp cut-off expression[3]:

$$\sigma_{CN} = \pi \lambdabar^2 (\ell_{cr}+1)^2 \ . \tag{3}$$

Now if we write $\sigma_R = \pi \lambdabar^2 \sum_0^\infty (2\ell+1) T_\ell$ and replace the diffuse edge region where T_ℓ decreases from 1 to 0 by a step function at a cut-off value

$\ell_{CT}(max)$, then $\sigma_R = \pi \lambdabar^2 (\ell_{CT}(max) + 1)^2$. For large ℓ-values, therefore, $\frac{\sigma_{CN}}{\sigma_R} \approx \frac{\ell_{cr}^2}{\ell_{CT}^2(max)}$. Then experimental measurements of σ_{CN} are expressed in terms of critical angular momenta $\ell_{cr}\hbar$. Although in itself, there is no evidence that channels leading to a compound nucleus are restricted to the low ℓ-part of the ℓ-population, such an assumption is the consequence of the idea that the main reason for the limitation has to be looked for in the angular momentum effects. Moreover, in addition to compound nucleus cross section measurements, there are other experimental results which support such a concept.

i) The study of the de-excitation characteristics of the compound nuclei (probability for alpha particle emission, angular distribution) show that the highest possible J values are not populated[5].

ii) New types of reactions have been observed where σ_{CN} was found smaller than σ_R. In addition to quasi-elastic transfer reactions, very inelastic exchanges of nucleons occurred. The process is believed to correspond to a "damped" grazing collision since the kinetic energies of both partners in the exit channel are much lower than expected from a quasi-elastic mechanism. However, such collisions should correspond to large impact parameters, even if they correspond to trajectories closer to the target nucleus than the tangential trajectory. In some cases, quantitative measurements[5,6] were made on all the non-complete fusion processes and σ_{CN} was deduced by subtraction from σ_R. A typical example of the separation between the three categories (quasi-elastic, damped collisions, compound nuclei) is shown on figure 1 where both σ_{CN} and the damped collision cross section were measured.

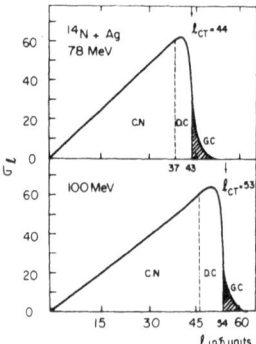

Fig. 1 - Angular momentum population for $^{14}N + Ag$ at two energies. ℓ_{CT} is the cut-off value equivalent to ℓ_{max}. DC is the area corresponding to «Damped collisions».

The total cross section may be expressed by a relation of the form:

$$\sigma_R = \pi \lambdabar^2 \left[\ell_{cr}^2 + (\ell_g^2 - \ell_{cr}^2) + \sum_{\ell_g}^{\infty} (2\ell + 1) T_\ell \right], \qquad (4)$$

where grazing collision waves are summed up above the grazing collision lower limit ℓ_g, damped collisions are summed up between ℓ_{cr} and ℓ_g with $T_\ell = 1$ and $P_\ell(CN) = 0$, and compound nucleus collisions are summed up between $\ell = 0$ and ℓ_{cr}. As we shall see later the very low part of the ℓ-population

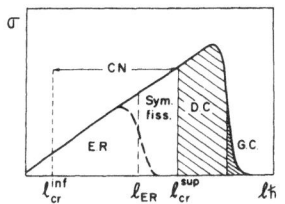

Fig. 2 - Typical separation between cross section for compound nuclei, damped collisions and grazing collision. In the CN part, ER corresponds to evaporation residues and sym. fiss to fission from compound nuclei.

might not contribute to the compound nucleus and a more general expression should take account of that effect as shown on figure 2.

The problem to exactly define what means complete fusion has been raised several times. When a medium kinetic energy proton or alpha particle (less than 50 MeV) is "absorbed" by a complex nucleus, it is generally assumed that this is equivalent to the formation of a compound nucleus. A dissipative process occurs inside the new composite species but there is no big change in the shape. With heavy ions and especially when projectiles and targets have equal masses, "absorption" does not lead automatically to a compound nucleus. Two incoming partners may stick together for a while, exchange energy and mass, and still may stay in a very deformed shape and may not reach the equilibrium shape of a compound nucleus. A composite deformed shape, similar to a fissioning nucleus at the saddle, may be formed, with a necking region. The system recoils with full momentum and then the Coulomb forces may produce a disruption. In our definition of the three categories, it would be a damped collision reaction. Then difficulties occur in order to distinguish experimentally between events resulting from a C.N. mechanism and events corresponding to damped collisions, and ambiguities exist on critical ℓ-values.

2. Criteria for measuring compound nucleus cross sections.

2.1- In principle, a compound nucleus recoils with the full momentum and de-excites during its time of flight. If light particles are emitted, neither the direction of emission, nor the kinetic energy are strongly modified except for highly excited light nuclei. Therefore any technique which is able to measure the momentum of the kinetic energy in a direction close to zero degree and for which a rough mass identification shows that the recoiling mass is in the vicinity of the total mass (projectile + target) is valuable for measuring the evaporation residue cross section. If there is a non negligible part of the de-excitation process through fission channels, measurements of fission fragments should be taken into account.

On this principle, track detectors have been used extensively for

measuring σ_{CN}. In some cases, particularly for light targets, the evaporation of alpha particles so much diminished the mass of the recoiling species that the residue had final characteristics below the detection threshold; many events were lost and the results give too low
cross sections. Also the method cannot be used with projectiles heavier
than argon.

By using a counter telescope consisting of a thin ΔE and an E-
detector, mass and Z identifications are in principle possible. However, for compound recoiling nuclei, very thin ΔE-detectors are necessary, and the telescope should be placed at small angles, close to the
beam direction. In the region of A = 60-100, the discrimination is sufficient to separate the mass and Z region of the evaporation residues
from lower Z values. Again when the fission process is becoming a
large fraction of the decay channels, the fission fragments should be
counted.

2.2- Another criterion for asserting that a compound nucleus has
been formed is to collect and to count all the residual nuclei at their
ground states after decay. By this activation method, one should be able
to measure the cross sections for (HI,xn), (HI,pxn), (HI,2pxn) etc ...
reactions, if the radioactive decays of the residual nuclei are not too
fast and if the decay schemes are known. There can be no ambiguity on
the final result, but the presicion is not good since the final σ_{CN} is
obtained by adding many partial cross sections, each of them having a
non negligible uncertainty. Also the fission fragments should be counted by some other method.

2.3- In principle, when a compound nucleus has been formed, it
should de- excite by particle emission or fission during a period of
time much longer than the collision time. For large excitation energies, there have been controversies about the lifetime of a compound
system as compared to a single rotation period or to a vibration period.
Since the energy must be dissipated in many degrees of freedom in order
to obtain an equilibrium C.N., there is no real significance if the statistical theory gives, for a high excitation energy, a lifetime shorter
than a rotational period. Furthermore, the well known angular distribution of emitted particles and fission fragments symmetric around 90°
could not be observed any more. Using the statistical theory, we have
estimated the lifetime of medium mass compound nuclei excited to energies around 100 MeV for several values of the total angular momentum[7].

It comes very close to the rotational period (10^{-20}s) at low J values, but for J > 50 ℏ, it is 10 times larger.

Therefore we can assert that part of the total reaction cross section which corresponds to <u>compound nucleus formation is to be searched in all reactions which exhibit angular distributions symmetric about 90 degrees in the center of mass system</u>. When the compound nucleus de-excites primarily by fission, the angular distribution of the fission fragments is a crucial point. In practice, it takes long to obtain a good angular distribution. For light particle emission, it is difficult to give absolute cross sections since one does not know very precisely how many particles are emitted per excited nucleus. However, there have been some attempts to estimate σ_{CN} by this method. Moreover, the analysis of the angular distribution can directly give information on the critical angular momentum[8].

2.4- <u>Fission fragment angular correlation</u>. Because of the influence of high angular momenta the fission probability is very strongly enhanced, even for compound nuclei which normally have high fission barriers. For example, the compound nucleus cross section is nearly equal to the fission cross section in the region of A = 200. Then, it has been suggested[9a] to obtain σ_{CN} by measuring the fission cross section. But the angular correlation should be studied carefully in order to check that events <u>following a full momentum transfer</u> are well separated from events following a partial momentum transfer after an incomplete fusion. By using this procedure Sikkeland[9b] has found that only 1/2 of the total reaction cross section corresponds to symmetric fission after full momentum transfer in the case of (Ar + ^{238}U) at 400 MeV. Another example of such a clear separation has been given in our Nashville paper[10]. An extensive discussion on the technical problems that had to be solved in order to separate true fission events (after full momentum transfer) from various other fission fragments is given in Tamain's thesis[11].

There is, however, a more fundamental question. The observation of two fission fragments at the true correlation angles might not be considered as a proof in favour of a long lived compound nucleus. The full momentum transfer is not a sufficient criterion for testing the complete equilibrium in a composite system. But extensive results by Hanappe et al.[12] on a variety of target nuclei (from Mo until U) bombarded by Ar ions have shown that mass and energy distributions of the

fission fragments behave just like those resulting from lighter projectiles for which fission channels are known as decay processes from a
compound nucleus. The fact that there is a symmetry around half-mass
of the complete fusion nucleus, means that there is no remembrance
either of the masses and of the ratio N/Z of projectile and target.
That does not prove an entire equilibrium for all degrees of freedom,
but it shows that at least the mixing time is much longer than the time
for a single nucleon-nucleon interaction. Very recently, angular distributions of the fission fragments have been obtained[13] for the system
^{40}Ar + ^{197}Au, with the help of X-ray measurements on catcher foils. The
well known shape of 1/sinθ predicted by Halpern and Strutinsky[14] has
been observed. This confirms that the compound nucleus ^{237}Bk has re-
coiled before fissioning and that this process accounts for a large
fraction of the total cross section. It has not been checked yet, that
the 1/sin θ distribution occurs for all fission fragments through the
entire Z range. If both heavy and light fragments present the same
distribution, then the compound nucleus formation will be definitively
proved.

3. Dependence of ℓ_{cr} on the entrance channel.

3.1- Experimental results.
A large survey of experimental data
on the compound nucleus cross sections has been given in Nashville[10].
Critical angular momenta were deduced using the expression

$$\ell_{cr} = \sqrt{\sigma_{CN}/\pi \lambda^2}.$$

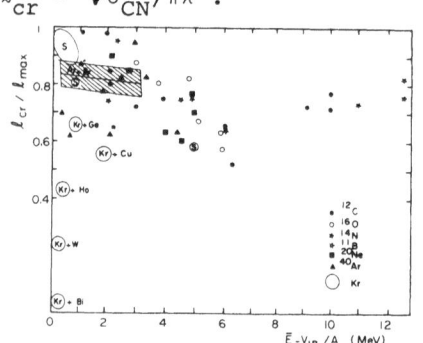

Fig. 3 - Plot of experimental results on ℓ_{cr}/ℓ_{max} for a
large number of reactions. S represents sulfur
induced reactions.

Figure 3 has been drawn in order to show that there is no real correlation between the ratio ℓ_{cr}/ℓ_{max}
and the energy per nucleon at the contact point (relative nucleon velocity).
It illustrates also that the ratio
might still be high for large velocities.

A more interesting presentation
is made on figure 4 in order to demon-
strate that at a given excitation energy ($E^x \sim$ 107 MeV) for the same
compound nucleus ^{170}Yb the entrance channel has a large effect on σ_{CN}
and then on $\ell_{cr}\hbar$. The experimental results have been obtained by Zebelman and Miller[15]. I have included in the figure results on ^{161}Tm pro-

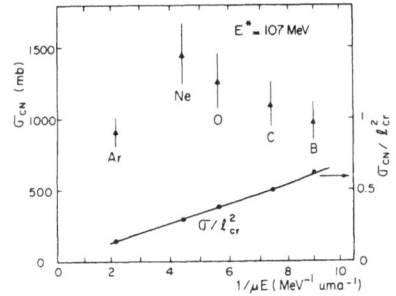

Fig. 4 - Plot of σ_{CN} versus $1/\mu E$ for rare earth compound nuclei at $E^* \approx 107$ MeV. On the left ordinates σ_{CN} has been devided by ℓ_{cr}^2 (see text and ref. 15).

duced with Ar ions, although it is not ex-actly the same excitation energy.

If the same J_{cr} would have been ob-tained, one should have $\sigma_{CN} = \frac{\pi\hbar^2}{2\mu\bar{E}} \cdot J_{cr}^2$. By plotting σ_{CN} versus $\frac{1}{\mu\bar{E}}$, where μ is the reduced mass, the result shows that the expected linear increase is not observed. On the contrary, ℓ_{cr} values can be extrac-ted from the product $\sigma_{CN} \times \mu\bar{E}$.

In this section we shall restrict ourself mainly on the argon induced reactions for which the highest cri-tical angular momenta have been obtained and where there are the largest disagreements with most of the theoretical models. In table I, a survey is made of the available data. As it was pointed out already, one can discuss the validity of some of the measured cross sections. In some cases the fission contribution has been neglected. In others, the full momentum transfer is the only criterion of the compound nucleus forma-tion. However, even if there are some uncertainties, the table shows clearly that ℓ_{cr} varies with the bombarding energy and may reach values larger than 100. Of course, it always stays at smaller values than the rigid rotor limit, defined by Grover[16] as the angular momentum above which there can be no levels for a given excitation energy (see table 1, column 8 in ref. 9). Then the origin of such critical values has to be explained on some other basis.

3.2- <u>Theoretical models</u>. Wilczynski[17] has attempted to explain the results by considering the approach between two nuclei and by assu-ming that complete fusion occurs when the two centers are closer than some particular distance at half densities of nuclear matter. The nuc-leus-nucleus force is derived from simple surface-energy considerations. The condition of force equilibrium

$$\frac{2\pi(\gamma_1+\gamma_2)R_1R_2}{R_{12}} = \frac{Z_1Z_2e^2}{R_{12}^2} + \frac{\ell_{cr}(\ell_{cr}+1)\hbar^2}{\mu R_{12}^3} \qquad (5)$$

gives the critical angular momentum at the distance R_{12} between the cen-ters, for which the nuclear force has its maximum value. The surface ten-sion coefficients, γ_1 and γ_2, are evaluated on the basis of the liquid

TABLE I - Compound nucleus cross sections and critical angular momenta deduced for Argon induced reactions

Target	CN	Ē	E*	σ_{CN} exp	l_{CR}(max)(calc)	l_{CR}(exp)	J(Bf = 0) see section 4	Experimental technique see section 2
58Ni	98Pd	112 / 170	97 / 155	900 ± 90 / 1 000 ± 100	72 / 110	64 ± 5 / 78 ± 6	75	Track Detectors ref. 27 / S.S. Detectors CN mass assignment ref. 28
63Cu	103Ag	115	100	1 563 ± 160	75	70 ± 5		(Fission neglected) ref. 28
77Se	117Te	95 / 132	71 / 107	—	70 / 110	52 ± 4 / 70 ± 4	90	J_{cr} from Angular distribution of emitted particles ref. 8
109Ag	149Tb	144 / 210	92 / 158	950 / 1 300	80 / 130	70 ± 5 / 115 ± 7	86	S.S. Detectors C.N. mass assignment ref. 27
118Sn (121Sb)	158Er (161Tm)	120 / 130 / 136 / 150 / 169 / 198 / 226	58 / 68 / 74 / 88 / 107 / 136 / 164	200 ± 40 / 500 ± 50 / 600 ± 50 / 900 ± 100 / 1 200 ± 200 / 1 460 ± 250 / 1 650 ± 250	53 / 70 / 75 / 93 / 109 / 132 / 150	30 ± 3 / 55 ± 5 / 60 ± 5 / 79 ± 6 / 95 ± 8 / 114 ± 10 / 130 ± 10	90	$\Sigma\sigma_{ER} + \sigma_{fiss}$ ref. 29
165Ho	205At	181 / 241	94 / 154	800 ± 90 / 1 350 ± 140	110 / 167	86 ± 5 / 129 ± 7	78	Fission Fragm. correlation ref. 12
197Au	237Bk	224	104	900 ± 200	120	100 ± 10	~ 0	Fission Fragm. correlation ref. 13
209Bi	249Mv	210	100	1 110 ± 200	123	108 ± 10	~ 0	Fission Fragm. correlation ref. 12
232Th	272 / 108	256 / 340	110 / 200	1 000 / 600	166 / 226	114 / 101		σ_R(calc) - σ_{DC} ref. 41
238U	278 / 110	214 / 256 / 346	82 / 118 / 204	760 ± 150 / 1 220 ± 120 / 1 441	117 / 165 / 240	91 ± 9 / 127 ± 7 / 162		Fission Fragm. correlation ref. 12 / Fission Fragm. correlation ref. 9b

drop model. Such a model does not take account of the entrance channel. It builds up a liquid drop potential for each ℓ-value and the criterion is that all ℓ-values which exhibit potential wells are assumed to lead to a compound nucleus. Those values of ℓ for which dV/dr ⩽ 0, for all values of r do not fuse. No energy is lost in the relative radial motion prior to the moment at which the projectile has reached the barrier.

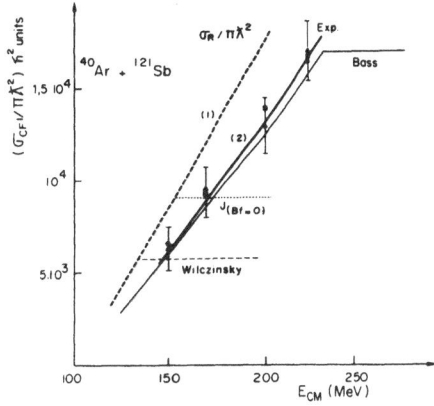

Fig. 5 - Plot of $\sigma_{CN}/\pi\lambda^2$ in $\hbar^2$ units versus energy for the system (Ar + Sb). (1) total reaction cross section (calculated). (2) and ■ calculated values assuming a critical distance with r_{cr} = 1 fm. ● Calculations by Gross et al. ▲ Experimental values. Also Wilczynski and Bass' models are shown [17,18].

The main characteristics of the model is that ℓ_{cr} is independent of bombarding energy, in contradiction to the experimental results, as is typically shown on figure 5 extracted from our Nashville report[10].

In the model of Bass[18], the assumption that each ℓ-wave which has a maximum outside a particular radius leads to a compound nucleus, is relaxed by a factor $f = \frac{5}{7}$, corresponding to the classical estimate of the conversion of angular momentum in the relative motion into internal angular momentum of the fragments:

$$V_\ell(r) = \frac{z_1 z_2 e^2}{r} + \frac{\hbar^2 f^2 \ell_{cr}(\ell_{cr}+1)}{2\mu r^2} - \frac{da_s A_1^{1/3} A_2^{1/3}}{R_{12}} e^{-\frac{r-R_{12}}{d}} \qquad (6)$$

where d = 1.35 fm, a_s = 17.0 MeV, R_{12} = 1.07 $(A_1^{1/3}+A_2^{1/3})$. Hence, the compound nucleus cross section is enhanced. However, above the barrier for ℓ_{cr} (maximum at which $\frac{dV}{dr}$ (r=R_{12})=0), the model predicts an energy independent value of ℓ_{cr}, and there are disagreements with some of our experimental data (figure 5). Such a model expresses the friction effects by putting a large barrier which plays the role of reducing the complete fusion cross section in a sharp way. As Wilczynski's model, it uses a single nuclear potential for the complete system and does not try to describe the approach of the two nuclei. Several years ago, Bjornholm and Swiatecki[19] have stressed the need of a full dynamical calculation for heavy ion reactions. An explanation of the limit to compound nucleus formation should not only be given on the basis of static shape of the fusioning systems, but should take account of the colli-

ding process. Along these lines, Beck and Gross[20] have considered a
potential which couples to internal degrees of freedom. This leads in
the classical limit, to a Newtonian equation of motion with frictional
forces proportional to the velocity. Then it becomes possible to explain both critical angular momenta and damped collision effects. Classical dynamical calculations have been performed by Gross and Kalinowski[21] and by Bondorf, Sobel and Sperber[22]. It is not our purpose to
discuss the theoretical aspects of damping effects. However, we would
like to support, in a qualitative statement, the idea that not a critical angular momentum but instead a <u>critical distance of approach</u> may be
the relevant quantity limiting complete fusion[23,24].

3.3- <u>The critical distance of approach</u>. The problem is to relate the repulsive effect of the angular momentum to the distance between the two centers at which it operates. For a given projectile-target system, the same partial ℓ-wave corresponds to a smaller and smaller
impact parameter when the bombarding energy increases. Therefore, a constant distance between the two colliding centers implies larger ℓ-values
for high energies than for low energies. The concept that a constant
distance of approach is the real clue for complete fusion limits comes
from the following consideration. The possibility for a large number
of intrinsic excitations ending up into a compound nucleus formation depends on two conflicting tendencies: attractive nuclear forces which
are more and more efficient when the distance between the two centers
diminishes; centrifugal forces which prevent the two nuclei to fuse into
a single composite system and to move towards a spherical shape. Then
it might be a good approximation to consider a potential energy as a
function of the distance between the two centers and to keep all degrees
of freedom frozen, except the relative motion. The potential energy is
calculated as usual:

$$V(r) = V_{nucl} + V_{coul} + \frac{\ell(\ell+1)\hbar^2}{2\mathcal{J}} \tag{7}$$

but for the nuclear part, instead of using a liquid drop potential, or
an optical model potential, we have used a <u>potential which conserves entirely the structure of each nucleus during the contact</u>. This implies
the <u>sudden approximation</u>. At the opposite, an adiabatic approach in
which there is a continuous exchange of energy between nucleons of both
partners would change continuously the nuclear potential. This would
irreversibly lead to a compound nucleus since no repulsive nuclear potential would appear. It seems to us that the sudden approximation,

even if it is not entirely correct, is a better way for building a model
which has the purpose to study the approach interaction and to estimate
at which distance between the two centers the kinetic energy has been decreased so much that the two body potential has to be abandonned and an
attractive well should describe the composite nucleus evolving into a
compound nucleus. In other words, we try to keep frozen all degrees of
freedom except relative motion and angular momentum, to build the corresponding potential, and to see at which distance one should unfreeze
other degrees of freedom in order to proceed to a compound nucleus. If
such a distance cannot be reached, then complete fusion does not occur.
The ion-ion potential has been calculated[24] using an energy-density
method as proposed by Bruckner et al.[25]. Recently, Ngô et al.[26] have
shown that such a potential reproduces remarkably well the experimentally known interaction barriers for all the heavy ion reactions under study
(from ^{12}C to ^{84}Kr on many targets). Such interaction barriers correspond to the value of the potential, where $(\frac{dV}{dr})_{RIB} = 0$, for the s-wave.
For a bombarding energy $\bar{E}$, the classical turning point for the partial
wave $\ell = \ell_{cr}(\bar{E})$ is found at a distance R_{cr}. All partial waves with $\ell <$
ℓ_{cr} contribute to complete fusion since the turning point is at a shorter distance than R_{cr}. For $\ell > \ell_{cr}$, the turning point is at a larger
distance than R_{cr} and the interpenetration is not sufficient to have
enough intrinsic excitations and energy dissipation leading to the compound nucleus. Since the first attempt to determine R_{cr}, we have tried to
use a folding nuclear potential and the results are not very different.
The interesting result is that for all systems studied, the critical
distance R_{cr} should be expressed in terms of a single radius parameter
r_{cr}, $R_{cr} = r_{cr}(A_1^{1/3}+A_2^{1/3})$ with $r_{cr} = 1.0 \pm 0.07$ fm. Three examples are
given on figures 6, 7 and 8 for argon induced reactions: on a light
target (Ni) where the fission cross section is probably negligible and
for which the experimental cross sections have been measured by Gutbrod
et al.[27] and more recently by Bingham et al.[28], on a medium target
(Sn-Sb) where σ_f is of the same order of magnitude as σ_{ER}[29], and on a
heavy target (Ho) where all σ_{CN} is going into the fission process[12].

The effect of the Coulomb potential is clearly seen, since $Z_1 Z_2$
is equal to 504 for Ni, to 900 for Sn and 1170 for Ho. While there is
still a pocket in the Ar + Ni potential curve at $\ell = 65$, such a pocket
has disappeared at $\ell = 80$ for Sn and at $\ell = 100$ for Ho. One should no-
tice that the critical angular momentum is not obtained at $(dV/dr = 0)$
as in Wilczynski's model. For a given bombarding energy $\bar{E}$, there is a
given $\ell_{max}\hbar$. The potential energy curve corresponding to that particu-

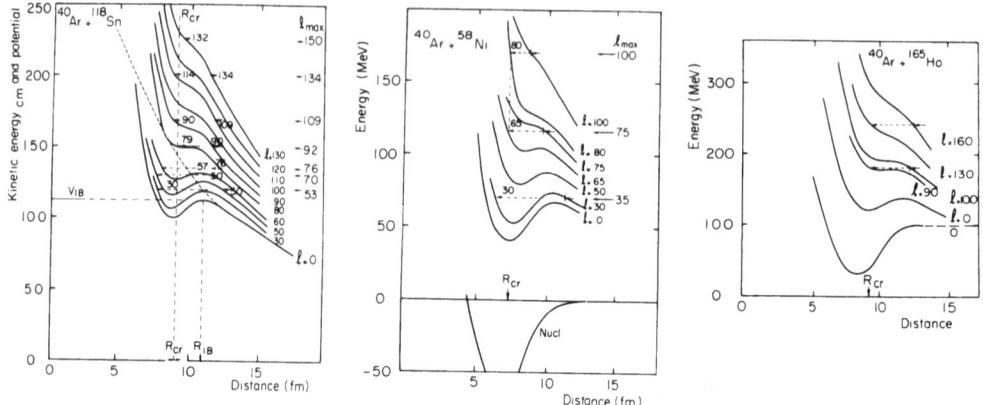

Fig. 6-7 and 8 - Potential energy curves calculated as the basis of expression (7) for three systems (Ar + ^{118}Sn), (Ar + ^{165}Ho) and (Ar + ^{58}Ni). For a given kinetic energy, ℓ_{max} is given on the right hand side. The turning point for such a ℓ_{max} is shown by the sign ⊢. The turning point which corresponds to R_{cr} is shown by ⇢ and the experimental critical ℓ is noted.

lar ℓ_{max} value may present a maximum at which $dV_{\ell}/dr = 0$. However, in the case of argon induced reaction, such a maximum $V_{IB}(\ell_{max})$ is always higher than $\bar{E}$. Therefore, σ_{CN} is always smaller than σ_{R} since there are necessarily ℓ-values for which the critical distance of approach cannot be reached. This is clearly shown on the three figures where the turning point for ℓ_{max} is marked. The situation is not the same for lighter ions. For example, figure 9 shows the potential curves for the reaction ^{12}C + ^{27}Al. The orbital angular momentum at a given energy is much smaller than for argon induced reactions. Therefore, at relatively low energies, the maximum $V_{IB}(\ell_{max})$ is lower than the bombarding energy. All ℓ-waves can penetrate the target nucleus and overcome the barrier. Hence, σ_{CN} should be very close to σ_{R}, since only soft grazing collisions might undergo direct reactions. The result is that $\ell_{cr} \approx \ell_{max}$ and there is not any more a well defined critical distance. All waves are trapped in the potential pocket, and energy can be dissipated in such a way

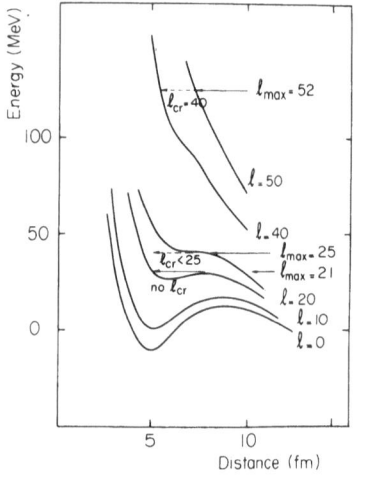

Fig. 9 - Potential energy curves for the system (^{12}C + ^{27}Al). At 30 MeV $\ell_{max} = 21$ and the turning point is already obtained for a deep penetration. It is only when $\ell_{max} = 25$ that the turning point for ℓ_{max} is found at large distances.

that a compound nucleus is always formed. Hence, the distance of approach for complete fusion which can be deduced, is the distance of the interaction barrier R_{IB} and corresponds to V_{IB} with large r_0 values, as they were found by Gutbrod et al.[1].

A consequence of the concept of critical distance is to deduce a simple approximative formula for calculating the compound nucleus cross section. There is obviously a relation between R_{cr} and ℓ_{cr}, which is expressed by the classical angular momentum conversation. Since at the contact, the distance is R_{cr}, the corresponding impact parameter is given by:

$$b_{cr}^2 \mu \bar{E} = \tfrac{1}{2} \ell_{cr}^2 \hbar^2 = R_{cr} \mu (\bar{E} - V_{cr}) , \qquad (8)$$

where V_{cr} is the potential at the distance R_{cr}. Such a potential should be deduced, in principle, from the potential curve $\ell = 0$, at the abscissa R_{cr}, since it corresponds to $V_{cr} = V_{nucl}(R_{cr}) + V_{coul}(R_{cr})$. Hence, the cross section for complete fusion is equal to $\sigma_{CN} = \pi b_{cr}^2 = \pi R_{cr}^2 (1 - \frac{V_{cr}}{E})$, a relation very similar to

$$\sigma_R = \pi R_{IB}^2 (1 - \frac{V_{IB}}{\bar{E}}) . \qquad (9)$$

Therefore, for light ions and low energies, $\sigma_R \approx \sigma_{CN}$ corresponds to a distance between centers R_{IB} and an interaction barrier V_{IB} which can be determined as the reaction threshold. This has been done by plotting σ_R versus $1/\bar{E}$ like in Gutbrod's experiments[1]. For heavy ions and high energies, $\sigma_{CN} < \sigma_R$ and one can apply $\sigma_{CN} = \pi R_{cr}^2 (1 - \frac{V_{cr}}{E})$, if V_{cr} is determined or calculated. Such a conclusion has been obtained by Mosel et al.[30] in a more elaborate calculation where they show that relations (8) and (9) are the asymptotic forms of the exact expression of σ_{CN}. In figure 10, experimental cross sections are compared to calculated values using relations (8) and (9) for the system Ar + Sn. It demonstrates that at high energies a good agreement is found with R_{cr} and V_{cr}, while for low energies experimental cross sections have a tendency to fit with R_{IB} and V_{IB} with the usual parameters.

Another evidence for the critical distance can be found in table II, where are reproduced experimental cross sections from Zebelman and Miller and from our work on Ar + Sb. They have been obtained at energies much above the barrier, and they fit with $\sigma_{CN} = R_{cr}^2 (1 - \frac{V_{cr}}{E})$, assuming nearly the same parameter r_{cr} whatever is the bombarding ion,

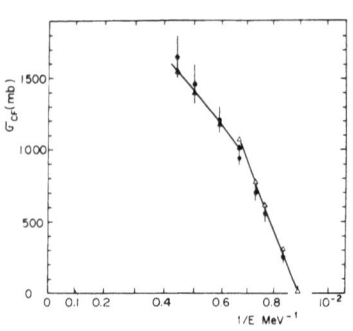

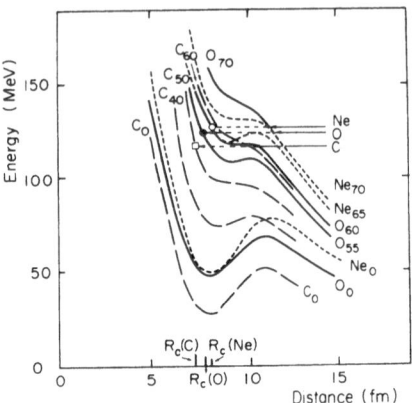

Fig. 10 - Plot of σ_{CF} versus 1/E for the system ($Ar + ^{118}Sn$) : Experimental points are shown by dark points with error bars. ▲ values calculated with expression (8), $r_{cr} = 1.05$ fm and $v_{cr} = 90$ MeV. △ Values calculated with expression (9), $r_B = 1.4$ fm and $V_{IB} = 112$ MeV.

Fig. 11 - Potential energy curves for the systems ($^{20}Ne + ^{150}Nd$) ($^{16}O + ^{144}Sm$) and ($^{12}C + ^{158}Gd$). Subscripts noted O correspond to $\ell = 0$.
○ Turning point for Ne for $r_{cr} = 1$fm ($\ell_{cr} = 65$) dotted line
● Turning point for O for $r_{cr} = 1$ fm ($\ell_{cr} = 55$) full line
□ Turning point for C for $r_{cr} = 1$ fm ($\ell_{cr} = 50$) dashed line.

Table II

Cross sections for the formation of compound nucleus Yb at $E^* = 107$ MeV

Projectile	$\bar{E}$ MeV	$\sigma_{CN}(exp)$ mb	R_{cr} fm	r_{cr} fm	V_{cr} MeV	$\sigma_{calc} = \pi R_{cr}^2 (1 - \dfrac{V_{cr}}{\bar{E}})$
^{11}B	107.5	980±150	7.2	0.95	35	1 000
^{12}C	117	1100±160	7.2	0.93	35	1 130
^{16}O	124	1260±200	7.9	1	48	1 310
^{20}Ne	127	1450±220	8.	1	49	1 380
^{40}Ar	150	900±100	8.1	1	90	820

and for the potential, values determined on the potential energy curves as shown on figure 11.

4. <u>Fission cross section σ_f, and evaporation residue cross section, σ_{ER}. Influence of the fission barrier</u>.

As was already pointed out in the first section, when a compound nucleus is formed it decays essentially through two main categories of channels, i) evaporation of particles which ends up with residual nuclei in the vicinity of A(CN), ii) fission which ends up with residual fragments with a symmetric mass distribution around $\frac{1}{2}$A(CN).

But the fission probability is strongly enhanced by high angular momenta in such a way that amongst the various ℓ-waves, those of high ℓ-values preferentially induce fission. A well known model for expressing this effect is the rotating liquid drop model of Plasil and Swiatecki[2] from which Plasil and Blann[31] have deduced fission barriers $B_f(J)$ as a function of J. The fission barrier of a rotating drop is given by the difference, $E_{sp}(J) - E(J)$, between the energy of the saddle shape with angular momentum J and the minimum energy of a rotating spherical drop with angular momentum J. It is smaller than $B_f(J=0)$, and $B_f(J)$ can be equal to zero even for medium mass nuclei when J is of the order of 90 $\hbar$. This concept has been explained in very comprehensive surveys by Blann[32] and Plasil[33]. In the ℓ-range, where $B_f(J)$ decreases between S_n, the neutron binding energy, and zero, the ratio Γ_f/Γ_n increases very sharply, and one may replace the decreasing function of σ_{ER} by a step function. Hence, Blann and Plasil have defined a critical angular momentum J_{cr}^{ER} above which the cross section for producing evaporation residues is negligible. Figure 12 illustrates how, inside the ℓ-population leading to a compound nucleus, a separation can be made between the two categories of decays.

A very spectacular evidence for such a behaviour has been given recently by Miller et al.[34], on the compound nucleus ^{170}Yb. They could show that it is only above J = 50 $\hbar$, a value which is not reached by carbon induced reactions, that σ_f starts to increase rapidly. Since they have been able to measure ℓ_{cr} values of 40, 46, 58 and 70 respectively for ^{11}B, ^{12}C, ^{16}O and ^{20}Ne ions, they could deduce how Γ_f/Γ_n sharply depends on $\ell\hbar$.

It has been argued more generally that since for $B_f(J) = 0$ there is not any more a maximum in the curve of potential energy versus distance, the J-value corres-

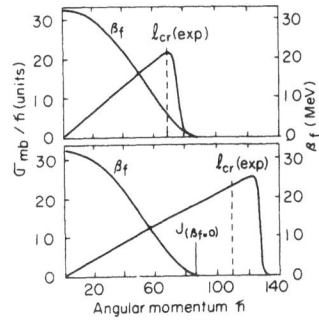

Fig. 12 · ℓ wave population for the system ^{40}Ar+Ag at two different c.m. energies (144 MeV) and (211 MeV). B_f is the rotating liquid drop barrier (31) (right side ordinate). Critical J from $B_f = 0$ is calculated at 85 $\hbar$. Experimental ℓ_{cr} are noted.

ponding to $B_f(J) = 0$ should be a critical value <u>also in the entrance channel</u>. Hence, no compound nucleus, whatever would be the decay process, could be formed for an orbital angular momentum $\ell_{cr}\hbar$ leading to J_{cr} so that $B_f(J)_{cr} = 0$. If the cross section for fission is higher than $\sigma = \pi\lambda^2 J_{cr}^2$ determined from such a J_{cr}, it would imply that a number of the fission events <u>are not</u> due to a de-excitation process from a compound nucleus (even if the full momentum transfer is observed). These would come from some sort of <u>composite system</u> not yet in equilibrium and has been classified as "preequilibrium" fission. However, the results shown in table I (column 8) seem to indicate that $\ell_{cr}\hbar$ values larger than $J_{cr}(B_f)$ calculated from $B_f(J_{cr}) = 0$, have been obtained especially in argon induced reactions. In those experiments, fission fragments have mass distributions similar to those of fissions issued from compound nuclei, and moreover in one case (Ar + Au) the angular distribution looks like a CN distribution[13] (see section 2-4). It seems, at least for the moment, that, either the $B_f(J_{cr}) = 0$ calculations give too small J_{cr} values, or that the concept of $B_f(J) = 0$ although clearly well established in the <u>exit channels</u> should not be applied without caution for the <u>entrance channel</u>.

5. <u>On the possible existence of an inferior critical angular momentum for very heavy ions (A > 40).</u>

In figure 2, the possibility that the part of the ℓ-population that induces compound nucleus reactions does not automatically start at $\ell = 0$ has been pointed out. It seems very obvious that for light projectiles, the best conditions for a total absorption by a heavy target correspond to s-waves, as far as the Coulomb barrier has been overcome. However, for two colliding ions of large masses, at energies slightly higher than the interaction barrier, it has been shown that complete fusion does not occur and that the reaction cross section is entirely due to other processes like very inelastic transfers of nucleons[35] or "quasi-fissions"[36]. A schematic sketch of excitation functions in the region of the threshold is presented on figure 13. For example, Péter et al.[37] have found recently that there is a shift of the order of 20 MeV between the quasi-fission and the symmetric fission threshold energies in the

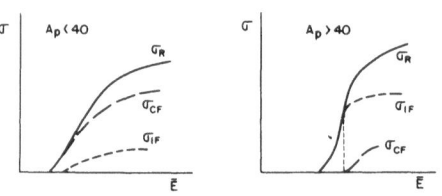

Fig. 13 - Schematic illustration of excitation functions for σ_R, σ_{CF} and σ_{IF} (incomplete fusion cross section) in two typical cases, for projectile mass A_p : smaller than 40, and higher than 40.

(^{65}Cu + ^{197}Au) reactions. Such an additional barrier for complete fusion might be attributed to many causes. A possibility could be that for the energy range under consideration, the angular momenta are too low, so that tangential frictional forces are not large enough to dissipate a sufficient amount of energy, and the repulsive Coulomb force can repulse the incoming ion after an inelastic scattering.

Another experimental result which is more convincing, is relative to the reaction (^{84}Kr + ^{74}Ge) which has been compared[38] with the reaction (^{40}Ar + ^{118}Sn) yielding the same compound nucleus. Excitation functions have been measured for the residual nuclei issued from (Ar,xn) and (Kr,xn) reactions with x = 4, 5 and 6. The threshold for a given (Kr,xn), for example (Kr,6n), is shifted towards higher excitation energies than (Ar,6n) by about 15 MeV, although the angular momentum populations are very similar in both cases in the same energy range (figure 14)

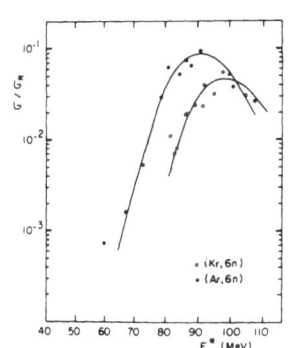

Fig. 14 - Excitation functions for the reactions (Ar,6n) and (Kr,6n) passing through the same compound nucleus ^{158}Er (After ref. 38).

Such a shift is not due to any Coulomb barrier effect because the energy is much higher than the Coulomb barrier. Moreover, the width for the (Kr,6n) excitation function is narrower than for the (Ar,6n). The same result was obtained for the (HI,5n) reaction. In principle, all (HI,5n) excitation functions should start at the same excitation energy, whatever is the heavy ion. This is indeed observed when comparing the (0,5n), (Ne,5n), (Ar,5n) reactions. Only the maximum of the excitation function and the descent on the high excitation energy side are known to be shifted to higher energies when larger angular momenta are shared by the compound nucleus.

The lack of events in the low energy side of the excitation function seems to indicate that low angular momenta are not contributing to the reaction. A quantitative code has been written in order to fit excitation functions, including the angular momentum. The best fit was obtained by assuming a distribution of J-values, for the argon induced reactions, between 0 and 80 ℏ at the lowest excitation energy and between 0 and 130 ℏ at the highest. For the krypton induced reactions, the fit was obtained with a population restricted to the interval 45 ℏ and 80 ℏ. This is not a real evidence of a cut-off on the low side of the angular momentum.

However, Tsang[39] has shown that such an inferior critical value has a theoretical meaning when the interaction is studied in terms of radial and tangential frictional forces. When the radial energy is zero (tangential collision), the system has too much angular momentum and nothing can be captured. As the radial energy is increased, the frictional force will transfer the initial orbital angular momentum into the spin angular momenta I_1 and I_2 by means of a sliding friction term and a smaller rolling friction term. Then, for some distance between the two centers which is critical, the system will be captured. Now, Tsang says that as one reaches a high radial energy and a low angular momentum, friction cannot dissipate sufficient energy and the system goes in and comes out over the barrier and no capture occurs. This effect implies indeed a lower cut off in critical angular momentum. Such theoretical considerations would be in very good agreement with our experimental findings, and for the moment the inferior critical angular momentum is only a very tentative assumption. It emphasizes the idea that the window open for compound nucleus formation might become very narrow between two limits of $\ell\hbar$ population.

Acknowledgements. - I should like to thank Drs. Beiner and Galin who have made the code for calculating   Bruckner's nuclear matter densities and the potential energy curves.

Note added during the conference: I very recently received a preprint from Kozub et al.[40] where the same equation $\sigma_{CN} = \pi R_{cr}^2 (1-V_{cr}/\bar{E})$ is derived. Experimental results were obtained on bombardments of ^{27}Al with ^{16}O, ^{20}Ne and ^{32}S. R_{cr} was derived from the ordinate intercepts of the straight line representing the dependence of σ_{CN} on $\bar{E}^{-1}$. The results indicate a smaller parameter r_{cr} than 1 fm, in disagreement with our results, and also very deep, attractive potentials V_{cr}.

References

1) H.H. Gutbrod, W.G. Winn, M. Blann, Nucl. Phys. A213 (1973) 267.
2) S. Cohen, F. Plasil, W. Swiatecki, Proceedings of the Third Conference on Reactions between Complex Nuclei (University of California Press, 1963) 325, and Ann. Phys. 82 (1974) 557.
3) L. Kowalski, J. Jodogne, J.M. Miller, Phys. Rev. 169 (1968) 894.
4) J.B. Natowitz, Phys. Rev. C1 (1970) 623 and 2157.
5) R. Basile, J. Galin, D. Guerreau, M. Lefort, X. Tarrago, J. Phys. 33 (1972) 9.
6) F. Pühlhofer, R.M. Diamond, Nucl. Phys. A191 (1972) 561.
7) M. Lefort, Lecture E. Fermi Summer School, Varenna (1974).
8) J. Galin, B. Gatty, D. Guerreau, C. Rousset, U. Schlotthauer-Voos, X. Tarrago, Phys. Rev. C9 (1974) 1126.
9a) T. Sikkeland, V.E. Viola, E.L. Haines, Phys. Rev. 125 (1962) 1350.
  b) T. Sikkeland, Phys. Let. 27B (1968) 277.
10) M. Lefort, Y. Le Beyec, J. Péter, Conference on Reactions between Complex Nuclei, Nashville (USA) June 1974.
11) B. Tamain, Thesis, Clermont-Ferrand, n° 182 (May 1974).
12) F. Hanappe, C. Ngô, J. Péter, B. Tamain, Int. Conference on Physics and Chemistry of Fission, IAEA/SM-174/42 (1973) 289.
13) R. Lucas, J. Poitou, H. Nifenecker, R. Bimbot, J. Péter, B. Tamain, comm. to this conf.
14) I. Halpern, V.M. Strutinsky, Proceedings U. N. Int. Conf. Peaceful Uses of Atomic Energy, 15, p. 408, p. 1513 (1958).
15) A.M. Zebelman, J.M. Miller, Phys. Rev. Lett. 30 (1973) 27.
16) J.R. Grover, Phys. Rev. 157 (1967) 832.
17) J. Wilczynski, Nucl. Phys. A216 (1973) 386.
18) R. Bass, Phys. Lett. 47B (1973) 139.
19) S. Björnholm, W.J. Swiatecki, Physics Reports, 4 (1972) 326.
20) R. Beck, D.H.E. Gross, Phys. Lett. B47 (1973) 143.
21) D.H.E. Gross, H. Kalinowski, Phys. Lett. B48 (1974) 302.
22) J.P. Bondorf, M.I. Sobel, D. Sperber, Preprint Nordita, 1974, Submitted to Physics Reports.
23) M. Lefort, J. Phys. A7 (1974) 107.
24) J. Galin, D. Guerreau, M. Lefort, X. Tarrago, Phys. Rev. C9 (1974) 1018.
25) K.A. Bruckner, J.R. Buchler, M.N. Kelly, Phys. Rev. 173 (1968) 944.
    K.A. Bruckner, J.R. Buchler, G. Jorna, R.J. Lombard, Phys. Rev. 171 (1968) 1188.
26) C. Ngô, B. Tamain, J. Galin, M. Beiner, R.J. Lombard, preprint IPNO/TH-74-19.
27) H.H. Gutbrod, H.C. Britt, B. Erkila, R.H. Stokes, F. Plasil, M. Blann, Conf. on Physics and Chemistry of Fission IAEA/SM Rochester 174/59 (1973).
28) H.G. Bingham, E.E. Gross, M.J. Saltmarsh, A. Zucker, C.R. Bingham, Bull. Am. Phys. Soc. 19 (1974) 428 and Communication to Conf. on Reactions between Complex Nuclei, Nashville, p. 130 (June 1974).
29) M. Lefort, Y. Le Beyec, J. Péter, Rivista Nuovo Cimento, 4 (1974) 79.
30) D. Glas, U. Mosel, Preprint University of Giessen (1974).
31) M. Blann, F. Plasil, Phys. Rev. Lett. 29 (1972) 303.
32) M. Blann, VI Summer School Warsaw University, Poland (1973).
33) F. Plasil, Conf. on Reactions between Complex Nuclei, Nashville (June 1974).
34) A.M. Zebelman, L. Kowalski, J. Miller, K. Beg, Y. Eyal, G. Jaffé, A. Kandel, D. Logan, Phys. Rev. C10 (1974) 200.
35) R. Bimbot, H. Gauvin, Y. Le Beyec, M. Lefort, N.T. Porile, B. Tamain, Nucl. Phys. A189 (1972) 539.
36) M. Lefort, C. Ngô, J. Péter, B. Tamain, Nucl. Phys. A216 (1973) 166.
    F. Hanappe, M. Lefort, C. Ngô, J. Péter, B. Tamain, Phys. Rev. Lett. 32 (1974) 738.

37) F. Hanappe, C. Ngô, J. Péter, B. Tamain, Conf. on Reactions between Complex Nuclei, Nashville (June 1974) p. 116.
38) H. Gauvin, R.L. Hahn, M. Lefort, Y. Le Beyec, Phys. Rev. C10 (1974) 722.
39) C.F. Tsang, Preprint LBL 2928 (1974).
40) R.L. Kozub, N.H. Lu, J.M. Miller, D. Logan, T.W. Debiak, L. Kowalski, Preprint Columbia University (1974).
41) A.G. Artukh, G.F. Gridnev, V.L. Mikheev, V.V. Volkov, J. Wilczynski, Nucl. Phys. A215 (1973) 91.

SUMMARY

by O. Hansen

Niels Bohr Institute
University of Copenhagen, Denmark
and
Los Alamos Scientific Laboratories
Los Alamos, New Mexico, USA

It is true that I have been sitting through all these talks and my revenge is to match in time the amount I have used listening to you now to speak to you. So be prepared for the next three or four hours, if you fall asleep it will not deter me in any way, because I have also slept and it did not deter any of the speakers in any way. Many of my friends have asked me during the day when they woke me up whether I had taken the whole thing in, whether I could remember the whole thing and were ready to give it back to you. So I went up and found an English-German dictionary to define what this whole thing was about: and it says that summary means "Hauptinhalt" and to summarize can be said to "zusammenfassen". So it leaves me some individuality. I don't have to replay every thing to you, I can actually just give back to you what I think were the most important things and I can even give them in my own distorted way which I am going to do. I think that yesterday night we all were touched upon the main thing. It taught me something about all the physicists who are present here. We learned a little bit about elephants and birds, classical elephants and quantum mechanical birds, and I cannot help noting that you must be a queer bunch because obviously you have the greatest difficulties in distinguishing a bird from an elephant. Well, last night about one o'clock when I walked home to the hotel, I began to understand why you people have so many difficulties knowing whether you deal with an elephant or a bird, because I was sure I saw an elephant fly by. So it occurred to me that the best advice I can offer in this situation is, that one should not look for the difference between elephants and birds. When you can see elephants flying, you can see birds with trunks. But I am sure one should listen, because even in my worst delirium I have never heard an elephant whistle like a nightingale. So that's the "Hauptinhalt" and the summary and I can go on to some malicious remarks.

The first thing I would like to say was that I have collected a list of things that had not been mentioned at this conference which were mentioned extensively at the monthly conferences on semiclassical and so on heavy ion things that we have had over the past year: Nashville, Brookhaven, Copenhagen - it goes even back to two Argonne meetings about the same thing. It is obvious that until the last speaker came on, words like recoil, spectroscopy, post-prior had not been mentioned and it tells you a little bit about the whole change of attitude. These are no longer so much things we discuss, they are things we use. We may not do it any better than we did two years ago, I don't know, but we got so accustomed to it, anyway, it doesn't bother us. But then there was another thing that hasn't been talked about at all in this conference and which I think is really a serious mistake. I would like to quote one of the theoretical speakers in an off-moment when he was trying to collect his thoughts for what should be coming next. He said something, and that's the only note I took, to the effect of writing a semiclassical program for fitting elastic scattering - then he paused and he kind of said to himself: "Well then the experimentalist would come running and asking me how long time every calculation would take and how much it would cost, and I don't want to bother with that". And it reminded me of something that I think is quite evident: the experimentalists are very subdued by the theoreticians now. I mean we have, we experimentalists, to sit quietly and listen to all these theoretical discussions. Many of them are actually technicalities, and we don't stand up and tell about the experiments, and I seriously think that the most exciting development that has happened over the past three years, has not happened on the theoretical front but on the experimental front: in the way we take data, in the quality of the data, in the new instrumentations that have come up. I am not going to describe them in technical detail, but I still think that physics is an empirical science, and one tends to forget that.

So I would therefore start to show you a couple of spectra of just honest data, that has not been fitted in the complex plane, that does not show Regge poles but simply plots, a number of counts versus channel or things like that. The first such thing is one that I got from Ben Zeidman and let's see how it looks like. This is fairly new Argonne data and it certainly shows an enormous improvement over the previous Argonne data. You can for example see several states isolated

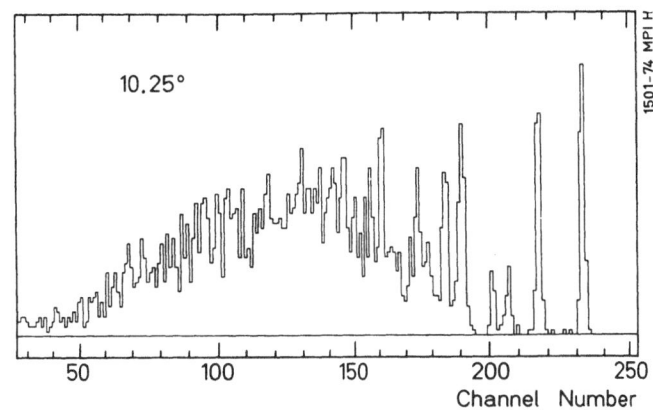

10.25°

50   100   150   200   250
Channel Number

Fig. 1: Argonne data showing improvement over previous Argonne data.

from the mud. This is counter data, so one can do counter data with 160 keV energy resolution, where we normally had 400 or 500, but I would say that this is still sloppy data. Ben, thank you for lending me this transparency. But here are some data that have not been shown here, have not been discussed, and I find it tremendously exciting. Look this is from the $^{186}W(^{12}C,^{14}C)^{184}W$ reaction. This is the ground state, and this is the rotational state at 111 keV. You

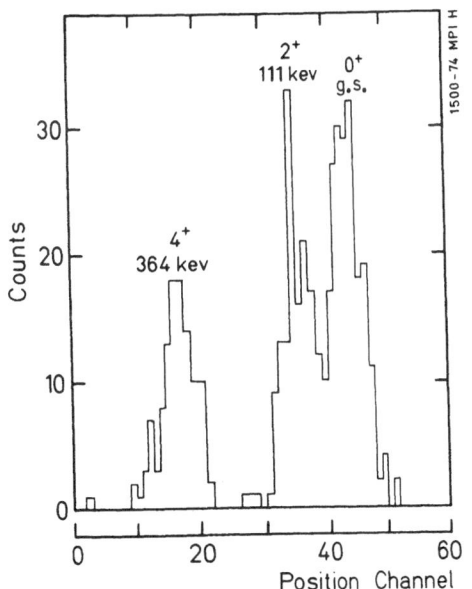

2⁺
111 kev   0⁺
g.s.

4⁺
364 kev

Fig. 2: Spectrum of the $^{186}W(^{12}C,^{14}C)$ ^{184}W reaction.

can judge for yourself, the energy resolution on that data is down to about 40 keV, and you have complete angular distributions! This shows one thing, it shows that new spectrographs, the multilens spectrographs, the Q3Ds or what you call them, they are there now. This data is from Brookhaven taken by a collaboration of Yale and Brookhaven. I think the leading authors are the wellknown experimentalists Bob Ascuitto and Ben Sørensen, but nonetheless that type of thing I find is really exciting. With the Q3D in Los Alamos (we have not done heavy ion reactions, but just to give you an idea) one can get energy resolutions of 1/2500 with a solid angle of 15 msterad. An old spectrograph would give us 1/1200 with a 50 times smaller solid angle. You can see this is a tremendous development and I would venture to say that this heavy ion data of $(^{12}C,^{14}C)$ is better than most of the (p,t) data you can find in the literature. So that is one thing. There have been similar developments in fast timing techniques. Some experimentalist flashed a mass spectrum and he knew that he was going to talk about the DWBA, but he couldn't resist saying "my mass resolution was .2 mass units". That has also come by. And I think, my dear theoretical colleagues, that there is a foundation for your future.

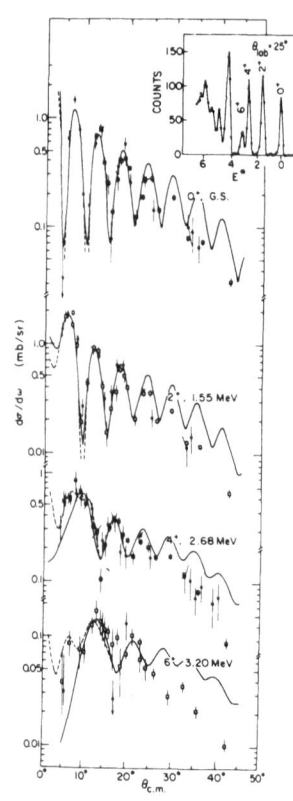

Because I have to live with Ben Zeidman for a while, this is another slide of his data, just to show you how heavy ion angular distributions are done. You have seen this slide before: they are done for every one degree, they start at angles around 5 or 6 degrees in the center of mass system, and if you remember most of the light ion data: there are 6 angles and 15 assignments. You see, the quality of data that you can take now is even above most of the light ion data. And I personally think, that this is an exciting development and I'll come back to one of the implications that I think this has.

I now tend to try to summarize the first two days. There was an obvious deviation between the first two days and today in contents. The first two days dealt very much with what we can call semiclassical and what we can call quantum mechanical, and I would like to summarize that in my own

Fig. 3: The $^{48}Ca(^{16}O,^{14}C)$ reaction at the incident energy of 56 MeV populating the lowest $0^+, 2^+, 4^+$, and 6^+ states in ^{50}Ti.

way, namely by showing you some selected data that show the various features that have been discussed. So apart from the semantics this should demonstrate to you that such features actually do exist in nature, because that is what the experimentalists are working with. This is again data from Argonne (fig. 4). What you can see is a big and fairly flat hill, which is a usual grazing peak. Then you can see the things break out in oscillations and if you actually follow the oscillations even into the peak then you will find that the angular difference between subsequent maxima is constant and if you look at your distorted wave calcula-

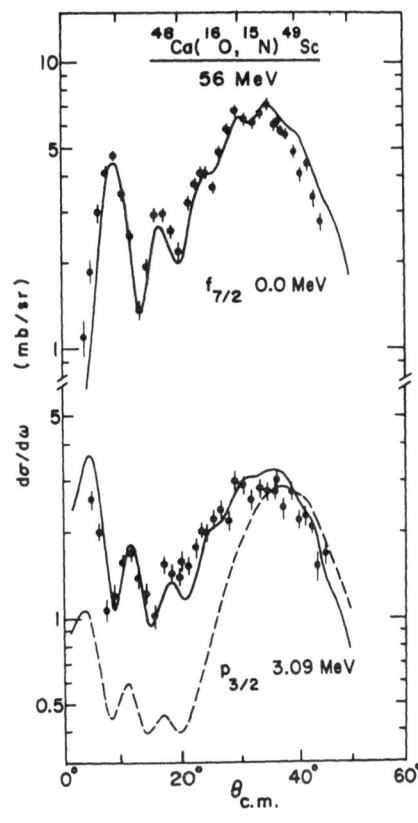

Fig. 4: The $^{48}Ca(^{16}O, ^{15}N)^{48}Sc$ reaction.

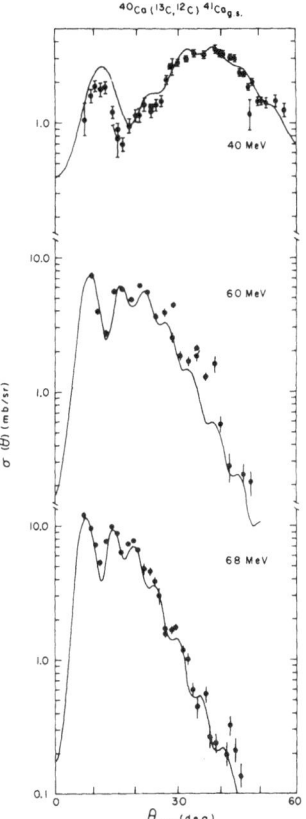

Fig. 5: The $^{40}Ca(^{13}C, ^{12}C)^{41}Ca_{g.s.}$ reaction at three different incident energies.

tions to fit them then you will find, that the angular difference between such maxima, $\Delta\Theta$ I call it, fulfils the $\Delta\Theta\cdot(\ell_{grazing}+1/2) = \pi$ rule. Or , if you are an experimentalist I cannot do it that way, then forget about the π and put the $\Delta\Theta$ in degrees and it works. Put 180 instead of π. It is a typical feature of these wiggles that they fulfil that simple relation. It corresponds to the ℓ-value, the partial wave that sits in the maximum of the S-matrix distribution in the ℓ-space.

The next slide is one that J. Garrett showed. This is Brookhaven data. The upper thing acutally shows more or less the same thing as the previous slide, maybe not as beautiful. But then one changes the energy and you can see how the smooth grazing peak moves forward and just like a child with measles these things break out all over. And we

also thought for a while that this meant it was sick, but it is not.
One can actually reproduce all these things by remeasuring and again
one will find that the Δθ for various oscillations is constant over
the thing and beautifully satisfies the relation $\Delta\theta \cdot (\ell_{grazing}+1/2) = \pi$,
where I define $\ell_{grazing}$ again from distorted wave calculations. It is
a true difficulty, that this is no number that comes out on any counter
when you do the experiment. So again, when you go up in energy this
pattern becomes strong and you can see that it does exist. But if you
want to talk about it to S. Kahana for example, who is now sleeping
away, you have to know the termonology which I learned at this meeting.
He calls this ±θ interference and he has a gaussian distribution at +θ
and a gaussian distribution at -θ and the S-matrix has interfered with
one another. If you want to stand on friendly terms with Frahn you draw
exactly the same picture but you now call it Fraunhofer scattering. If
you want to talk to K. McVoy remember not to show him any pictures what
so ever because he is a true Regge poles freak - back to back in the complex plane. The last way of getting these wiggles, consists in changing
the ℓ-window. The numbers of partial waves that contribute have been
shown by both Mermaz and Jerry Garrett in a comparison between one nuc-

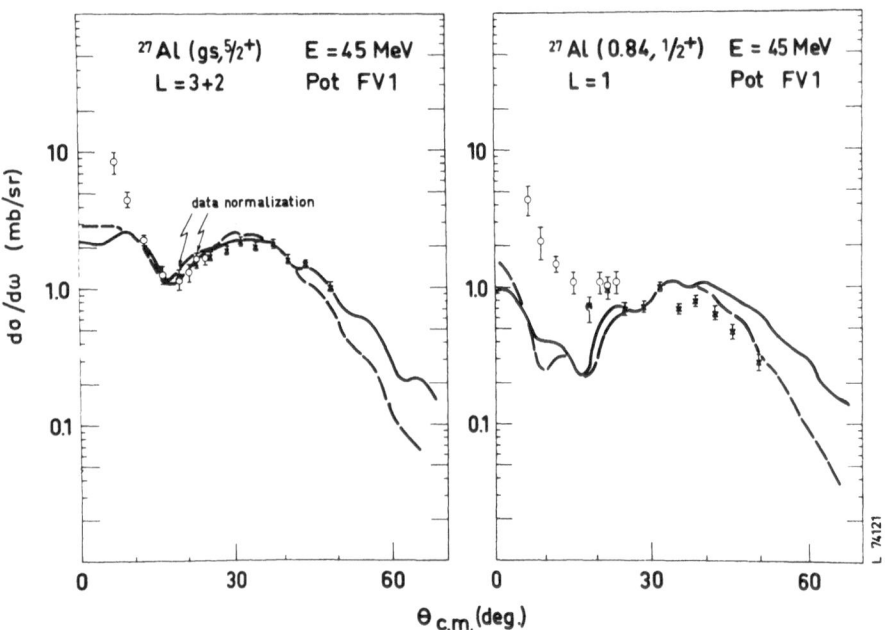

Fig. 6: The $^{26}Mg(^{16}O,^{15}N)^{27}Al$ reaction at the incident energy of
45 MeV.

leon and two nucleon transfer reactions. There you see, one is smooth and the other one wiggles. It has been said several times that this has not much to do with the form factor falling off more sharply, which you would imagine to cut out a little fewer contributing partial waves but has much more to do with changes in ℓ-values and so on. So I will venture to present this third view (fig. 6) of how measles can break out in high energy or heavy ion or what it is that you are doing physics on: this is an angular distribution from $^{26}Mg(^{16}O,^{15}N)^{27}Al_{g.s.}$ done at 45 MeV. You can see you have this fairly smooth distribution, the ground state of Al is $5/2^+$ and the ground state of ^{26}Mg is 0^+. Then we - i.e. the Copenhagen research group - did the $^{27}Al(^{16}O,^{15}N)^{28}Si$ reac-

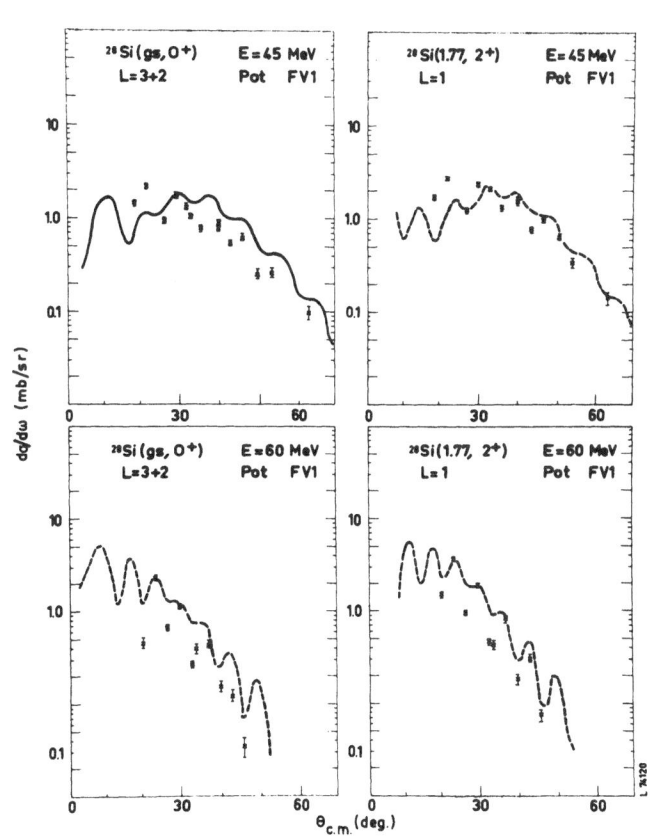

tion (fig. 7). There we started from the $5/2^+$ ground state and went to the ground state of ^{28}Si, so again you are forced to transfer $5/2^+$, in other words the ℓ-transfer is exactly the same in those two reactions, and the form factor is the same, but the Q-values are very different. Here, you go from even Z do an odd Z, whereas in the other case you go from an odd Z to even Z, so in the other case you gain binding energy by the pairing and thereby (as I have learned) the form factor becomes steeper in a stripping reaction and I would expect to see some oscillations if the old argument is true.

Fig.7: The $^{27}Al(^{16}O,^{15}N)^{28}Si$ reaction at the incident energy of 45 MeV.

Yes, it can be seen, (don't look at the theoretical curves, they don't fit), but you can see sure enough here are oscillations. This is 45 MeV in both reactions, and there is no ℓ-difference between those two that is ensured by angular momentum conservation which also holds in the classical limit. So you can see that one can make these things break out by essentially changing the binding energy and thereby the slope of the form factor, of course with that goes also that you change the energy in the outgoing channel. So these are the three ways to make typical oscillations which satisfy $\Delta\Theta\cdot(\ell+1/2) = \pi$. And that has been one big part of the discussions. I don't care whether you call it semiclassical or quantum mechanical. It is obvious that if you want to oscillate you must have some interference, you must use a superposition of two amplitudes somewhere and in my look that stands for quantum mechanical, but I am absolutely willing to learn.

Now, there are other types of interference, which had not been discussed too much but have been touched upon, namely, the type where you have the interference from two orbits. I venture to draw one of these deflection functions and if I remember correctly, I put ℓ that way

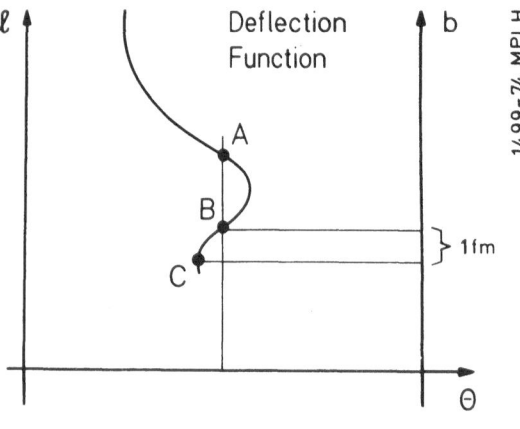

Fig. 8

and Θ that way not to get confused. The curve is something I remember by heart. Here, near the rainbow you can in the elastic or inelastic scattering have contributions from two orbits, (this was also one of the limits that Frahn talked about), and the characteristic feature of that is that the interference pattern does not satisfy $\Delta\Theta\cdot(\ell+1/2) = \pi$, actually the $\Delta\Theta$ changes with the angle from bump to bump as it depends on which two orbitals are interfering. In the inelastic scattering, (that has been mentioned by R. Malfliet), you get across the same bumps but out of phase because you have completely changed the weighting of the forces that are responsible for the scattering at these two points A and B. In the elastic scattering the trajectories are Rutherford dominated but in the inelastic the Coulomb scattering is much weaker and actually the nuclear contribution from B becomes more important and since one force is pushing

the trajectories away, the other pulling them in, you get them out of
phase. Again just look at a set of real data that show this effect.
These data are taken from F. Videbaek's thesis. It is hard to see the

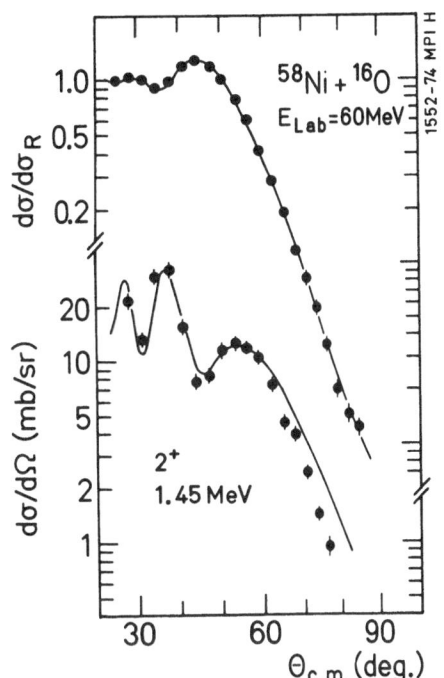

oscillations in the elastic channel. You see the first one, but
the next ones get damped out very
fast. In the inelastic scattering
this is quite a nice example: here
comes the first oscillation down
and up again, they are out of
phase with the elastic, and this
is what has been interpreted by
Landowne or Malfliet in such a
truely semiclassical approximation, where you use an orbit picture. I don't want to say more
about it and I think these pictures or data summarize the discussion pretty well. You can see
the phenomenon directly in the data.

Fig. 9: Elastic and inelastic
scattering of ^{16}O on ^{58}Ni.

(There's the first one that
fell asleep, at least that I noticed. Well, we only have three,
two more hours to go, so it's not
too bad). Okay, so much for that. I found that the various theoreticians, who entertained us for about 8 hours discussing these things actually agreed, except that they didn't agree that they agreed. I did
not really see any phantastic differences. People use slightly different languages to talk about it, but I think we understand it.

Now, the rest of the morning today, when I was almost awake,
seemed to cover completely different items - oh sorry, I wanted to do
one very simple remark. Many people have in a sense the idea, that all
these things we have been talking about, wiggles, non-wiggles, semiclassical and so on is old hat, is things we went through with the light
ions many years ago and we have learned absolutely nothing new, we are
just reinventing everything - and I think, honestly, that point of view
is wrong. I think we have learned quite a lot and I think that the reason for that lies in the very nature of the heavy ion reactions, namely

that when you do inelastic scattering the amount of energy ΔE that you give away really fulfills $\Delta E \ll E$, i.e. it is much less than the energy available. When you do a transfer reaction like $(^{16}O, ^{15}O)$ or even $(^{32}S, ^{33}S)$ or something, you really fulfill that the transferred mass Δm is much less than the mass available, $\Delta m \ll m$. That means that the outgoing particle and the incoming particle are very close to one another. Apart from symmetrization things, they really look alike, whereas when you do a (d,p) reaction, there is a very violent thing happening, the guy is loosing half his mass and forgets his identity completely. You also fulfill that the transferred angular momentum $\Delta \ell$ is much less than the partial wave ℓ_o responsible for the main transfer $\Delta \ell \ll \ell_o$. So, that's the reason that you can talk about both elastic and inelastic transfer reactions within the same language, within the same approximation, and you will learn about the same things from looking at them; whereas with light ion reactions, you would be a little bit out of your mind to postulate a (d,p)-reaction to look like elastic scattering or to be very closely related to it. Here, the difference between the potentials in the in- and outgoing channels amounts to a factor of 2, while in heavy ion reactions they are so small that we usually use the same potentials in and out. I think this is a very essential feature of the heavy ion reactions. We have learned a lot about describing these reactions, we have learned about the picture of quantum mechanical and semiclassical effects and the transitions between them in a very clean cut way, and I think we should say, it's been fun.

Now, what happened in the morning was, people put a lot of energy both in their talks and also in their particles and we could look at events where you could loose 100 MeV in an inelastic scattering event. That is not possible with a 60 MeV O-beam from a tandem. So, immediately the language changed and people were talking about friction, critical angular momenta that they definitely think are different from ours, the ones we call critical or grazing. I found myself in a situation that I had a hard time understanding what people were talking about, because they use another language. On the other side I find the whole thing was very exciting, because when the two nuclei come together you have in the intersection between them a situation that I don't think that we ever have been able to study in quantum physics. You have a manybody problem and you can study these nuclei where two different balls are slightly intersecting. The particles that sit there and have to go forward or backward, they are in a situation that no nucleon has ever been in since the big explosions of stars. To me some of the most exciting physics that I think we can ever learn from heavy ion reactions must be in that field. I have a very hard time digesting that in such situation of two overlapping nuclei, where maybe ten nucleons are involved, you can apply a simple macroscopic classical idea and hope to get away with it. If you look at such a deflection function (as in fig. 8), then we have studied the thing down to about here (point B in fig. 8). If instead of having a scale of ℓ you put in the distance of closest approach b, you would find that the part that comes into play now (between B and C in fig. 8) is probably beyond what we probe in the low energy reactions. I think it is just about a Fermi beyond what we have looked at until now. In that situation you obviously have the possibility to transfer many nucleons and you see that people now start to use statistical approaches to it, but the question that arises is: are we at all in a situation, where you can think you have statistical equilibrium and can really apply these concepts with confidence? I don't think that these concepts were given to us with that aspect, I think they were given to us as pioneering first simple ideas and we should see what we can do with them.

But what I would like to do in the next five minutes (and then I am done) is to try to indicate that the first two days which I call the days of the spectroscopists, can be connected with the days of the macromafia who is doing these "Klumpen-Physik" type experiments, where they smash things together on a 100 MeV scale and are happy afterwards (terrible waste of energy by the way). One simple thing that we think that we have learned from studying these transfer reactions, elastic and inelastic scattering is about the ion-ion deflection function close to where the two nuclei overlap. I don't think you can in a unique way derive the potential from the data. Experiments have the irritating habit of integrating over some distances and give you an answer which is not the potential at any single point. But, there is at least one connection to the ion-ion potential that is a serious, a simple one. From the elastic scattering we know about the amplitudes $\eta\ell$ and that means that we can predict the absolute reaction cross section. That comes out of every little optical model code! It is printed (if you haven't seen it) always down to the left! And it says something like σ_{reac} = ·· and if you don't use DWUCK it is always in millibarns, if you use DWUCK it is in Fermi squared. But apart from that, I seriously say that these absolute reaction cross sections are containing an important test of the potentials. They are not (as for example Körner showed) dependent on which of the many potentials you are using, because the potentials that are useful, are the same in the region of interest and therefore invariably lead to the same total reaction cross sections (within a few percent). Therefore, if you measure the total reaction cross section, it must come out close to what you predict from the potentials we think we have derived from our work. And that was the one single thing that I do not think was discussed extensively and maybe should have been. I saw a hint of it in Oeschler's statement, I wasn't awake enough to find out whether he agreed or didn't agree, but I remember vaguely that he does not agree. He finds in his measurements that the measured total reaction cross section is substantially smaller than predicted. We also know how these potentials behave with energy, so again we should be able to apply the method at the higher energies and this should hang together. I think it is extremely important, if we claim to understand the potentials, to make sure that simple properties like the total reaction cross section come out right, and that means that we probably should discuss the methods of carefully measuring to-tal reaction cross sections, maybe instead of measuring the next (^{16}O, ^{15}O) reaction.

Another place where we can meet grounds with the high energy heavy ion physics is the theory. I really feel a little bit like speaking my mind on these matters. I feel that in the situation where you talk about friction and where you go just beyond the point where the simple transfer reactions happen you have the opportunity of trans-ferring more and more particles. They can go forth and back, and you could have a succession (if you want to put it that way) of doorways, as you came closer and closer. This is a terribly difficult problem, if you look at it in a detailed description, because you now have very strong coupling and you have many many channels. I think one would nor-mally just have shrugged one's shoulders and said, you can't do it - but what R. Malfliet showed us is that you can formulate the classical, semiclassical, semiquantal (I don't know what to call it) approach in such a way that you can get very accurate answers in any of these situ-ations. Of course, in that language you can formulate the coupled equa-tions that describe very complicated situations, where you transfer five particles forth and back successively, or simultaneously. Maybe it leads to 500 coupled equations or even 1000, or what do I know, but in that formulation the coupled channel equations can be solved with the compu-ter techniques we have available. It is no longer that staggering prob-lem as it would be quantum mechanically and I therefore think an approach is opened to attack these problems.

At the same time I would strongly advocate that we start applying all the beautiful techniques we have learned to the experiments, so that we can make them much more detailed and may be really measure those deep inelastic things with a mass resolution, who knows, of .2 mass units instead of 15 mass units. I don't know if it can be done, but it seems to me at least that that's the way we are moving. The accelerators that are being built have these capabilities in them. I mean Darmstadt will open up with all the energy and an intense beam (may be not with the energy resolution you would want). Berlin is coming on. Oak Ridge is building a high resolution machine that will not match these others in energies but nonetheless will have enough energy to get up (at least in lighter systems) to this very interesting region near the critical, ja, what Wilcynski and these people call the critical angular momentum. So, if I should summarize what I have learned from this then I see the very exciting possibility of applying the microscopic methods, that we took over from spectroscopy and that we have experimentally refined, to heavy ion reactions. They can lead into a completely new field where we can study the nuclear matter, the nucleons or nuclei in a situation where they are about to comit nuclicide, where they are about to loose their identity. You therefore have a very unusual situation in that you can treat these things as quantum mechanical to some extent, statistical to some extent, but I don't think they satisfy any of our standard approaches. Understanding them would require that the two groups of people who sometimes have difficulties understanding one another's language - the spectroscopists and the macromafia - get together and start working together. It also requires that we stop having conferences and go down into the laboratories - it takes about a month to recover from a conference, about a month to think of the next one and there it is. Okay, that's all I have to say.

LIST OF PARTICIPANTS

A.Y. **Abul-Magd**, Dept. of Physics Faculty of Science, Cairo University, Giza, Egypt

D. **Agassi**, Max-Planck-Institut für Kernphysik, Heidelberg, Germany

K. **Alder**, Institut für Theoretische Physik der Univeristät Basel, Basel, Switzerland

P. **Armbruster**, Gesellschaft für Schwerionenforschung, Darmstadt, Germany

Y. **Avishai**, Ben Gurion Negev University, Beer Sheva, Israel

H. **Backe**, Institut für Kernphysik der Technischen Hochschule, Darmstadt, Germany

J. **Barrette**, Max-Planck-Institut für Kernphysik, Heidelberg, Germany

R. **Bass**, Institut für Kernphysik der Universität Frankfurt, Frankfurt, Germany

D. **Berndt**, Max-Planck-Institut für Kernphysik, Heidelberg, Germany

K. **Bethge**, Institut für Kernphysik der Universität Frankfurt, Frankfurt, Germany

R. **Bock**, Gesellschaft für Schwerionenforschung, Darmstadt, Germany

J. de **Boer**, Sektion Physik der Universität München, Garching, Germany

H.G. **Bohlen**, Hahn-Meitner-Institut für Kernforschung, Berlin, Germany

W. **Bohne**, Hahn-Meitner-Institut für Kernforschung, Berlin, Germany

H. **Bokemeyer**, Gesellschaft für Schwerionenforschung, Darmstadt, Germany

W.J. **Braithwaite**, Center for Nuclear Studies, University of Texas, Austin, Texas, U.S.A.

P. **Braun-Munzinger**, Max-Planck-Institut für Kernphysik, Heidelberg, Germany

P. **Brix**, Max-Planck-Institut für Kernphysik, Heidelberg, Germany

B. **Buck**, Dept. of Theoretical Physics, University of Oxford, Oxford, United Kingdom

S. **Buhl**, II. Physikalisches Institut der Universität, Heidelberg, Germany

P.J.A. **Buttle**, Sektion Physik der Universität München, Garching, Germany

N. **Carjan**, Institut für Kernphysik der Technischen Hochschule, Darmstadt, Germany

B. **Cauvin**, CEN Saclay, Gif-sur-Yvette, France

M.E. **Cobern**, CEN Saclay, Gif-sur-Yvette, France

J.G. **Cramer jr.**, Nuclear Physics Lab., University of Washington, Seattle, Washington, U.S.A.

H. **Damjantschitsch**, Max-Planck-Institut für Kernphysik, Heidelberg, Germany

K. **Dietrich**, Physik Department der Technischen Universität München, Garching, Germany

H. **Doubre**, Institut de Physique Nucléaire, Orsay, France

C. **Dover**, Physics Department, Brookhaven National Lab., Upton, Long Island, N.Y., U.S.A.

J.L. **Durell**, Schuster Laboratory, University of Manchester, Manchester, United Kingdom

K.A. **Eberhard**, Sektion Physik der Universität München, Garching, Germany

Y. **Eyal**, The Weizmann Institute of Science, Rehovot, Israel

D. **Fick**, Max-Planck-Institut für Kernphysik, Heidelberg, Germany

E. **Flynn**, Los Alamos Scientific Lab., Los Alamos, New Mexico, U.S.A.

W.E. **Frahn**, University of Cape Town, Physics Department, Rondebosch, South Africa

H. **Freiesleben**, Physikalisches Institut der Universität, Marburg, Germany

H. **Fuchs**, Hahn-Meitner-Institut für Kernforschung, Berlin, Germany

H.W. **Fulbright**, Laboratoire de Physique Nucléaire, Strasbourg-Cronenbourg, France

A. **Gamp**, Max-Planck-Institut für Kernphysik, Heidelberg, Germany

J.D. **Garrett**, Brookhaven National Lab., Dept. of Physics, Upton, Long Island, N.Y., U.S.A.

B. **Gebauer**, Hahn-Meitner-Institut für Kernforschung, Berlin, Germany

C.K. **Gelbke**, Max-Planck-Institut für Kernphysik, Heidelberg, Germany

N.K. **Glendenning**, Lawrence Berkeley Laboratory, Berkeley, California, U.S.A.

A. **Gobbi**, Gesellschaft für Schwerionenforschung, Darmstadt, Germany

U. **Götz**, Institut für Theoretische Physik der Universität, Basel, Switzerland

P.A. **Gottschalk**, The Weizmann Institut of Science, Rehovot, Israel

G. **Graw**, Max-Planck-Institut für Kernphysik, Heidelberg, Germany

D.H.E. **Gross**, Hahn-Meitner-Institut für Kernforschung, Berlin, Germany

E. **Grosse**, Max-Planck-Institut für Kernphysik, Heidelberg, Germany

H.H. **Gutbrod**, Gesellschaft für Schwerionenforschung, Darmstadt, Germany

E.C. **Halbert**, Oak Ridge National Lab., Dept. of Physics, Oak Ridge, Tennessee, U.S.A.

M.L. **Halbert**, Oak Ridge National Lab., Dept. of Physics, Oak Ridge, Tennessee, U.S.A.

O. **Hansen**, Niels Bohr Institut, Copenhagen, Denmark

H.L. **Harney**, Max-Planck-Institut für Kernphysik, Heidelberg, Germany

K.D. **Hildenbrand**, Max-Planck-Institut für Kernphysik, Heidelberg, Germany

E.R. **Hilf**, Institut für Kernphysik der Technischen Hochschule, Darmstadt Germany

H. **Homeyer**, Hahn-Meitner-Institut für Kernforschung, Berlin, Germany

M. **Ivanovich**, AERE Harwell, Nuclear Physics Div., Oxfordshire, United Kingdom

M. **Ivascu**, Institute of Atomic Physics, Bukarest, Romania

J.C. **Jacmart**, Institut de Physique Nucléaire, Orsay, France

R. **Jährling**, Max-Planck-Institut für Kernphysik, Heidelberg, Germany

S. **Kahana**, Oxford University, Department of Theoretical Physics, Oxford, United Kingdom

H. **Kalinowski**, Hahn-Meitner-Institut für Kernforschung, Berlin, Germany

D. **Kamke**, Institut für Exp.-Physik I der Ruhr-Universität, Bochum, Germany

E. **Kankeleit**, Institut für Kernphysik der Technischen Hochschule, Darmstadt, Germany

J. **Knoll**, CEN Saclay, Gif-sur-Yvette, France

C.M. **Ko**, Max-Planck-Institut für Kernphysik, Heidelberg, Germany

H.J. **Körner**, Physik Department E12, Technische Universität München, Garching, Germany

W. **Kohl**, Gesellschaft für Schwerionenforschung, Darmstadt, Germany

B. **Kohlmeyer**, Physikalisches Institut der Universität, Marburg, Germany

D. **Kolb**, Gesellschaft für Schwerionenforschung, Darmstadt, Germany

H. **Krappe**, Hahn-Meitner-Institut für Kernforschung, Berlin, Germany

E. **Kuphal**, Institut für Kernphysik der Technischen Hochschule, Darmstadt, Germany

K. **Kusterer**, Max-Planck-Institut für Kernphysik, Heidelberg, Germany

J. **Kuzminski**, Max-Planck-Institut für Kernphysik, Heidelberg, Germany

S. **Landowne**, Hahn-Meitner-Institut für Kernforschung, Berlin, Germany

Ch. **Leclercq-Villain**, Université Libre de Bruxelles, Physique Théorique, Brüssel, Belgium

Th. **Ledergerber**, The Weizmann Institute of Science, Rehovot, Israel

S.M. **Lee**, Institut für Kernphysik der Universität, Köln, Germany

M. **Lefort**, Institut de Physique Nucléaire, Orsay, France

K.H. **Lindenberger**, Hahn-Meitner-Institut für Kernforschung, Berlin, Germany

P. **Lucas**, CEN Saclay, Gif-sur-Yvette, France

U. **Lynen**, Gesellschaft für Schwerionenforschung, Darmstadt, Germany

M.H. **Macfarlane**, Argonne National Lab., Physics Div., Argonne, Illinois, U.S.A.

R. **Malfliet**, Kernfysisch Versneller Instituut, Universiteitscomplex Paddepoel, Groningen, Netherlands

M.C. **Mallet-Lemaire**, CEN Saclay, Gif-sur-Yvette, France

R. **Malmin**, State University of New York, Physics Dept., Stony Brook, N.Y., U.S.A.

P. **Manakos**, Institut für Kernphysik der Technischen Hochschule, Darmstadt, Germany

N. **Marquardt**, Laboratoire de Physique Nucléaire, Université de Montréal, Montréal, Canada

K.W. **Mc Voy**, University of Wisconsin, Dept. of Physics, Madison, Wisconsin, U.S.A.

M.C. **Mermaz**, CEN Saclay, Gif-sur-Yvette, France

V. **Metag**, Max-Planck-Institut für Kernphysik, Heidelberg, Germany

K. **Möhring**, Hahn-Meitner-Institut für Kernforschung, Berlin, Germany

U. **Mosel**, Institut für Theoretische Physik der Universität, Giessen, Germany

O. **Nathan**, Niels-Bohr-Institut, Copenhagen, Denmark

W. **Nörenberg**, Max-Planck-Institut für Kernphysik, Heidelberg, Germany

W.J. **Ockels**, Kernfysisch Versneller Instituut, Universiteitscomplex, Paddepoel, Groningen, Netherlands

W. von **Oertzen**, Hahn-Meitner-Institut für Kernforschung, Berlin, Germany

H. **Oeschler**, Niels-Bohr-Institut, RISØ, Roskilde, Denmark

L. **Papineau**, CEN Saclay, Gif-sur-Yvette, France

H.C. **Pauli**, Max-Planck-Institut für Kernphysik, Heidelberg, Germany

M. **Pauli, jr.**, Institut für Theoretische Physik der Universität, Basel, Switzerland

D. **Pelte**, Max-Planck-Institut für Kernphysik, Heidelberg, Germany

F. **Pühlhofer**, Gesellschaft für Schwerionenforschung, Darmstadt, Germany

W. **Reiter**, Max-Planck-Institut für Kernphysik, Heidelberg, Germany

A. **Richter**, Institut für Experimentalphysik der Ruhr-Universität, Bochum, Germany

E. **Roeckl**, Gesellschaft für Schwerionenforschung, Darmstadt, Germany

G. **Rufenach**, Max-Planck-Institut für Kernphysik, Heidelberg, Germany

N. **Sanderson**, Department of Physics, University of Brimingham, Birmingham, United Kingdom

R. **Santo**, Institut für Kernphysik der Universität, Münster, Germany

D.K. **Scott**, University of California, Lawrence Berkeley Lab., Berkeley, California, U.S.A.

R.H. **Siemssen**, Kernfysisch Versneller Instituut, Universiteitscomplex Paddepoel, Groningen, Netherlands

F. **Siller**, II. Physikalisches Institut der Universität, Erlangen, Germany

R. da **Silvera**, Institut de Physique Nucléaire, Orsay, France

H.J. **Specht**, II. Physikalisches Institut der Universität, Heidelberg, Germany

R. **Schaeffer**, CEN Saclay, Gif-sur-Yvette, France

U.C. **Schlotthauer**, Institut de Physique Nucléaire, Orsay, France

Ch. **Schmelzer**, Gesellschaft für Schwerionenforschung, Darmstadt, Germany

U. **Schmidt-Rohr**, Max-Planck-Institut für Kernphysik, Heidelberg, Germany

W.F.W. **Schneider**, Gesellschaft für Schwerionenforschung, Darmstadt, Germany

G. **Schrieder**, Institut für Experimentalphysik der Ruhr-Universität, Bochum, Germany

W.U. **Schröder**, Institut für Kernphysik der Technischen Hochschule, Darmstadt, Germany

D. **Schwalm**, Max-Planck-Institut für Kernphysik, Heidelberg, Germany

St. **Steadman**, Max-Planck-Institut für Kernphysik, Heidelberg, Germany

R. **Stock**, Physikalisches Institut der Universität, Marburg, Germany

U. **Strohbusch**, Fakultät für Physik, Universität Freiburg, Freiburg, Germany

N. **Takigawa**, Hahn-Meitner-Institut für Kernforschung, Berlin, Germany

J. **Theobald**, Institut für Kernphysik der Technischen Hochschule, Darmstadt, Germany

C. **Toepffer,** University of the Witwatersrand, Johannesburg, South Africa
D. **Trautmann,** Institut für Theoretische Physik der Universität, Basel, Switzerland
I. **Tserruya,** Max-Planck-Institut für Kernphysik, Heidelberg, Germany
Chr. **Uhlhorn,** Ruhr-Universität, Bochum, Germany
R. **Vandenbosch,** University of Washington, Dept. of Physics, Seattle, Washington, U.S.A.
H. **Voit,** Physikalisches Institut der Universität, Erlangen, Germany
Th. **Walcher,** Max-Planck-Institut für Kernphysik, Heidelberg, Germany
H.A. **Weidenmüller,** Max-Planck-Institut für Kernphysik, Heidelberg, Germany
A. **Weiguny,** Institut für Theoretische Physik der Universität, Münster, Germany
H. **Weiss,** Physik Department der Technischen Universität München, Garching, Germany
H. **Wolter,** Kernforschungsanlage Jülich, Institut für Kernphysik, Jülich, Germany
J.P. **Wurm,** Max-Planck-Institut für Kernphysik, Heidelberg, Germany
Y. **Yariv,** Max-Planck-Institut für Kernphysik, Heidelberg, Germany
B. **Zeidman,** Physics Department, Argonne National Lab., Argonne, Illinois, U.S.A.

Lecture Notes in Physics